Cognitive Intelligence and Robotics

Cognitive Intelligence refers to the natural intelligence of humans/animals involving the brain to serve the necessary biological functioning to perform an intelligent activity. Although tracing a hard boundary to distinguish intelligent activities from others remains controversial, most of the common behaviors/activities of living organisms that cannot be fully synthesized by artificial means are regarded as intelligent. Thus the act of natural sensing and perception, understanding of the environment and voluntary control of muscles, blood-flow rate, respiration rate, heartbeat, and sweating rate, which can be performed by lower level mammals, indeed, are intelligent. Besides the above, advanced mammals can perform more sophisticated cognitive tasks, including logical reasoning, learning and recognition and complex planning/coordination, none of which could be realized artificially to the level of a baby, and thus are regarded as cognitively intelligent.

The series aims at covering two important aspects of the brain science. First, it would attempt to uncover the mystery behind the biological basis of cognition with special emphasis on the decoding of stimulated brain signals/images. The coverage in this area includes neural basis of sensory perception, motor control, sensory-motor coordination and also understanding the biological basis of higher-level cognition, such as memory and learning, reasoning and complex planning. The second objective of the series is to publish brain-inspired models of learning, perception, memory and coordination for realization on robots to enable them to mimic the cognitive activities performed by the living creatures. These brain-inspired models of machine intelligence would supplement the behavioral counterparts, studied in traditional AI.

The series includes textbooks, monographs, contributed volumes and even selected conference proceedings.

More information about this series at http://www.springer.com/series/15488

Kaushik Das Sharma · Amitava Chatterjee
Anjan Rakshit

Intelligent Control

A Stochastic Optimization Based Adaptive Fuzzy Approach

Kaushik Das Sharma
Department of Applied Physics
University of Calcutta
Kolkata, West Bengal, India

Anjan Rakshit
Department of Electrical Engineering
Jadavpur University
Kolkata, West Bengal, India

Amitava Chatterjee
Department of Electrical Engineering
Jadavpur University
Kolkata, West Bengal, India

ISSN 2520-1956 ISSN 2520-1964 (electronic)
Cognitive Intelligence and Robotics
ISBN 978-981-13-4603-3 ISBN 978-981-13-1298-4 (eBook)
https://doi.org/10.1007/978-981-13-1298-4

This Springer imprint is published by the registered company Springer Nature Singapore Pte Ltd.
The registered company address is: 152 Beach Road, #21-01/04 Gateway East, Singapore 189721, Singapore

Dedicated to
our beloved students

Preface

Fuzzy control evolved as a superior alternative to the conventional classical control strategies mainly for nonlinear, complex, and mathematically ill-defined systems. The mathematical formulation of a fuzzy controller inherently contains the approximation capability, and the human expert knowledge can easily be incorporated to infer a decision directly to the real world. In the last decade of the twentieth century, the Lyapunov theory was successfully employed to study the stability of the fuzzy controllers and subsequently the concept of adaptation in the domain of stable fuzzy control was introduced. But in the early phases of adaptive fuzzy controller design procedures, the basic structure of a stable controller was usually assumed to be static. As time elapsed, gradual improvements over the early designs were proposed. Hybridization of genetic algorithm-based strategy with a Lyapunov theory-based strategy was proposed to adapt both controller structures and free parameters, containing center locations of input membership functions, scaling gains, learning rate, etc., in addition to the positions of the singletons describing output fuzzy sets of the fuzzy controller. But the main problem associated with such hybrid models is that the genetic algorithm-based Lyapunov method needed several thousands of fitness evaluations to achieve parameter convergence and to guarantee the closed-loop stability.

This book presents systematic designs of stable adaptive fuzzy logic controllers employing hybridizations of the Lyapunov strategy-based approach (LSBA) and contemporary stochastic optimization techniques. The book also presents another class of systematic hybrid adaptive fuzzy controller design techniques employing H^{∞} strategy-based robust approach (HSBRA) and contemporary stochastic optimization techniques. Particle swarm optimization (PSO), harmony search (HS), and covariance matrix adaptation (CMA) have been considered here as the candidate stochastic optimization techniques utilized in conjunction with the Lyapunov theory or H^{∞} control theory, to develop the hybrid models discussed in this book. The hybridization process attempts to combine strong points of both the Lyapunov theory or H^{∞} control theory-based local search methods and the stochastic optimization-based global search methods to evolve superior methods. The objectives of these design strategies are to perform simultaneous adaptations of both the structure of the fuzzy controller and its free parameters so that two competing requirements can be fulfilled: (i) to guarantee stability of the controllers designed and (ii) to achieve very high degrees of automation in the process of these controller designs, by employing global search methods. The LSBA/HSBRA helps to guarantee closed-loop stability with improved speed of convergence. On the other hand, the stochastic optimization algorithms help to achieve desired automation for the design methodologies of the adaptive fuzzy controller by getting rid of many manually tuned parameters, which are typical in LSBA/HSBRA design procedure. Hence, the objective is to design stable adaptive fuzzy controllers which can simultaneously provide a high degree of automation, guarantee asymptotic stability, and also achieve satisfactory transient performance.

In the present book, several novel variants of hybridization of locally operative LSBA/HSBRA and globally operative stochastic optimization techniques have been presented. These approaches have generally been named as hybrid adaptation strategy-based approaches (HASBA). Three modes of hybridizations, namely cascade, concurrent, and preferential modes of hybridizations, have been presented and discussed in detail here. To demonstrate the usefulness of these hybrid methodologies, several nonlinear benchmark problems available in the literature have been simulated and exhaustive performance studies have been carried out. Several real implementations in the areas of process control, vision-based robot navigation domains, and manipulator end-effector tracking have also been illustrated to show the effectiveness of the hybrid techniques presented in this book.

The following list articulates the key features of this book:

- Stochastic optimization techniques are hybridized with the Lyapunov theory/ H^{∞} control theory to design stable adaptive fuzzy controllers, in both simulation case studies and real implementations.
- The stable fuzzy adaptive controllers are designed for both fixed structure controller configurations and variable structure controller configurations.
- Three different variants of hybridizations of the Lyapunov theory/H^{∞} control theory and the stochastic optimization techniques are presented, namely cascade model, concurrent model, and preferential model.
- Stochastic optimization-based system identification methodology is developed, and temperature of an air heater process is controlled by means of the Lyapunov theory and stochastic optimization-based hybrid controllers, with the estimated process parameters.
- A novel idea for vision-based mobile robot navigation utilizing the stable adaptive fuzzy tracking controllers, designed by the Lyapunov theory and stochastic optimization-based hybrid methodologies, is implemented in real experiments.
- A novel idea for two-link manipulator control utilizing the stable adaptive fuzzy tracking controllers, designed by the H^{∞} control theory and stochastic optimization-based hybrid methodologies, is implemented in real experiments.

The materials presented in this book are divided into five parts, and these are composed as follows:

Part I: Prologue

This part of the book introduces the idea of adaptive fuzzy control and gives a preview of few very prominent contemporary stochastic optimization techniques applied in different fields of engineering. A non-adaptive fuzzy controller design technique is also presented employing a stochastic optimization algorithms.

- Chapter 1 introduces the adaptive fuzzy control with its journey from modern control theories. This chapter also elaborates the overall literature available for the interested readers in this domain.
- Chapter 2 describes the glimpses of some contemporary stochastic optimization techniques. GA, PSO, HS, and CMA are introduced in a nutshell.
- Chapter 3 describes the design techniques adopted for stable fuzzy controllers, employing stochastic optimization algorithm-based approach. A non-fuzzy supervisory control-based stable operation of fuzzy controllers has been evaluated for some benchmark nonlinear systems.

Part II: Lyapunov Strategy-Based Design Methodologies

This part of the book presents a detailed discussion about the Lyapunov theory-based adaptation procedure and different hybrid adaptive fuzzy controller design strategies.

- Chapter 4 describes the formulation of the control law for a class of nonlinear systems and the application of the Lyapunov theory to ensure the stability of the closed-loop system. The Lyapunov strategy-based approach of designing fuzzy controllers is developed and analyzed in detail.
- Chapter 5 describes the stochastic optimization algorithm-aided Lyapunov theory-based hybrid stable adaptive fuzzy controller design methodologies considered in this book. As mentioned before, three variants of the hybrid design schemes are discussed in this chapter.

Part III: H^{∞} Strategy-Based Design Methodologies

This part of the book presents another avenue with a discussion about the H^{∞} strategy-based robust control-based adaptation procedure and its hybrid variants.

- Chapter 6 describes the formulation of the H^{∞} strategy-based robust control law for a class of nonlinear systems with external disturbances. Here, the H^{∞} control-based robust direct adaptive fuzzy controller design methodology is discussed and analyzed in detail. The Riccati-like equation has been utilized here to ensure the stability of the closed-loop system with H^{∞} strategy-based control, with disturbance rejection phenomenon.
- Chapter 7 describes the stochastic optimization algorithm-aided H^{∞} strategy-based hybrid robust stable adaptive fuzzy controller design methodologies, considered in this book. Three variants of the hybrid design schemes, as discussed in Chap. 5, are discussed in this chapter.

Part IV: Applications

It is the most interesting part of the book where three real-life applications from diverse domains are presented. The first two chapters in this part discuss the applications developed utilizing the Lyapunov theory-based hybrid models of fuzzy controller design, whereas the third chapter utilizes the H^{∞} strategy-based robust control-based hybrid models to develop a real application.

- Chapter 8 describes the real implementation of temperature control of an air heater system, employing the hybrid controllers presented in Chaps. 4 and 5.
- Chapter 9 describes the real implementation of vision-based navigation of a mobile robot, employing the Lyapunov theory-based controller and hybrid controllers presented in Chaps. 4 and 5.
- Chapter 10 describes the real implementation of tracking control-based robot manipulator end-effector movement, employing the H^{∞} strategy-based controller and hybrid controllers presented in Chaps. 6 and 7.

Part V: Epilogue

- Chapter 11 concludes the book. This chapter describes some emerging areas of intelligent fuzzy control along with the future scope of research in this domain.

The book can be studied in two ways. For the readers interested only in the Lyapunov theory-based design, they can go through Chaps. 1–5 and 8, 9. On the other hand, if H^{∞} control theory is of prime interest, then the route will be Chaps. 1–3, 6, 7, and finally Chap. 10. Practicing engineers and senior researchers may skip Chaps. 1–4 and Chap. 6 as these chapters mostly deal with the basic theories. Chap. 11 may be beneficial to the new researchers as they can find some exciting ideas for their future research.

We are grateful to many people for their openhearted support and encouragement during the course of writing of this book. In particular, we thank Mr. Debasish Biswas, Research Scholar, Department of Applied Physics, University of Calcutta, for preparing the experimental arrangements utilized in Chap. 10. Finally, on a personal note, we would like to thank our families for their support and inspiration during this endeavor.

Kolkata, India
April 2018

Kaushik Das Sharma
Amitava Chatterjee
Anjan Rakshit

Contents

About the Authors

Dr. Kaushik Das Sharma received his B.Tech. and M.Tech. degrees in electrical engineering from the Department of Applied Physics, University of Calcutta, Kolkata, India, in 2001 and 2004, respectively, and Ph.D. degree from Jadavpur University, Kolkata, India, in 2012. He is currently serving as Associate Professor in Electrical Engineering Section of Department of Applied Physics, University of Calcutta, Kolkata, India. He is the recipient of the Kanodia Research Scholarship in 2002 and **University Gold Medal** in 2004 from the University of Calcutta. His key research interests include *fuzzy control, stochastic optimization, machine learning, robotics*, and *systems biology*. He has authored/co-authored about 45 technical articles, including 20 international journal papers. He is Senior Member of IEEE (USA), Member of IET (UK), and Life Member of Indian Science Congress Association. He has served/is serving in important positions in several international conference committees all over the world. Presently, he also serves as Secretary of IEEE Joint CSS-IMS Kolkata Chapter and Executive Committee Member of IEEE Kolkata Section and IET (UK) Kolkata Local Network.

Amitava Chatterjee did his BEE, MEE, and Ph.D. from the Department of Electrical Engineering, Jadavpur University, Kolkata, in 1991, 1994, and 2002, respectively. He is currently serving as Professor in electrical engineering, Department of Jadavpur University, Kolkata, India. He is the recipient of the Japanese Government (Monbukagakusho) Scholarship in 2003 and the recipient of the Japan Society for the Promotion of Science (JSPS) Post-Doctoral Fellowship in 2004. In 2004 and 2009, he visited the University of Paris XII, Val de Marne, France, and, in 2017, he visited the University of Paris-Est, France, as Invited Teacher. His key research interests include *nonlinear control, intelligent instrumentation, signal processing*, *computer vision, robotics, image processing*, and *pattern recognition*. He has co-authored a book and co-edited two books all published by the Springer-Verlag, Germany. He presently serves as Editor of *IEEE Transactions on Vehicular Technology* and *Engineering Applications of Artificial Intelligence Journal* (Elsevier) and as Associate Editor of *IEEE Transactions on Instrumentation and Measurement* and *IEEE Sensors Journal*. He has authored/co-authored more than 145 technical articles, including 90 international journal papers. He is Member of the Technical Committee on "Imaging Measurements and Systems" of IEEE Instrumentation and Measurement Society, USA. He is Fellow of IETE (India), Fellow of the Institution of Engineers (India), and Senior Member of IEEE (USA). He has served/is serving in important positions in several international conference committees all over the world, including several flagship IEEE conferences.

Anjan Rakshit received his ME and Ph.D. degrees in electrical engineering from Jadavpur University, Kolkata, India, in 1977 and 1987, respectively. He is a retired professor of the Department of Electrical Engineering, Jadavpur University. His research interests include digital signal and image processing, Internet-based instrumentation, smart instrumentation, intelligent control, and design of real-time systems. He has published about 75 technical papers in his areas of research interest.

Part I
Prologue

Chapter 1
Intelligent Adaptive Fuzzy Control

1.1 Way to Modern Control Theory

Automatic control theory, predominantly the concept of feedback, has played a fundamental role to the development of automation [1]. A journey from classical control theory to modern control strategies spans over the time frame of few centuries. Classical control theory was applied for designing the controllers for the systems those are principally linear time invariant in nature [2]. The need for advanced control strategies arose when accurate control laws were required to be developed for detail mathematical modeling of the systems. Most of the practical systems that we encounter in real life are nonlinear in nature. The development of control laws for nonlinear systems necessitated the need for adopting modern approaches, e.g., state variable theory [3–6], which evolved as a general solution for the controller design problem. For the systems, which can be well described with a linear second-order mathematical model, there are a number of conventional procedures for designing the controllers [2, 7], while for the systems modeled with higher-order linear models, pole placement design technique [8, 9] or the frequency domain design procedures can be utilized [6, 7, 10]. Many of the current control applications involve complex, nonlinear single-input–single-output (SISO), multi-input–single-output (MISO), or multi-input–multi-output (MIMO) models, for which controller(s) is(are) required to deliver high-end performance specifications in the presence of plant uncertainty, input uncertainty (including noise or disturbances), and actuator constraints. Since classical control theories are not able to deal with such problems, several modern methodologies emerged in the past three decades that systematically address the fundamental limitations and capabilities of linear state feedback controllers, for such challenging systems, both in continuous-time case [6] and in digital implementations [11, 12]. Examples of such methodologies are the adaptive control [13, 14], H^{∞}-based robust control [15, 16], sliding mode control [17–19], optimal control [20, 21], etc., in conjunction with different intelligent techniques, such as neural networks [22, 23], fuzzy logic [24–26], machine

K. Das Sharma et al., *Intelligent Control*, Cognitive Intelligence and Robotics,
https://doi.org/10.1007/978-981-13-1298-4_1

learning [27, 28], evolutionary computations [29, 30], genetic algorithms [30, 31], etc. In some recent controller design strategies, intelligent control techniques have been utilized individually or in combinations to enhance the performance of the designed controllers.

Among the intelligent control techniques, "fuzzy logic controllers" (FLCs) are a class of nonlinear controllers that make use of human expert knowledge and an implicit imprecision to apply control to the systems which can be mathematically ill-defined. These controllers evolved as a popular domain of control system and control engineering more than three decades back. Although the original interest was initiated by the urge to design controllers for those systems which are very difficult to be characterized mathematically, with time, study on FLCs expanded extensively for both model-free and model-based systems.

1.2 Overview of Fuzzy Control

Fuzzy control emerged on the foundations of L. A. Zadeh's fuzzy set theory, introduced in 1965 [32, 33]. Fuzzy logic controllers have proven to be a successful control option for many complex and nonlinear systems or even for non-analytic systems too [34]. It has been suggested as an alternate approach to the conventional PID-type control techniques in many cases, even for linear systems where fuzzy controllers were exploited to obtain better performances [35].

One of the earliest fuzzy controllers was developed by E. H. Mamdani and S. Assilian in 1975, where control of a small laboratory prototype steam engine was considered [36, 37]. The primal fuzzy control algorithm consisted of fuzzy sets, a set of heuristic control rules and some methods to evaluate the rules. Since its emergence, fuzzy logic control has received huge attention from both the academic researchers and industrial practitioners. Many researchers have dedicated a great deal of time and effort equally in theoretical research and application techniques of fuzzy controllers in various fields like power systems [38–40], telecommunications [41–44], robotic systems [45–52], automobile [53–55], process control [56–59], medical services [56, 60–62], electronics [63–65], chaos control [66, 67] and nuclear reactors [68, 69], and many more.

In a nut shell, the basic structure of a fuzzy controller consists of four important components: (i) knowledge base (fuzzy rule base), (ii) fuzzification interface, (iii) inference engine, and (iv) defuzzification interface [70]. The block diagram of a generic fuzzy control system is shown in Fig. 1.1. The knowledge base encompasses all the knowledge about the controller (might be obtained from the input–output data set of the system to be controlled or may be obtained from a domain expert), and it contains a rule base for the fuzzy controller and a database. The description about the definition of membership functions (MFs) utilized in the fuzzy control rules, scaling gains, etc., is to be written in the knowledge base, which is the declarative part of the database. Whereas, the rule base of the fuzzy controller is the procedural part of the knowledge base. The rule base contains the information on

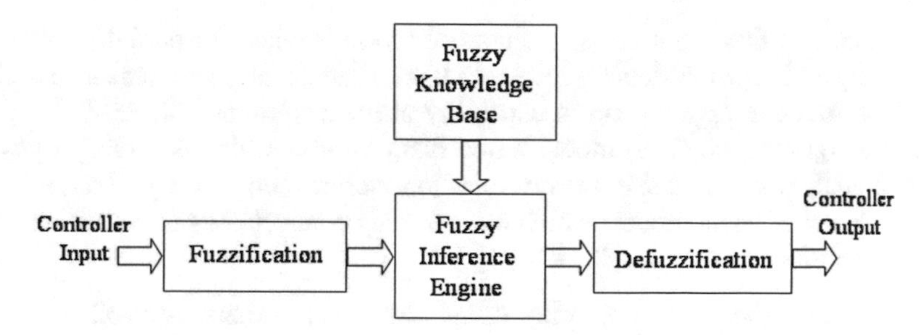

Fig. 1.1 Generic block diagram of a fuzzy logic controller

how this knowledge can be employed to infer new control actions in various situations. The inference procedure is performed by the inference engine part, which provides the reasoning mechanism to the control action under the physical constraints. It is the most essential part of a fuzzy control system. The fuzzifier, i.e., the fuzzification interface delineates a mapping from a real (crisp) space to a fuzzy space. Similarly, the defuzzifier, i.e., the defuzzification interface, delineates a mapping from a fuzzy space, guided by a universe of discourse for output variables, to a real (crisp) space [26].

The fuzzy logic controllers can be approximately categorized, according to their methods of rule generation and the construction of the inference engine, into the following classes:

(i) Conventional fuzzy control which is a classical Mamdani-type design [34, 37] and early version of Sugeno-type design [36] of fuzzy controller.
(ii) Fuzzy proportional–integral–derivative (PID) control where a fuzzy controller is designed to generate control actions like a conventional PID controller [71], while some other alternate definitions are also there in few literatures [72, 73].
(iii) Neuro-fuzzy control where the complementary features of artificial neural network-based control, providing learning abilities and high efficacy in terms of computational burden in parallel implementation, and fuzzy control, providing a efficient framework for expert knowledge representation, are combined together to enhance the flexibility, data processing capability, and adaptability of the designed controller [58, 69, 74, 75].
(iv) Fuzzy sliding mode control for controlling processes with parameter uncertainties and/or load disturbances, where fuzzy control and sliding mode control are combined in such a way that can eliminate the chattering in sliding mode control by fuzzy-based switching and the robustness of the designed controller is achieved with the traditional philosophy of sliding mode control [17, 18, 47, 55, 76–78]. In this context, hybridization of robust stabilization concept with fuzzy control method has given rise to another active branch of fuzzy control in recent times, termed as robust fuzzy control [79–82].

(v) Adaptive fuzzy control where control action is taken for partially known or fully unknown systems with some kind of adaptation mechanism based on the function approximation capability of fuzzy systems [24, 83–93].

(vi) Takagi–Sugeno (T-S) model-based fuzzy control which is a fuzzy dynamic model, based on a set of fuzzy rules to describe a nonlinear system, in terms of a set of linear models which are smoothly connected by fuzzy membership functions [24–26, 87, 88, 92–94].

Several researchers are exploring different aspects of fuzzy controllers both in theoretical domain and in application arena in different categories as stated before [70, 95]. As the present book mainly deals with the domain of stable adaptive fuzzy control, thus, the development of this category of adaptive fuzzy controllers is scrutinized in brief further.

1.3 Overview of Adaptive Fuzzy Control

The non-adaptive fuzzy control [24, 25, 27] were initially in use in some applications long time back. However, it is sometimes difficult to specify the configurations of the fuzzy controllers, i.e., controller structure, scaling gains, rule base parameters for some plants, or it might be needed to tune those nominal parameters with the variation of the plant dynamics to provide a minimum desired performance. In some cases, to cope with the temporal variations in the reference input signal applied to the plant, the rigorous adjustments of the fuzzy controller configurations are required. This provides the motivation for adaptive fuzzy control, where the focus is on the automatic synthesis and tuning of fuzzy controller parameters, fulfilling the requirements of control objectives.

The very first adaptive fuzzy controller was introduced by Procyk and Mamdani in [96], called the linguistic self-organizing controller (SOC). Later Layne and Passino have studied several applications of this method in [97], and further, they introduced the "fuzzy model reference learning controller" (FMRLC) in [98–100]. Kwong and Passino developed some extensions of FMRLC in [101]. The FMRLC had also been successfully implemented for rotational inverted pendulum, a single-link flexible robot, and an induction machine [100, 102] and also been applied to the multiple-input–multiple-output (MIMO) problem of a two-link rigid robot [103, 104].

Many other adaptive fuzzy control techniques have also been employed for several applications [83, 105–108]. In [94], Ordonez et al. provided a comparative analysis of FMRLC with conventional adaptive and/or non-adaptive nonlinear control methods for experimental studies on rotational inverted pendulum system. The main problem with the SOC and the FMRLC that were developed in [101, 102], by Passino et al., was that although the developed approaches emerged as adaptive fuzzy control strategies, the closed-loop stability of the system was not guaranteed.

Su and Stepanenko [86], Wang [87], Johansen [108], and other researchers in [84, 85, 90, 93, 109] explored the ideas derived from conventional adaptive control to establish stability conditions for a variety of adaptive fuzzy control techniques. On the whole, these techniques can be split into two categories: direct and indirect adaptive fuzzy control [24, 25, 84, 87, 91, 95, 109–112]. In indirect adaptive fuzzy control, there is an estimation mechanism that produces a model of the plant which is then used to specify the controller configurations. In direct adaptive control, on the contrary, the model of the plant is not estimated; instead, the controller parameters are directly tuned by using input–output data of the plant.

Along with the property of adaptability of a fuzzy controller, it is always desired that the system should be stable in behavior. The notion of stability in fuzzy controller, and in control theory in general, should be a major concern during the design process of the controller. Thus, the next section briefly describes the different approaches available in the literature for stability analysis of a fuzzy controller.

1.4 Stability Issues in Adaptive Fuzzy Control

The area of stability analysis of fuzzy control systems is receiving an ever increasing attention from many researchers for the last two decades, primarily due to the advancements made in the theoretical perspectives of fuzzy control and their several possible applications. A brief overview of some popular approaches adopted for stability analysis of fuzzy controller is summarized as follows:

Kalman and Bertram in [113], Chand and Hansen in [114] and Chiu and Chand in [115] presented some earlier studies based on Lyapunov methods for analyzing stability of fuzzy control systems. Langari et al. in [116, 117] also used Lyapunov's direct method and generalized theorem of Popov [118] to provide sufficient conditions for stability of fuzzy control systems. In [119], Aracil, Ollero, and Garcia-Cerezo introduced stability indices for fuzzy control systems using phase portrait for characterizing and understanding the dynamical behavior of low-order fuzzy control systems [120, 121]. The circle criterion [69] is used by Ray et al. in [122, 123] to provide sufficient conditions for stability of fuzzy control systems. The stability analysis of fuzzy control systems, developed using sliding mode control technique, was detailed in [76, 124]. Some earlier treatments of stabilizing a gain scheduling type of T-S fuzzy control systems [125] were presented in [26]. In [110, 125, 126, 128] Tanaka et al. explored the Lyapunov theory-based stability analysis and implementation of stable fuzzy controllers in different applications like model car control, backing up control of a track trailer, chaos control, and the robustness of the designed controllers were also studied in [129, 130]. Nonlinearities in control systems, such as the systems with fuzzy controllers, sometimes cause limit cycle oscillations in the system response [131]. The circle criterion can be used to prove the absolute stability of the system and therefore the absence of limit cycles. Similarly, using a steady-state error prediction procedure [131], one can prove the absence of limit cycles, if the steady-state error is zero. However, neither of these methods is able to calculate the amplitude or frequency of limit cycles when they do exist in practice. In [132], Jenkins presented how describing function analysis [133] can be used to predict the existence of limit cycles, thus, the stability of the fuzzy control systems. Jenkins and Passino also presented a method in [134] for calculating frequency and amplitude of limit cycle in fuzzy control systems [135, 136]. Some other methods for stability analysis of fuzzy control systems using linear matrix inequalities (LMIs) method were presented in [110, 126, 127, 130, 137]. Similarly, Chen in [138, 139] and Fei and Isik in [140] provided another alternate approach, called "cell-to-cell mapping" [141, 142], for analysis of stability in fuzzy dynamical systems.

Among the different stability analysis methodologies for fuzzy control systems, Lyapunov theory -based analyses are most popular in fuzzy control research fraternity. The possible reason can be that it can provide a systematic technique which can easily infer the stability of a closed-loop system. There are many variations of Lyapunov theory applied to the fuzzy controller stability analysis, such as (i) stability analysis based on common quadratic Lyapunov function [128, 143, 144], (ii) stability analysis based on piecewise quadratic Lyapunov function [145–147], and (iii) stability analysis based on fuzzy Lyapunov function [148–150]. In [151], Curtis and Beard developed a graphical method for understanding the Lyapunov-based nonlinear control.

Robustness is another issue, along with stability of the adaptive fuzzy controller that was also rigorously investigated by the researchers for last two decades [42, 89, 112, 152–166]. Chen et al. in [89] presented the earlier most work on H^{∞} theory-based robust tracking control applied on fuzzy controller design. The adaptation rule and robust control law, augmented with the fuzzy control action, were developed utilizing the Riccati-like equation. The attenuation of tracking error, when the non-linear plant is subjected to the external disturbances, was ensured by the use of H^{∞} theory-based optimal control.

Assawinchaichote et al. in [152, 154] described design strategies of output feedback fuzzy controllers, based on H^{∞} control scheme, for a class of nonlinear systems that may be singularly perturbed employing T-S-type fuzzy modeling. Linear matrix inequality (LMI) approach had been utilized to develop H^{∞} output feedback fuzzy controllers that regulated the L_2-gain of the system in such a manner that the overall stability of the system was guaranteed. Moreover, in [154], authors successfully applied their formulation of LMI-based H^{∞} robust tracking control scheme for the Markovian jump nonlinear systems. In [155], the authors also presented a similar methodology for LMI-based H^{∞} robust tracking control scheme and it was validated by applying it for a real-time non-minimum phase system.

Wu et al. in [156] studied the T-S-type fuzzy controller design problem by minimizing the mixed H^2/H^{∞} norm. The concept of fuzzy region clustering in connection with fuzzy rule base design and the H^{∞}-based robust control technique was employed to stabilize all clustered plant rules, for each fuzzy region.

Wu et al. in [157] proposed a H^{∞} fuzzy observer-based controller design technique for a class of nonlinear parabolic partial differential equation (PDE) systems. The characteristics of the system, in terms of eigenspectrum, was divided into a finite-dimensional slower one and an infinite-dimensional stable fast component. A T-S-type fuzzy observer-based indirect controller was developed to stabilize the nonlinear PDE system, and H^{∞} disturbance attenuation was also achieved in an optimal manner, considering the control constraints.

Guan et al. in [158] formulated a feedback H^{∞} control problem for discrete time T-S-type fuzzy systems utilizing the output measured by a memoryless logarithmic quantizer. The asymptotical stability of the closed-loop system was guaranteed by the full-order dynamic output feedback controller while the prescribed H^{∞} performance level was also ensured to attenuate the disturbances. On the other hand, Pan et al. in [159] focused on the H^{∞} control-based robust stabilization problem of fuzzy controller for a class of continuous-time T-S-type systems.

1.5 Intelligent Adaptive Fuzzy Control

Computational intelligence (CI), termed as soft computing in many cases, provides a framework to address design, analysis, and modeling in control problems, in the presence of uncertain and imprecise information [29, 30, 167]. The main components of intelligent computing technologies are fuzzy logic, neural networks, and stochastic optimization, and these are considered as complementary and synergistic to each other. CI components, individually or in a hybrid mode, have been applied to improve the performances of the design strategies in the domains of parameter estimation and nonlinear control systems, which are characterized by imprecise data and incomplete knowledge [53]. Neuro-fuzzy systems [168] are the most well-known representative of hybrid systems in terms of number of applications in the field of control system design. Compared to neuro-fuzzy systems, genetic fuzzy systems, i.e. genetic algorithm-based design of fuzzy controllers, are not put into practice until the end of the twentieth century, in the industrial applications. The earliest genetic algorithm-based fuzzy control systems were introduced in the following four pioneering works described next.

Karr in his seminal work in [169] introduced genetic tuning of linguistic fuzzy rule base systems (FRBSs). The definition of the fuzzy database was encoded in the chromosome, which contained the concatenated parameters of the input and output fuzzy sets.

In [170], Valenzuela-Rendon presented the first genetic fuzzy system for learning fuzzy rule bases with disjunctive normal form (DNF) of fuzzy rules, employing a supervised learning algorithm. In [171], the author further extended his original proposal to incorporate the concept of reinforcement learning.

In [172], Thrift presented another pioneering work for learning fuzzy rule bases. This method works by using a complete decision table that represents a special case of crisp relation, defined over the collections of fuzzy sets corresponding to the input and output variables. In this work, genetic algorithm employed an integer coding.

Pham and Karaboga, in [173], presented a quite different approach for designing fuzzy controllers that uses a fuzzy relation, instead of the decision table-based classical crisp relation. The genetic algorithm (GA) was used to modify the fuzzy relational matrix of a SISO fuzzy model.

After the initiation, the application of genetic algorithm for designing the fuzzy controllers and neuro-fuzzy controllers gained tremendous attention from the researchers during 1990s. A brief overview of some prominent applications of genetic algorithm for designing fuzzy or neuro-fuzzy controllers is presented here next:

Bonissone et al. in [54] and Hwang in [174] described genetic tuning schemes for optimization of a fuzzy controller applied in transportation systems. The design goal that was successfully achieved in [54] was a controller that could accurately track a desired velocity profile while at the same time could maintain a smooth ride in order to minimize the stress load on the couplers connecting the rail cars. In [174], fuzzy *c*-means clustering and genetic algorithm were applied in conjunction to identify the structure and parameters of a linguistic fuzzy model, which was used by the railway operators to build efficient and economic control strategies.

In [77], Huang et al. designed a fuzzy sliding mode controller using a real-coded GA in a position control application for an industrial XY-table. By using laser as a sensor, the table could be positioned with submicron-level accuracy. This work also showed that the evolutionary optimization of fuzzy controllers is feasible for hardware-in-the-loop systems and can help to reduce the costs for controller design and tuning.

In [175], Shim et al. reported an application of evolutionary computation, in combination with neural networks and fuzzy systems, for intelligent consumer products. Here, the role of the evolutionary algorithm was to adapt the number of rules and to fine-tune the membership functions to improve the performance of fuzzy systems for estimation and control. This work also mentioned evolutionary computation for fuzzy rule generation applied to process control.

In [176], Alcala et al. considered GA to develop a smart tuning strategy for fuzzy logic controllers applied to the control of heating, ventilating and air conditioning systems, considering energy performance and indoor comfort requirements. To solve this problem, a quick convergence GA was proposed based on an aggregated fitness function, considering trustable weights provided by the experts.

Homaifar and McCormick in [177] proposed a method for knowledge base learning using GA for control problems, where both membership functions and fuzzy rules were determined simultaneously in order to address their codependency. This method considered the simple GA, with integer coding, for both rule consequents [172] and for membership function support amplitude, in the same chromosome.

In [178] Ishibuchi, Nozaki, Yamamoto and Tanaka utilized GAs for selecting a small number of fuzzy IF-THEN rules with high classification performance. The proposed algorithm was based on a simple GA with binary coding, representing whether a rule should be selected or not from an initial set of candidate rules.

In [179], Setnes and Roubos proposed a two-step approach for function approximation, dynamic systems modeling, and data classification problems, by learning approximate T-S rules. In this work, first a compact initial knowledge base was obtained using fuzzy clustering technique and then this model was optimized by a real-coded GA, subjected to constraints in order to maintain the semantic properties of the rules. Each chromosome represents the parameters defining each candidate fuzzy model containing membership functions of the antecedents and coefficients of the consequents. To obtain a more interpretable approximate T-S model using GA, this work contributed significantly in this domain.

In [180], Park et al. used a new fuzzy reasoning method to enhance the performance of fuzzy controllers designed from prior knowledge provided by an expert. As the design method mostly relied on the expert's knowledge about the system, so to avoid the subjective selection of fuzzy reasoning models, genetic algorithm was used to find simultaneously the optimal fuzzy relation matrix and the fuzzy membership functions, extending the fuzzy controller design model proposed by Pham and Karaboga in [173]. In this way, each chromosome is divided into two parts, one part representing the fuzzy relation matrix and the other part representing the fuzzy membership functions.

Similar works for designing fuzzy controllers utilizing genetic algorithm are also reported in [175, 181–190].

The survey of the genetic algorithm- based hybrid design methodologies for fuzzy controllers reveals that it can successfully solve industrial and commercial problems. The chief motivation behind this development, during last two decades, is the need for low-cost solutions that utilize intelligent methodologies for information processing, design and optimization, where the controller designs are essentially performed off-line.

1.6 State of the Art

Lyapunov theory-based strategies, in many forms, are utilized in conventional control theory for long to design controllers that can guarantee closed-loop stability of the system. Wang in [87] and Fischle and Schroder in [88] described two of the earliest design methods of Lyapunov-based direct adaptive fuzzy controllers.

In [87], Wang developed a direct adaptive fuzzy controller which guaranteed the global stability of the resulting closed-loop system in the sense that all signals involved were uniformly bounded and the convergence of the plant output towards a given reference signal is guaranteed. He also showed how a supervisory control forced the states to be within the constraint set. Despite that the controller in [87] had a drawback that it needed adaptation of controller parameters for every variation in reference signal, which implied that for temporally varying reference signals the controller parameters would never converge.

In [88], Fischle and Schroder demonstrated that a modified form of stable AFLC can be proposed where the controller parameters would converge even in presence of a temporally varying reference signal. The main problem in these methods is that it requires a considerable amount of time to track the reference input and also the inherent approximation error in striving to achieve an unknown theoretical control law leads the system towards larger integral absolute error (*IAE*). Another major drawback of many existing designs of adaptive fuzzy logic controllers is that they are developed for systems with unlimited actuation authority. It was also commented in [88] that some further developments are necessary to reduce the learning time, which tend to become unacceptably large for systems of higher order. If it is possible to overcome this drawback, stable adaptive fuzzy control has the potential to provide uncomplicated, reliable and easy-to-understand solutions for a large variety of nonlinear control objectives. However, those designs of controllers were proposed based on control laws adapting a parameter vector containing the positions of the fuzzy sets of the controller outputs only.

As mentioned before, several approaches have also been developed to utilize stochastic optimization techniques, specially genetic algorithm, to automatically design the fuzzy controllers. In [190], Giordano et al. proposed and successfully implemented stable adaptive fuzzy controller where they combined conventional Lyapunov theory and real-coded genetic algorithm to design variable structure fuzzy controller. In this seminal paper, they utilized Lyapunov theory for local adaptation of consequent parts of the T-S fuzzy system and genetic algorithm was used to determine the structure and scaling gains of the fuzzy controller. They had successfully implemented the proposed hybrid stable adaptive fuzzy controller for controlling the speed of a real DC motor. In [190], it was also presented that the stability of closed-loop system and the convergence of the plant output to a desired reference signal were precisely guaranteed.

In the last few decades, optimal control theory-based on robust control strategy has been well developed and applied extensively to cater to the robust stabilization and external disturbance rejection problems [42, 89, 112]. However, in the classical H^{∞} control strategy, the model of the plant must be known a priori. In [89], Chen et al. extended the conventional H^{∞} control strategy towards the nonlinear systems with unknown or uncertain models employing adaptive fuzzy logic concept. In [165], an adaptive fuzzy controller was introduced to guarantee the stability of the adaptive fuzzy controller. In that study the H^{∞} tracking design technique has also been utilized to attenuate the effect of the matching error and external disturbance on the tracking error. The design method, for the first time, showed the way to combine the H^{∞}-based attenuation scheme, T-S-type fuzzy approximation, and adaptive control law for the robust tracking controller design for nonlinear plants with uncertainty and/or unknown variation in plant parameters.

Over the last decade, designs of optimized adaptive fuzzy controllers for various types of nonlinear plants are of focal interest of the researchers around the globe [92, 93]. Various population-based optimization techniques are very popular for hybridization with adaptive fuzzy control strategies based on Lyapunov theory [191]. Further, H^{∞} control-based robust adaptive fuzzy controllers are also hybridized with the contemporary stochastic optimization techniques and successfully implemented for many real-life applications [165].

1.7 Summary

This chapter has introduced the concepts of basic fuzzy controllers and their different forms in a nutshell. The adaptive fuzzy controller and the stability issues, especially in light of Lyapunov theory, has also been scrutinized in brief. The H^{∞} strategy-based robust fuzzy control schemes are also discussed in brief. The developmental introduction of hybrid fuzzy controller design methodology, particularly GA-based design, has been explored too. Finally, the inspiration for the present book is put forward, the concepts of designing hybrid stable adaptive fuzzy controllers employing Lyapunov theory or H^{∞} strategy-based robust control theory and stochastic optimization techniques are introduced, and the works carried out within the scope of this book are put forward, in brief.

References

1. O. Mayr, *The Origins of Feedback Control* (MIT Press, Cambridge, 1970)
2. C.-T. Chen, *Control System Design, Orlando* (SaundersCollege Publishing, Florida, 1993)
3. L.A. Zadeh, C.A. Desoer, *Linear System Theory: A State Space Approach* (McGraw-Hill Book Company, New York, 1963)
4. K. Ogata, *State Space Analysis of Control Systems, Englewood Cliffs* (Prentice Hall, New Jersey, 1967)
5. R.R. Mohler, *Nonlinear Systems: Dynamics and Control*, vol. I (Prentice Hall, Englewood Cliffs, New Jersey, 1991)
6. B. Friedland, *Control System Design: An Introduction to State-Space Methods* (McGraw-Hill, New York, 1986)
7. G. Ellis, in *Control System Design Guide: A Practical Guide*, 3rd Ed. (Elsevier Academic Press, 2004)
8. E.J. Devison, On pole assignment in multivariable linear systems. *IEEE Trans. Autom. Control*, AC-**13**, 747–748 (Dec 1968)
9. W.M. Wonham, On pole assignment in multi-input controllable linear systems. *IEEE Trans. Autom. Control*, AC-**12**, 660–665 (Dec 1967)
10. A.G.J. MacFarlane, I. Postlethwaite, The generalized Nyquist Stability Criterion and Multivariable Root Loci. Int. J. Contr. **25**, 81–127 (1977)
11. J.R. Ragazzini, L.A. Zadeh, The Analysis of Sampled-Data Systems, *Trans. AIEE*, **71**, part II, 225–234 (1952)
12. B.C. Kuo, *Digital Control Systems*, 2nd edn. (SaundersCollege Pub, Orlando, Florida, 1992)
13. K.J. Astrom, B. Wittenmark, *Adaptive Control, Reading* (Addison-Wesley Pub. Co., Massachusetts, 1989)
14. K.S. Narendra, R.V. Monopoli (eds.), *Applications of Adaptive Control* (Academic Press, New York, 1980)
15. K. Zhou, J.C. Doyle, K. Glover, *Robust and Optimal Control* (Prentice-Hall, New Jersey, 1996)
16. B.A. Francis, in *A Course in H-infinity Control Theory, Lecture notes in Control and Information Sciences*, no. 88 (Springer-Verlag, Berlin, 1987)
17. V.I. Utkin, Variable structure systems with sliding modes, *IEEE Trans. Autom. Control* AC-**22**(2), 212–222 (Apr 1977)
18. V.I. Utkin, *Sliding Modes in Control and Optimization* (Springer-Verlag, Berlin, 1992)
19. S.H. Zak, *Systems and Control* (Oxford University Press, New York, 2003)

20. H. Kwakernaak, R. Sivan, *Linear Optimal Control Systems* (Wiley Interscience Pub, New York, 1972)
21. H.M. Power, R.J. Simpson, *Introduction to Dynamics and Control* (McGraw-Hill, London, 1978)
22. S. Haykin, in *Neural Networks: A Comprehensive Foundation*, 2nd Ed. (Prentice-Hall, 1998)
23. M. Norgaard, O. Ravn, N.K. Poulsen, L.K. Hansen, *Neural Network for Modelling and Control of Dynamic Systems* (Springer-Verlag, London, 2000)
24. L.-X. Wang, *Adaptive Fuzzy Systems and Control: Design and Stability Analysis, Englewood Cliffs* (Prentice-Hall, NJ, 1994)
25. K.M. Passino, S. Yurkovich, in *Fuzzy control*, in Handbook on Control, ed. by W. Levine (CRC, Boca Raton, FL, 1996), pp. 1001–1017
26. D. Driankov, H. Hellendoorn, M.M. Reinfrank, *An Introduction to Fuzzy Control* (Germany, Springer-Verlag, Heidelberg, 1993)
27. N.J. Nilsson, in *Artificial Intelligence: A New Synthesis* (Morgan Kaufmann Pub., 1998)
28. B. Coppin, in *Artificial Intelligence Illuminated* (Jones and Bartlett Pub., 2004)
29. A. Konar, *Computational Intelligence: Principles, Techniques and Applications* (Springer-Verlag, Berlin, 2005)
30. A.P. Engelbrecht, in *Fundamentals of Computational Swarm Intelligence* (Wiley, 2006)
31. D.E. Goldberg, *Genetic Algorithms in Search, Optimization and Machine Learning* (Kluwer Academic Publishers, Boston, 1989)
32. L.A. Zadeh, Fuzzy sets. Informat. Contr. **8**, 338–353 (1965)
33. L.A. Zadeh, Outline of a new approach to the analysis of a complex system and decision processes. IEEE Trans. Syst., Man, Cybern. **3**, 28 (1973)
34. E.H. Mamdani, Twenty years of fuzzy control: Experiences gained and lessons learnt, in *Proceedings of the 2nd IEEE International Conference on Fuzzy System* (San Francisco, USA, 1993), pp. 339–344
35. D. Chakroborty, K.D. Sharma, A study on the design methodologies of fuzzy logic controller, in *Proceedings of 17th State Science & Technology Congress, WBUAFS* (Kolkata, India, March 2010), pp. 69
36. M. Sugeno, An Introductory Survey of Fuzzy Control. Inf. Sci. **36**, 59–83 (1985)
37. E.H. Mamdani, S. Assilian, An experiment with in linguistic synthesis with a fuzzy logic controller. Int. J. Man-Machine Studies **7**, 1–13 (1975)
38. T. Abdelazim, O.P. Malik, An adaptive power system stabilizer using on-line self-learning fuzzy systems, in *Proceedings of IEEE Power Engineering Society General Meeting* (Toronto, ON, Canada, 2003), pp. 1715–1720
39. A. Flores, D. Saez, J. Araya, M. Berenguel, A. Cipriano, Fuzzypredictive control of a solar power plant. IEEE Trans. Fuzzy Syst. **13**(1), 58–68 (2005)
40. T. Guesmi, H.H. Adballah, A. Toumi, Transient stability fuzzy control approach for power systems, in *Proceedings of IEEE International Conference on Industrial Technology* (Hammamet, Tunisia, 2004), pp. 1676–1681
41. Y.H. Aoul, A. Nafaa, D. Negru, A. Mehaoua, FAFC: Fast adaptive fuzzy AQM controller for TCP/IP networks, in *Proceedings of IEEE Global Telecommunication Conference* (Dallas, TX, 2004), pp. 1319–1323
42. B.S. Chen, C.L. Tsai, D.S. Chen, Robust H-infinity and mixed H_2/ H-infinity filters for equalization designs of nonlinear communication systems: Fuzzy interpolation approach. IEEE Trans. Fuzzy Syst. **11**(3), 384–398 (2003)
43. B.S. Chen, Y.S. Yang, B.K. Lee, T.H. Lee, Fuzzy adaptive predictive flow control of ATM network traffic. IEEE Trans. Fuzzy Syst. **11**(4), 568–581 (2003)
44. A. Kandel, O. Manor, Y. Klein, S. Fluss, ATM-traffic management and congestion control using fuzzy logic. IEEE Trans. Syst., Man, Cybern. C, Appl. Rev. **29**(3), 474–480 (1999)
45. T.-H.S. Li, Y. Ying-Chieh, W. Jyun-Da, H. Ming-Ying, C. Chih-Yang, Multifunctional intelligent autonomous parking controllers for carlike mobile robots. IEEE Trans. Indust. Electron. **57**(5), 1687–1700 (2010)

46. R. Gutierrez-Osuna, J.A. Janet, R.C. Luo, Modeling of ultrasonic range sensors for localization of autonomous mobile robots. IEEE Trans. Indust. Electron. **45**(4), 654–662 (1998)
47. Z. Jinhui, X. Yuanqing, Design of static output feedback sliding mode control for uncertain linear systems. IEEE Trans. Indust. Electron. **57**(6), 2161–2170 (2010)
48. D Santosh, S. Achar, C.V, Jawahar, Autonomous image-based exploration for mobile robot navigation, in *Proceedings of IEEE International Conference on Robotics Automation ICRA 2008*, pp. 2717–2722
49. T. Das, I.N. Kar, Design and implementation of an adaptive fuzzy logic-based controller for wheeled mobile robots. IEEE Trans. Contr. Syst. Tech. **14**(3), 501–510 (2006)
50. U. Larsson, J. Forsberg, A. Wernersson, Mobile robot localization: integrating measurements from a time-of-flight laser. IEEE Trans. Indust. Electron. **43**(3), 422–431 (1996)
51. N.N. Singh, Vision Based Autonomous Navigation of Mobile Robots Ph.D. dissertation, Jadavpur University, 2010
52. S.Y. Hwang, J.B. Song, *Monocular Vision-based SLAM in indoor environment using corner, lamp, and door features from upward-looking camera* (IEEE Trans. Indust. Electron, Early Access, 2011)
53. P.P. Bonissone, Y.-T. Chen, K. Goebel, P.S. Khedkar, Hybrid soft computing systems: industrial and commercial applications. Proc. IEEE **87**(9), 1641–1667 (1999)
54. P.P. Bonissone, P.S. Khedkar, Y.-T. Chen, Genetic algorithms for automated tuning of fuzzy controllers: a transportation application, in *Proceedings of 5th IEEE International Conference on Fuzzy System* (New Orleans, USA, 1996), pp. 674–680
55. S.J. Huang, W.C. Lin, Adaptive fuzzy controller with sliding surface for vehicle suspension control. IEEE Trans. Fuzzy Syst. **11**(4), 550–559 (2003)
56. R.J.G.B. Campello, L.A.C. Meleiro, W.C. Amaral, Control of a bioprocess using orthonormal basis function fuzzy models, in *Proceedings of IEEE International Conference on Fuzzy System* (Budapest, Hungary, 2004), pp. 801–806
57. B. Chen, X. Liu, Fuzzy approximate disturbance decoupling of MIMO nonlinear systems by back stepping and application to chemical processes. IEEE Trans. Fuzzy Syst. **13**(6), 832–847 (2005)
58. C.W. Frey, H.B. Kuntze, A neuro-fuzzy supervisory control system for industrial batch processes. IEEE Trans. Fuzzy Syst. **9**(4), 570–577 (2001)
59. J.I. Horiuchi, M. Kishimoto, Application of fuzzy control to industrial bioprocesses in Japan. Fuzzy Sets Syst. **128**(1), 117–124 (2002)
60. H. Zheng, K.Y. Zhu, A fuzzy controller based multiple model adaptive control system for blood pressure control, in *Proceedings of 8th Conference Control, Automation, Robotics and Vision* (Kunming, China, 2004), pp. 1353–1358
61. H.F. Kwok, D.A. Linkens, M. Mahfouf, G.H. Mills, SIVA: a hybrid knowledge and model based advisory system for intensive care ventilators. IEEE Trans. Inform. Technol. Biomed. **8**(2), 161–172 (2004)
62. H. Seker, M.O. Odetayo, D. Petrovic, R.N.G. Naguib, A fuzzy logic based-method for prognostic decision making in breast and prostate cancers. IEEE Trans. Inform. Technol. Biomed. **7**(2), 114–122 (2003)
63. T. Haruki, K. Kikuchi, Video camera system using fuzzy logic. IEEE Trans. Consumer Electron. **38**(3), 624–634 (1992)
64. S. Kumar, A review of smart volume controllers for consumer electronics. IEEE Trans. Consumer Electron. **51**(2), 600–605 (2005)
65. S.H. Lee, Z. Bien, Design of expandable fuzzy inference processor. IEEE Trans. Consumer Electron. **40**(2), 171–175 (1994)
66. C.L. Chen, G. Feng, D. Sun, Y. Zhu, H-infinity output feedback control of discrete-time fuzzy systems with application to chaos control. IEEE Trans. Fuzzy Syst. **13**(4), 531–543 (2005)
67. K.Y. Lian, C.S. Chiu, T.S. Chiang, P. Liu, LMI-based fuzzy chaotic synchronization and communications. IEEE Trans. Fuzzy Syst. **9**(4), 539–553 (2001)

68. M. Boroushaki, M.B. Ghofrani, C. Lucas, M.J. Yazdanpanah, Identification and control of a nuclear reactor core (VVER) using recurrent neural networks and fuzzy systems. IEEE Trans. Nucl. Sci. **50**(1), 159–174 (2003)
69. S.R. Munasinghe, M.S. Kim, J.J. Lee, Adaptive neuro fuzzy controller to regulate UTSG water level in nuclear power plants. IEEE Trans. Nucl. Sci. **52**(1), 421–429 (2005)
70. G. Feng, A survey on analysis and design of model-based fuzzy control systems. IEEE Trans. Fuzzy Syst. **14**(5), 676–697 (2006)
71. B.G. Hu, G.K.I. Mann, R.G. Gosine, A systematic study of fuzzy PID controllers—Function based evaluation approach. IEEE Trans. Fuzzy Syst. **9**(5), 699–712 (2001)
72. S.Z. He, S. Tan, F.L. Xu, P.Z. Wang, Fuzzy self-tuning of PID controllers. Fuzzy Sets Syst. **56**, 37–46 (1993)
73. Z.Y. Zhao, M. Tomizuka, S. Isaka, Fuzzy gain-scheduling of PID controllers. IEEE Trans. Syst., Man, Cybern. **23**(5), 1392–1398 (1993)
74. J.S.R. Jang, ANFIS: adaptive network-based fuzzy inference system. IEEE Trans. Syst., Man, Cybern. **23**, 665–685 (1993)
75. J. Zhang, Modeling and optimal control of batch processes using recurrent neuro-fuzzy networks. IEEE Trans. Fuzzy Syst. **13**(4), 417–426 (2005)
76. R. Palm, Sliding mode fuzzy control, in *Proceedings of IEEE International Conference on Fuzzy System* (San Diego, CA, March 1992), pp. 519–526
77. P.-Y. Huang, S.-C. Lin, Y.-Y. Chen, Real-coded genetic algorithm based fuzzy sliding-mode control design for precision positioning, in *Proceedings of IEEE World Congress on Computational Intelligence* (Anchorage, AK, USA, 1998), pp. 1247–1252
78. T. Orlowska-Kowalska, M. Dybkowski, K. Szabat, Adaptive sliding-mode neuro-fuzzy control of the two-mass induction motor drive without mechanical sensors. IEEE Trans. Indust. Electron. **57**(2), 553–564 (2010)
79. H.J. Lee, J.B. Park, G. Chen, Robust fuzzy control of nonlinear systems with parametric uncertainties. IEEE Trans. Fuzzy Syst. **9**(2), 369–379 (2001)
80. H. Han, C.-Y. Su, Robust fuzzy control of nonlinear systems using shape-adaptive radial basis functions, *Fuzzy Sets & Syst.* **125**, 23–28 (2002)
81. N. Sadati, A. Talasaz, A robust fuzzy sliding mode control for uncertain dynamic systems, in *Proceedings of IEEE International Conference on System, Man Cybernetics, Oct 2004*, pp. 2225–2230
82. R. Palm, Sliding mode fuzzy control, in *Proceedings of IEEE International Conference on Fuzzy System* (1992), pp. 519–526
83. S. Isaka, A. Sebald, A. Karimi, N. Smith, M. Quinn, On the design and performance evaluation of adaptive fuzzy controllers, in *Proceedings of IEEE Conference on Decision and Control* (Austin, TX, Dec 1988), pp. 1068–1069
84. L.-X. Wang, Stable adaptive fuzzy control of nonlinear systems, in *Proceedings of 31st Conference on Decision Control* (Tucson, AZ, 1992), pp. 2511–2516
85. Y.P. Glorennec, Adaptive fuzzy control, in *Proceedings of 4th IFSA Congress* (Brussels, 1991), vol. Eng., pp. 33–36
86. C.-Y. Su, Y. Stepanenko, Adaptive control of a class of nonlinear systems with fuzzy logic. IEEE Trans. Fuzzy Syst. **2**, 285–294 (1994)
87. L.X. Wang, Stable adaptive fuzzy control of nonlinear system, *IEEE Trans. Fuzzy Syst.* **1**(2), 146–155 (May 1993)
88. K. Fischle, D. Schroder, An improved stable adaptive fuzzy control method. IEEE Trans. Fuzzy Syst. **7**(1), 27–40 (1999)
89. B.S. Chen, C.H. Lee, Y.C. Chang, H-infinity tracking design of uncertain nonlinear SISO systems: adaptive fuzzy approach. IEEE Trans. Fuzzy Syst. **4**(1), 32–43 (1996)
90. D.L. Tsay, H.Y. Chung, C.J. Lee, The adaptive control of nonlinear systems using the Sugeno-type of fuzzy logic. IEEE Trans. Fuzzy Syst. **7**(2), 225–229 (1999)
91. T. Wang, S. Tong, Adaptive fuzzy output feedback control for SISO nonlinear systems. Int. J. Innovative Comput. Info Control **2**(1), 51–60 (2006)

92. K.D. Sharma, A. Chatterjee, F. Matsuno, A Lyapunov theory and stochastic optimization based stable adaptive fuzzy control methodology, in *Proceedings of SICE Annual Conference 2008*, Japan, pp. 1839–1844, August 20–22
93. K. Das Sharma, A. Chatterjee, A. Rakshit, A hybrid approach for design of stable adaptive fuzzy controllers employing Lyapunov theory and particle swarm optimization. IEEE Trans. Fuzzy Syst. **17**(2), 329–342 (2009)
94. R. Ordonez, J. Zumberge, J.T. Spooner, K.M. Passino, Adaptive fuzzy control: experiments and comparative analyses. IEEE Trans. Fuzzy Syst. **5**(2), 167–188 (1997)
95. L.-X. Wang, in *A course in fuzzy systems and control* (Englewood Cliffs, Prentice Hall, NJ, 1997)
96. T. Procyk, E. Mamdani, A linguistic self-organizing process controller. Automatica **15**(1), 15–30 (1979)
97. J.R. Layne, K.M. Passino, Fuzzy model reference learning control. J. Intell. Fuzzy Syst. **4** (1), 33–47 (1996)
98. J.R. Layne, K.M. Passino, Fuzzy model reference learning control, in *Proceedings of 1st IEEE Conference on Control Applications* (Dayton, OH, Sept 1992), pp. 686–691
99. J.R. Laync, K.M. Passino, Fuzzy model reference learning control for cargo ship steering. IEEE Contr. Syst. Mag. **13**, 23–34 (1993)
100. J. Layne, K. Passino, S. Yurkovich, Fuzzy learning control for antiskid braking systems, in *Proceedings of IEEE Conference on Decision Control* (Tucson, AZ, Dec 1992), pp. 2523–2528
101. W.A. Kwong, K.M. Passino, Dynamically focused fuzzy learning control. IEEE Trans. Syst., Man, Cybern. **26**, 53–74 (1996)
102. W. Lennon, K. Passino, Intelligent control for brake systems, in *Proceedings of IEEE International Symposium on Intelligent Control* (Monterey, CA, Aug 1995), pp. 499–504
103. J. Zumberge, K.M. Passino, A case study in intelligent control for a process control experiment, in *Proceedings of IEEE International Symposium on Intelligent Control* (Dearborn, MI, Sept 1996)
104. V.G. Moudgal, W.A. Kwong, K.M. Passino, S. Yurkovich, Fuzzy learning control for a flexible-link robot. IEEE Trans. Fuzzy Syst. **3**, 199–210 (1995)
105. S. Shao, Fuzzy self-organizing controller and its application for dynamic processes. Fuzzy Sets Syst. **26**, 151–164 (1988)
106. R. Tanscheit, E. Scharf, Experiments with the use of a rule-based self-organizing controller for robotics applications. Fuzzy Sets Syst. **26**, 195–214 (1988)
107. T. Yamazaki, An improved algorithm for a self-organizing controller and its experimental analysis, Ph.D. Thesis, London University, 1982
108. T.A. Johansen, Fuzzy model based control: stability, robustness, and performance issues. IEEE Trans. Fuzzy Syst. **2**, 221–234 (1994)
109. R. Chekkouri, L. Romeral, M. Moreno, Direct adaptive fuzzy controller by changing the output membership functions for nonlinear motion applications, *IEEE ICIT 2003-Maribor*, (Slovenia, 2000), pp. 22–26
110. K. Tanaka, Design of model-based fuzzy controller using Lyapunov's stability approach and its application to trajectory stabilization of a model car, in *Theoretical Aspects of Fuzzy Control*, ed. by H.T. Nguyen, M. Sugeno, R. Tong, R.R. Yager (Wiley, New York, 1995), pp. 31–50
111. S.S. Chen, J.L. Wu, S.F. Su, J.C. Chang, Y.C. Chang, T.T. Lee, On the analysis of direct adaptive fuzzy controllers, in *Proceedings of 3rd International Conference on Machine Learning. Cybernetics* (Shanghai, 2004), pp. 4368–4373
112. H.-X. Li, S. Tong, A hybrid adaptive fuzzy control for a class of nonlinear MIMO system. IEEE Trans. Fuzzy Syst. **11**(1), 24–34 (2003)
113. R.E. Kalman, J.E. Bertram, Control System Analysis and Design via the 'Second Method' of Lyapunov. I. Continuous-time Systems, *Trans. ASME J. Basic Eng.*, pp. 371–393, June 1960

114. S. Chand, S. Hansen, Energy based stability analysis of a fuzzy roll controller design for a flexible aircraft wing, in *Proceedings of IEEE Conference on Decision and Control* (Tampa, FL, Dec 1989), pp. 705–709
115. S. Chiu, S. Chand, Fuzzy controller design and stability analysis for an aircraft model, in *Proceedings of the American Control Conference* (Boston, June 1991), pp. 821–826
116. G. Langari, A Framework for Analysis and Synthesis of Fuzzy Linguistic Control Systems, Ph.D. thesis, University of California, Berkeley, 1990
117. G. Langari, M. Tomizuka, Stability of fuzzy linguistic control systems, in *Proceedings of IEEE Conference on Decision and Control* (Honolulu, HI, Dec 1990), pp. 2185–2190
118. K. Narendra, J. Taylor, *Frequency Domain Criteria for Absolute Stability* (Academic Press, New York, 1973)
119. J. Aracil, A. Ollero, A. Garcia-Cerezo, Stability indices for the global analysis of expert control systems, *IEEE Trans. Syst., Man, Cybern.* **19**(5), 998–1007 (Sep/Oct 1989)
120. S.S. Farinwata, G. Vachtsevanos, Stability analysis of the fuzzy controller designed by the phase portrait assignment algorithm, in *Proceedings of 2nd IEEE International Conference on Fuzzy Systems* (San Francisco, CA, Mar 1993), pp. 1377–1382
121. C.J. Harris, C.G. Moore, Phase plane analysis tools for a class of fuzzy control systems, in *Proceedings of IEEE International Conference on Fuzzy Systems* (San Diego, CA, Mar 1992), pp. 511–518
122. K.S. Ray, A.M. Ghosh, D.D. Majumder, L_2-stability and the related design concept for SISO linear systems associated with fuzzy logic controllers, *IEEE Trans. Syst., Man, Cybern.* **14** (6), 932–939 (Nov/Dec 1984)
123. K.S. Ray, D.D. Majumder, Application of circle criteria for stability analysis of linear SISO and MIMO systems associated with fuzzy logic controllers, *IEEE Trans. Syst., Man, Cybern.* **14**(2), 345–349 (Mar/Apr 1984)
124. R. Palm, D. Driankov, H. Hellendoorn, *Model Based Fuzzy Control* (Springer-Verlag, New York, 1997)
125. T. Takagi, M. Sugeno, Fuzzy identification of systems and its applications to modeling and control. IEEE Trans. Syst., Man, Cybern. **15**(1), 116–132 (1985)
126. K. Tanaka, T. Ikeda, H.O. Wang, Robust stabilization of a class of uncertain nonlinear systems via fuzzy control: Quadratic stabilizability, H-infinity control theory, and linear matrix inequalities. IEEE Trans. Fuzzy Syst. **4**(1), 1–13 (1996)
127. K. Tanaka, M. Sano, Concept of stability margin for fuzzy systems and design of robust fuzzy controllers, in *Proceedings of IEEE International Conference on Fuzzy Systems* (San Francisco, CA, Mar 1993), pp. 29–34
128. K. Tanaka, M. Sugeno, Stability analysis and design of fuzzy control systems. Fuzzy Sets Syst. **45**, 135–156 (1992)
129. K. Tanaka, M. Sano, A robust stabilization problem of fuzzy control systems and its application to backing up control of a truck-trailer. IEEE Trans. Fuzzy Syst. **2**(2), 119–134 (1994)
130. H.O. Wang, K. Tanaka, An LMI-based stable fuzzy control of nonlinear systems and its application to the control of chaos, in *Proceedings of IEEE International Conference on Fuzzy Systems* (New Orleans, LA, Sept, 1996), pp. 1433–1439
131. H.K. Khalil, *Nonlinear Systems*, 2nd edn. (Prentice-Hall, Englewood Cliffs, NJ, 1996)
132. D.F. Jenkins, Nonlinear analysis of fuzzy control systems, Master's Thesis, Department of Electrical Engineering, The Ohio State University, 1993
133. D.P. Atherton, *Nonlinear Control Engineering: Describing Function Analysis and Design* (Berkshire, England, 1982)
134. D. Jenkins, K.M. Passino, An introduction to nonlinear analysis of fuzzy control systems. *J. Intell. Fuzzy Systems* **7**(1), 75–103 (1999)
135. D.P. Atherton, A describing function approach for the evaluation of fuzzy logic control, in *Proceedings of American Control Conference* (San Francisco, CA, June 1993), pp. 765–766
136. W.J.M. Kickert, E.H. Mamdani, Analysis of a fuzzy logic controller. Fuzzy Sets Syst. **1**, 29–44 (1978)

137. H.O. Wang, K. Tanaka, M.F. Griffin, An approach to fuzzy control of nonlinear systems: stability and design issues. IEEE Trans. Fuzzy Syst. **4**(1), 14–23 (1996)
138. Y.-Y. Chen, The Global Analysis of Fuzzy Dynamical Systems, Ph.D. thesis, University of California, Berkeley, 1989
139. Y.Y. Chen, T.C. Tsao, A description of the dynamical behavior of fuzzy systems, *IEEE Trans. Syst., Man, Cybern.* **19**(4), 745–755 (Jul/Aug 1989)
140. J. Fei, C. Isik, The analysis of fuzzy knowledge-based systems using cell-to-cell mapping, in *Proceedings of IEEE International Symposium on Intelligent Control* (Philadelphia, Sept 1990), pp. 633–637
141. C.S. Hsu, A theory of cell-to-cell mapping dynamical systems. *Trans. ASME, J. Applied Mech.* **47**, 931–939 (Dec 1980)
142. C.S. Hsu, R.S. Guttalu, An unraveling algorithm for global analysis of dynamical systems: an application of cell-to-cell mapping. *Trans. ASME, J. Applied Mech.* **47**, 940–948 (Dec 1980)
143. G. Feng, Stability analysis of discrete time fuzzy dynamic systems based on piecewise Lyapunov functions. IEEE Trans. Fuzzy Syst. **12**(1), 22–28 (2004)
144. M. Johansson, A. Rantzer, K.E. Arzen, Piecewise quadratic stability of fuzzy systems. IEEE Trans. Fuzzy Syst. **7**(6), 713–722 (1999)
145. S.G. Cao, N.W. Rees, G. Feng, Stability analysis and design for a class of continuous-time fuzzy control systems. Int. J. Contr. **64**, 1069–1087 (1996)
146. S.G. Cao, N.W. Rees, G. Feng, H-infinity control of nonlinear continuous-time systems based on dynamic fuzzy models. Int. J. Syst. Sci. **27**, 821–830 (1996)
147. S.G. Cao, N.W. Rees, G. Feng, Analysis and design for a class of complex control systems-Part I: fuzzy modeling and identification. Automatica **33**, 1017–1028 (1997)
148. D.J. Choi, P.G. Park, H-infinity state-feedback controller design for discrete-time fuzzy systems using fuzzy weighting-dependent Lyapunov functions. IEEE Trans. Fuzzy Syst. **11** (2), 271–278 (2003)
149. T.M. Guerra, L. Vermeiren, LMI-based relaxed nonquadratic stabilization conditions for nonlinear systems in the Takagi–Sugeno's form. Automatica **40**(5), 823–829 (2004)
150. K. Tanaka, T. Hori, H.O. Wang, A multiple Lyapunov function approach to stabilization of fuzzy control systems. IEEE Trans. Fuzzy Syst. **11**(4), 582–589 (2003)
151. J.W. Curtis, R.W. Beard, A graphical understanding of Lyapunov based nonlinear control, in *Proceedings of 41st IEEE Conference on Decision and Control* (Las Vegas, Nevada USA, Dec 2002)
152. W. Assawinchaichote, S.K. Nguang, Fuzzy H^{∞} output feedback control design for singularly perturbed systems: an LMI approach, in *Proceedings of 42nd IEEE Conference on Decision and Control* (Hawaii, USA, December 2003), pp. 863–868
153. X. Zhengrogn, C. Qingwei, H. Weili, Robust H control for fuzzy descriptor systems with state delay. IEEE J. Syst. Eng. Electron. **16**(1), 98–101 (2005)
154. S.K. Nguang, W. Assawinchaichote, P. Shi, Y. Shi, Robust H^{∞} control design for uncertain fuzzy systems with Markovian jumps: an LMI approach, in *Proceedings of American Control Conference* (Portland USA, June 2005), pp. 1805–1810
155. T. Agustinah, A. Jazidie, Md Nuh, Fuzzy tracking control based on H^{∞} performance for nonlinear systems. WSEAS Trans. Syst. Control **6**(11), 393–403 (2011)
156. S.-M. Wu, C.-C. Sun, H.-Y. Chung, W.-J. Chang, Mixed H_2/H_{∞} region-based fuzzy controller design for continuous-time fuzzy systems. J. Intelligent Fuzzy Syst. **18**, 19–30 (2007)
157. H.-N. Wu, H.-X Li, H^{∞} fuzzy observer-based control for a class of nonlinear distributed parameter systems with control constraints, *IEEE Trans. Fuzzy Syst.* **16**(2), 502–516 (April 2008)
158. Y. Guan, S. Zhou, W.X. Zheng, H^{∞} dynamic output feedback control for fuzzy systems with quantized measurements, in *Proceedings of Joint 48th IEEE Conference on Decision and Control and 28th Chinese Control Conference* (Shanghai, P.R. China, Dec 2009), pp. 7765–7770

159. J. Pan, S. Fei, Y. Ni, M. Xue, New approaches to relaxed stabilization conditions and H^∞ control designs for T-S fuzzy systems. J Control Theory Appl. **10**(1), 82–91 (2012)
160. H. Yishao, Z. Dequn, C. Xiaoxui, D. Lin, H^∞ tracking design for a class of decentralized nonlinear systems via fuzzy adaptive observer. IEEE J. Syst. Eng. Electron. **20**(4), 790–799 (2009)
161. W. Miaoxin, L. Jizhen, L. Xianqjie, L. Juncheng, Robust H^∞ controller design for a class of nonlinear fuzzy time-delay systems. IEEE J. Syst. Eng. Electron. **20**(6), 1286–1289 (2009)
162. B.-S. Chen, Y.-P. Lin, Y.-J. Chuang, Robust H^∞ observer-based tracking control of stochastic immune systems under environmental disturbances and measurement noises. Asian J. Control **13**(5), 667–690 (2011)
163. G. Chao, L. Xiao-Geng, W. Jun-Wei, W. Fei, Robust H^∞ decentralized fuzzy tracking control for bank-to-turn missiles, in *Proceedings of 32nd Chinese Control Conference*, July 26–28, 2013, China, pp. 3498–3503
164. S. Park, H. Ahn, Robust T-S fuzzy H^∞ control of interior permanent magnet motor using weighted integral action, in *Proceedings of 2014 14th International Conference on Control, Automation and Systems (ICCAS 2014)*, Oct. 22–25, 2014, Korea, pp. 1504–1507
165. K.D. Sharma, A. Chatterjee, P. Siarry, A. Rakshit, CMA—H^∞ hybrid design of robust stable adaptive fuzzy controllers for non-linear systems, in *Proceedings of 1st International Conference on Frontiers in Optimization: Theory and Applications (FOTA-2016)*, November 2016, India
166. J. Xiong, X.-H. Chang, Observer-based robust non-fragile H^∞ control for uncertain T-S fuzzy singular systems, in *Proceedings 36th Chinese Control Conference*, July 26–28, 2017, China, pp. 4233–4238
167. S. Sumathi, P. Surekha, *Computational Intelligence Paradigms* (CRC Press, 2010)
168. J.-S.R. Jang, C.-T. Sun, E. Mizutani, *Neuro-Fuzzy and Soft Computing* (Prentice-Hall, Englewood Cliffs, NJ, 1997)
169. C. Karr, Genetic algorithms for fuzzy controllers, *AI Expert*, pp. 26–33, Feb. 1991
170. M. Valenzuela-Rendon, The fuzzy classifier system: a classifier system for continuously varying variables, in *Proceedings of 4th International Conference Genetic Algo.*, 1991, pp 346–353
171. M. Valenzuela-Rendon, Reinforcement learning in the fuzzy classifier system. Expert Syst. Appl. **14**, 237–247 (1998)
172. P. Thrift, Fuzzy logic synthesis with genetic algorithms, in *Proceedings 4th International Conference on Genetic Algo.*, 1991, pp 509–513
173. D.T. Pham, D. Karaboga, Optimum design of fuzzy logic controllers using genetic algorithms. J. Syst. Eng. **1**, 114–118 (1991)
174. H.-S. Hwang, Control strategy for optimal compromise between trip time and energy consumption in a high-speed railway. IEEE Trans. Syst., Man, Cybern. **28**(6), 791–802 (1998)
175. M. Shim, S. Seong, B. Ko, M. So, Application of evolutionary computation at LG electronics, in *Proceedings of IEEE International Conference Fuzzy Systems*, Seoul, South Korea, 1999, pp. 1802–1806
176. R. Alcala, J.M. Benittez, J. Casillas, O. Cordon, R. Perez, Fuzzy control of HVAC systems optimised by genetic algorithms. Appl Intelligence **18**, 155–177 (2003)
177. A. Homaifar, E. McCormick, Simultaneous design of membership functions and rule sets for fuzzy controllers using genetic algorithms. IEEE Trans. Fuzzy Syst. **3**(2), 129–139 (1995)
178. H. Ishibuchi, K. Nozaki, N. Yamamoto, H. Tanaka, Selection fuzzy IF-THEN rules for classification problems using genetic algorithms. IEEE Trans. Fuzzy Syst. **3**(3), 260–270 (1995)
179. M. Setnes, H. Roubos, GA-fuzzy modeling and classification: complexity and performance. IEEE Trans. Fuzzy Syst. **8**(5), 509–522 (2000)
180. D. Park, A. Kandel, G. Langholz, Genetic-based new fuzzy reasoning models with applications to fuzzy control. IEEE Trans. Syst., Man, Cybern. **24**(1), 39–47 (1994)

181. F. Herrera, M. Lozano, J.L. Verdegay, Tuning Fuzzy Logic Controllers by Genetic Algorithms. *Int. J. Approx. Reasoning* **12**, 299–315 (1995)
182. K. Kropp, U.G. Baitinger, Optimization of fuzzy logic controller inference rules using a genetic algorithm, in *Proceedings 1st European Congress Fuzzy and Intelligent Technology*, Aachen, 1993, pp. 1090–1096
183. T. Hessburg, M. Lee, H. Takagi, M. Tomizuka, Automatic design of fuzzy systems using genetic algorithms and its application to lateral vehicle guidance, *SPIE's Int. Symp. Optical Tools Manufac. and Adv. Automation*, vol. 2061, Boston, 1993
184. A. Varsek, T. Urbancic, B. Filipic, Genetic algorithms in controller design and tuning, *IEEE Trans. Syst., Man, Cybern.* **23**, 1330–1339 (1993)
185. S. Voget, M. Kolonko, Multidimensional optimization with a fuzzy genetic algorithm. J. Heuristics **4**(3), 221–244 (1998)
186. F. Herrera, M. Lozano, Adaptation of genetic algorithm parameters based on fuzzy logic controllers, in *Genetic Algo. and Soft Comp.*, Physica-Verlag, Wurzburg, pp. 95–125, 1996
187. I.G. Damousis, P. Dokopoulos, A fuzzy expert system for the forecasting of wind speed and power generation in wind farms, in *Proceedings of International Conference on Power Industry Computer Application, 2001* (Sydney, Australia, 2001) pp. 63–69
188. F. Cupertino, E. Mininno, D. Naso, B. Turchiano, L. Salvatore, On-line genetic design of anti-windup unstructured controllers for electric drives with variable load. IEEE Trans. Evol. Comput. **8**(4), 347–364 (2004)
189. C.L. Chiang, C.T. Sub, Tracking control of induction motor using fuzzy phase plane controller with improved genetic algorithm. *Elect. Power Syst. Res., Elsevier*, **73**, 239–247 (2005)
190. V. Giordano, D. Naso, B. Turchiano, Combining genetic algorithm and Lyapunov-based adaptation for online design of fuzzy controllers, *IEEE Trans. Syst., Man, Cybern. B Cybern.* **36**(5), 1118–1127 (Oct 2006)
191. K.D. Sharma, *Hybrid methodologies for stable adaptive fuzzy control*, Doctoral Dissertation, Jadavpur University, India, 2012

Chapter 2
Some Contemporary Stochastic Optimization Algorithms: A Glimpse

2.1 Introduction

Over the last decade, the applications of population-based stochastic optimization methods, like GA [1, 2], PSO [3, 4], CMA [5], HS algorithm [6], GSA [7] in the area of control system design have displayed very promising outcomes. Stochastic optimization methods are general-purpose procedures, conceptualized mostly as empirical evolutionary laws and often mimic the nature-inspired strategies [8, 9]. Moreover, such optimization algorithms are very flexible and they can be efficaciously applied to several varieties of function optimizations with constraints [10]. Though these algorithms generally find a solution which is near optimal, but the solution usually suffices the approximations required for that optimization problem. Again, these population-based stochastic algorithms can be utilized to solve the problems with n number of parameters to be optimized, and the algorithms are not much sensitive to the initial parameter guesses. Sometimes, the dimension of n may be quite large, but that does not pose any negative impact on the performance of the algorithms except the computational cost. One of the key advantages of these multi-agent, stochastic, metaheuristic algorithms is that these algorithms do not require to calculate the derivatives of the fitness functions to proceed and these are theoretically capable of finding the global optimum [10, 11]. In this chapter, the most widely used population-based algorithm, i.e. GA, and other three algorithms utilized in this book, i.e., PSO, CMA, and HSA, are described in brief. Readers may refer [9, 10] for more similar kinds of optimization methods.

2.2 Genetic Algorithm (GA)

The principle of genetic algorithm (GA) is based on Darwinian's theory of survival of the fittest and mimics the metaphor of natural biological evolution. It was first proposed by John Holland in early 1960s [1, 2]. GA is a stochastic global search

K. Das Sharma et al., *Intelligent Control*, Cognitive Intelligence and Robotics,
https://doi.org/10.1007/978-981-13-1298-4_2

technique and operates on a population of potential candidate solutions, termed as gene pool, applying the principle of survival of the fittest to produce optimal/ near-optimal solutions. At each generation, a new set of candidate solutions is generated by selecting individuals according to their fitness values in the problem domain, as set by the designer. Then, the reproduction process is initiated among the selected candidate solutions, employing the biological operators motivated from natural genetics. This process directs to the evolution of populations of individuals that are better suited to the environment [10].

According to the GA taxonomy, the individuals, or candidate solutions, are encoded as strings, termed as chromosomes, and the chromosome values, i.e., the genotypes, are mapped onto a corresponding decision variable, i.e., phenotypic domain. Traditionally, the chromosomes were represented in binary values such as "0" or "1". Later, other types of representations of chromosome have also been extensively used, like ternary, integer, real-valued. The decision variables are effectively generated by decoding the chromosome, and then, they are employed to construct the potential candidate solutions. A candidate solution is utilized to evaluate the fitness level for it, and the fitness levels of all individual candidate solutions are used to select the parents. The parents are then employed to reproduce the offspring for their evaluation in the next generation. During the reproduction phase, each individual candidate solution has got a fitness value evaluated using the objective function set by the designer. These fitness values play an important role in selection of the parents for mating purpose. The candidate solutions with high fitness values compared to the entire gene pool have a greater probability of being selected for mating, whereas less fit candidate solutions have a low probability of being selected accordingly. This feature of GA is in full conformity with the survival of the fittest theory proposed by Charles Darwin [1].

Once a section of candidate solutions gets selected as parents, the genetic operators are applied on them to obtain the offspring. Genetic operators are able to manipulate the chromosomes in conformation with the philosophy conceived from the nature. The genetic or recombination operators are used to exchange genetic information between pairs, or larger groups, of individuals. The simplest recombination operator is that of single-point crossover [2, 10].

Let us consider the two-parent candidate solutions represented as binary strings of 8 bit [2, 10]:

$$\begin{array}{lc|cccccccc} P1 & = & 0 & 1 & 1 & 0 & 1 & 1 & 0 & 0 \\ P2 & = & 1 & 0 & 1 & 1 & 0 & 0 & 0 & 1 \end{array}.$$

Let i be a randomly selected integer, indicating a position in the string of length l (here it is 8), and the genetic information is exchanged between the individual candidate solutions about this point. Then, the two offspring candidate solutions are produced. Now, if $i = 3$, then the two child chromosome will be generated as follows [2, 10]:

$$\begin{array}{lc|cccccccc} C1 & = & 0 & 1 & 1 & 1 & 0 & 0 & 0 & 1 \\ C2 & = & 1 & 0 & 1 & 0 & 1 & 1 & 0 & 0 \end{array}.$$

Algorithm 1: GA

BEGIN

Set fitness function $fit(\underline{z})$, $z_d \in Z_d$, $d = 1,2,\ldots,n$, $\underline{z}$ is a candidate solution vector comprising the set of decision variable z_d, $\underline{z} = [z_1, z_2, \cdots, z_n]^T \in R^n$, Z_d is the universe of discourse of the dth decision variable z_d i.e. $z_{d-\min} \le Z_d \le z_{d-\max}$ and n is the number of decision variable.

Create a mating pool of N chromosomes and make randomized initialization of their n dimensional positions Z;

FOR $i = 1$ to N

  Calculate fitness: $fit_i(k)$;

END FOR

Initialize iteration number $k = 1$;

REPEAT:

  FOR $i = 1$ to N

    Select two parents from mating pool for reproduction with probability of selection: $\frac{fit_i(k)}{\sum_i fit_i(k)}$;

    Carry out crossover operation to recombine parents for two offspring using crossover probability C_r;

    Carry out mutation operation to using mutation probability M_r;

    Calculate fitness of offspring: $fit_{i0}(k)$;

    Insert offspring in new generation as: $fit_{i0}(k) \leftarrow \max_{i \in \{1,2,\cdots,N\}} fit_i(k)$, if $fit_{i0}(k) < \max_{i \in \{1,2,\cdots,N\}} fit_i(k)$;

  END FOR

UNTIL termination criterion is satisfied

END

The crossover operation is performed with crossover probability C_r i.e. this genetic operation is not applied to the entire population or gene pool. Depending on the crossover probability, the pairs are chosen for mating to produce offspring. The crossover operation thus introduces stochasticity in the optimization process. Once the crossover operation is over, another genetic operator, called mutation, is applied to the new offspring chromosomes. The mutation operator is also applied with a probability M_r. Mutation operation changes the individual genetic representation of a chromosome according to a probabilistic rule. In binary GA, the mutation operates on a single bit to toggle its state, i.e., 0 to 1 or vice versa. For example, if the offspring $C1$ is chosen for mutation operation considering the probability of mutation M_r, and the second bit of $C1$ has to toggle its state, then the mutated chromosome $C1m$ is given as follows [2, 10]: $C1m$ = 0 **0** 1 1 0 0 0 1.

The mutation operation produces a new candidate solution, and this genetic operator is meant for fine movement of the candidate solution near the optimal solution point. However, this may have the effect of tending the candidate solution to converge to a local optimum, instead of global optimum.

After the application of genetic operators, like crossover and mutation, each candidate solution once again undergoes evaluation of the objective function and a fitness value is assigned to that individual candidate solution. Subsequently, the individuals are selected for reproduction, based on their fitness values and the procedure continues in the subsequent generations. Thus, the performance of the entire gene pool is likely to get improved, because population of good solutions will be increased and they are utilized for mating with each other and thus creating better offspring in future generations. Along with this feature, GA also intends to reject such candidate solutions which have poor fitness values, over the generations. The GA will stop searching the solution space when some stopping criteria are satisfied, such as reaching to a maximum iteration $iter_{\max}$ or the solution reaches within pre-specified error criteria [2, 10–12]. The basic GA, conceptualized as a minimization problem, is presented in a concise form in Algorithm 1.

2.3 Particle Swarm Optimization (PSO)

Particle swarm optimization (PSO), originally developed by Kennedy and Eberhart in 1995, is a population-based swarm algorithm [3, 4]. In the PSO computational algorithm, population dynamics simulates bio-inspired behavior, i.e., a "bird flock's" behavior which involves sharing of information and allows particles to take profit from the discoveries and previous experience of all other particles during the search of food. Each particle in PSO has a randomized velocity associated with it, which moves through the search space. Each particle in PSO keeps track of its coordinates in the solution space, which are associated with the best solution (fitness) it has achieved so far. This value is called *pBest* (personal best) of a particle. Another best value that is tracked by the global version of the particle swarm optimizer is the overall best value (fitness). Its location, called *gBest* (global best), is obtained from all the particles in the population. The past best position of the particle itself and the best overall position in the entire swarm are employed to obtain new position for the particle in quest to minimize (or maximize) the fitness. The PSO concept consists, in each time step, of changing the velocity of each particle flying towards its *pBest* and *gBest* location. The velocity is weighted by random terms, with separate random numbers being generated for velocities towards *pBest* and *gBest* locations respectively [10, 13–21].

Now, let the swarm size be given by N, each particle i ($1 \le i \le N$) represent a trial solution with $d = 1,2,\ldots,n$ parameters. Each particle, at kth instant, has a current

---

Algorithm 2: PSO (global version)

BEGIN

Setfitness function $fit(\underline{z})$, $z_d \in Z_d$, $d = 1,2,\ldots,n$, $\underline{z}$ is a candidate solution vector comprising the set of decision variable z_d, $\underline{z} = [z_1, z_2, \cdots, z_n]^T \in R^n$, Z_d is the universe of discourse of the dth decision variable z_d i.e. $z_{d-\min} \le Z_d \le z_{d-\max}$ and n is the number of decision variable.

Create N particles and make randomized initialization of their n dimensional positions Z.

Initialize iteration number $k = 1$;

REPEAT:

FOR $i = 1$ to N

Calculate fitness: $fit_i(k)$;

Calculate personal best fitness for each particle: $p_i(k) = \min_{i \in \{1,2,\cdots,N\}} fit_i(k)$;

Calculate global best fitness: $g(k) = \min_{i \in \{1,2,\cdots,N\}} p_i(k)$;

Calculate the inertia weight: $w(k) = \dfrac{(iter_{\max} - k) * (w_{start} - w_{end})}{iter_{\max}} + w_{end}$;

FOR $d = 1$ to n

Calculate the velocity update:

$v_{i,d}(k+1) = w(k)v_{i,d}(k) + c_1 r_1 (p_{i,d}(k) - z_{i,d}(k)) + \ldots$
$c_2 r_2 (g(k) - z_{i,d}(k))$

Calculate the position update: $z_{i,d}(k+1) = z_{i,d}(k) + v_{i,d}(k+1)$;

END FOR

END FOR

UNTIL termination criterion is satisfied;

END

---

position $z_i(k) = (z_{i,1}(k), z_{i,2}(k), \ldots, z_{i,j}(k), \ldots, z_{i,n}(k))$ in the search space, a current velocity $v_i(k) = (v_{i,1}(k), v_{i,2}(k), \ldots, v_{i,j}(k), \ldots, v_{i,n}(k))$ and a personal best position $p_i(k) = (p_{i,1}(k), p_{i,2}(k), \ldots, p_{i,j}(k), \ldots, p_{i,n}(k))$ in the search space. Let us assume that, to solve a given problem, a fitness function fit_i for the ith particle need be minimized, and then, the velocities of all particles in a swarm are updated, at each iteration, as follows [10]:

$$v_{i,d}(k+1) = w(k)v_{i,d}(k) + c_1 r_1 (p_{i,d}(k) - z_{i,d}(k)) + c_2 r_2 (g(k) - z_{i,d}(k)) \quad (2.1)$$

where $g(k)$ denotes the global best particle, c_1 and c_2 are the "trust" parameters and are usually set to 2, which control the velocity change of a particle in a single iteration, and $r_1 \to U(0,1)$, and $r_2 \to U(0,1)$ are elements from two uniform random sequences in the interval [0,1]. The term $w(k)$ is the inertia weight of the particle at kth iteration, as introduced by Shi and Eberhart [16] and is given by

$$w(k) = \frac{(iter_{\max} - k) * (w_{\text{start}} - w_{\text{end}})}{iter_{\max}} + w_{\text{end}} \tag{2.2}$$

where $iter_{\max}$ is the maximum number of iterations of PSO. Normally, $w(k)$ is reduced linearly from w_{start} to w_{end} throughout the PSO generation. A typical combination is $w_{\text{start}} = 0.9$ and $w_{\text{end}} = 0.4$.

Now the update formulation for each particle's position vector is given as follows [8, 10]:

$$z_i(k+1) = z_i(k) + v_i(k+1) \tag{2.3}$$

The personal best particle position $p_i(k)$ and the global best particle swarm position $g(k)$ are updated by the following equations, respectively, as follows:

$$p_i(k+1) = \begin{cases} p_i(k) & \text{if } f(z_i(k+1)) \geq fit(p_i(k)) \\ z_i(k+1) & \text{otherwise} \end{cases} \tag{2.4}$$

$$g(k+1) = \arg\min\,(fit(p_i(k+1))) \tag{2.5}$$

where $\arg\min(\cdot)$ denotes the minimum argument and $fit(\cdot)$ is a fitness function.

The values of each component in every v_i vector can be clamped to the range $[-v_{\max}; v_{\max}]$ in order to reduce the likelihood of particles leaving the search space. This mechanism does not restrict the values of z_i in the range of v_i, it only limits the maximum distance that a particle will move during each iteration [18–20]. The PSO algorithm (global version), conceptualized as a minimization problem, is given in Algorithm 2.

2.4 Covariance Matrix Adaptation (CMA)

Covariance matrix adaptation was introduced by Hansen, Ostermeier, and Gawelczyk [5] in 1995, and further improvements were developed by Hansen et al. in 2001 [22] and 2003 [23]. Originally, CMA was designed for small population sizes and was interpreted as a robust local search strategy [5, 22]. The drawbacks of the basic CMA were the complexity of the adaptation process and the dependence on the design parameter formulae for which very little theoretical knowledge is available. In [23], the rank-μ-update version of CMA was proposed and it exploits the advantage of large population size without affecting the performance for small population size. In 2005, Hansen et al. proposed restart version [24] of CMA which arguably proved the best-performing evolutionary algorithm for continuous optimization for a set of test functions [25]. The present book considers the basic rank-μ-update and weighted recombination version CMA [22].

It is considered that there are N number of potential candidate solutions and each of the N individuals is generated at $(k + 1)$th generation as follows [22, 27]:

$$z_i(k+1) = \aleph\left(\langle z(k)\rangle_w, \sigma(k)^2 C(k)\right) = \langle z(k)\rangle_w + \sigma(k)B(k)D(k)N(0,I), \quad i = 1,\ldots,N \tag{2.6}$$

where $\aleph(m, C)$ = normally distributed random vector with mean m and covariance matrix C. $\langle z(k)\rangle_w = \sum_{j=1}^{\mu} \omega_j z_{j:\lambda}(k)$ = weighted mean of the selected individuals based on the fitness function $fit_i(k)$, $\omega_j > 0$, $\forall\, j = 1, \ldots, \mu$ and $\sum_{j=1}^{\mu} \omega_j = 1$. The adaptation of the covariance matrix $C(k)$ is calculated along the evolution path $p_c(k+1)$ and by the μ weighted difference vectors between the recent parents and $\langle z(k)\rangle_w$ as follows [5, 22, 27]:

$$p_c(k+1) = (1 - c_c).p_c(k) + H_\sigma(k+1)\sqrt{c_c(2 - c_c)}.\frac{\sqrt{\mu_{eff}}}{\sigma(k)}\left(\langle z(k+1)\rangle_w - \langle z(k)\rangle_w\right) \tag{2.7}$$

$$C(k+1) = (1 - c_{\text{cov}}).C(k) + c_{\text{cov}}\frac{1}{\mu_{\text{cov}}}p_c(k+1)(p_c(k+1))^T \ldots + c_{\text{cov}}.\left(1 - \frac{1}{\mu_{\text{cov}}}\right)\sum_{j=1}^{\mu}\frac{\omega_j}{\sigma(k)^2}\left(z_{i:\lambda}(k+1) - \langle z(k)\rangle_w\right)\left(z_{i:\lambda}(k+1) - \langle z(k)\rangle_w\right)^T \tag{2.8}$$

where,

$$H_\sigma(k+1) = 1, \text{ if } \frac{\|p_\sigma(k+1)\|}{\sqrt{1 - (1 - c_\sigma)^{2(k+1)}}} < \left(1.5 + \frac{1}{n - 0.5}\right)E(\|\aleph(0, I)\|), \quad = 0, \text{ otherwise}$$

$\mu_{\text{eff}} = 1/\sum_{i=1}^{\mu}\omega_i^2$ = variance effective selection mass and $\mu_{\text{eff}} = \mu$, if $\omega_i = 1/\mu$ i.e. the condition of intermediate recombination [22, 25, 27], The weights ω_i is a matrix with rank equal to $\min(\mu, n)$. $c_{\text{cov}} \approx \min(1, 2\mu_{\text{eff}} \sim /n^2)$ = learning rate for the covariance matrix C.

The adaptation of the global step size $\sigma(k)$ is calculated from the conjugate evolution path $p_\sigma(k+1)$ as follows [22, 27]:

$$p_\sigma(k+1) = (1-c_\sigma).p_\sigma(k) \\ +\ldots\sqrt{c_\sigma(2-c_\sigma)}.B(k)D(k)^{-1}B(k)^T\frac{\sqrt{\mu_{\text{eff}}}}{\sigma(k)}\left(\langle z(k+1)\rangle_w - \langle z(k)\rangle_w\right) \tag{2.9}$$

$B(k)$, the orthogonal matrix and $D(k)$, the diagonal matrix are both obtained from the principal component analysis of $C(k)$. Thus, the global step size is calculated as follows [22, 27]:

$$\sigma(k+1) = \sigma(k).\exp\left(\frac{c_\sigma}{d_\sigma}\left(\frac{\|p_\sigma(k+1)\|}{E(\|\aleph(0,I)\|)} - 1\right)\right) \tag{2.10}$$

where $E(\|\aleph(0,I)\|) = \sqrt{2}\Gamma\left(\frac{n+1}{2}\right)/\Gamma\left(\frac{n}{2}\right) \approx \sqrt{n}\left(1-\frac{1}{4n}+\frac{1}{21n^2}\right)$ = the expected length of p_σ under random selection.

The initial values are to be selected as $p_\sigma(0) = p_c(0) = 0$ and $C(0) = I$, whereas $x(0)$ and $\sigma(0)$ are problem dependent. The default strategy parameters are given as follows [25–27]:

$$\lambda = 4 + \lfloor 3.\ln(n)\rfloor,\ \mu = \lfloor \lambda/2 \rfloor,\ \omega_{i=1\ldots\mu} = \frac{\ln(\mu+1) - \ln(i)}{\sum_{j=1}^{\mu}\ln(\mu+1) - \ln(i)}$$

$$c_\sigma = \frac{\mu_{\text{eff}}+2}{n+\mu_{\text{eff}}+3},\ d_\sigma = 1 + 2.\max\left(0, \sqrt{\frac{\mu_{\text{eff}}-1}{n+1}} - 1\right) + c_\sigma,\ c_c = \frac{4}{n+4}$$

$$\mu_{\text{cov}} = \mu_{\text{eff}},\ c_{\text{cov}} = \frac{1}{\mu_{\text{cov}}}.\frac{2}{\left(n+\sqrt{2}\right)^2} + \left(1-\frac{1}{\mu_{\text{cov}}}\right)\min\left(1, \frac{2\mu_{\text{eff}}-1}{(n+2)^2+\mu_{\text{eff}}}\right)$$

In this evolutionary strategy, $1/c_\sigma$ and $1/c_c$ are considered as memory time constants and d_σ as damping parameter. Further details about these parameters are discussed in [5, 22]. The CMA algorithm (rank-μ-update and weighted recombination version), conceptualized as a minimization problem, is given in Algorithm 3.

Algorithm 3: CMA (rank-μ-update and weighted recombination version)

BEGIN

Set fitness function $fit(\underline{z})$, $z_d \in Z_d$, $d = 1,2,\ldots,n$, $\underline{z}$ is a candidate solution vector comprising the set of decision variable z_d, $\underline{z} = [z_1, z_2, \cdots, z_n]^T \in R^n$, Z_d is the universe of discourse of the dth decision variable z_d i.e. $z_{d-\min} \le Z_d \le z_{d-\max}$ and n is the number of decision variable.

Create N number of potential candidate solutions and make randomized initialization of their n dimensional position $\underline{z}$.

Initialize iteration number $k = 1$;

REPEAT:

FOR $i = 1$ to N

Calculate fitness: $fit_i(k)$;

Calculate the weighted mean of the selected individuals:

$\langle z(k)\rangle_w = \sum_{j=1}^{\mu} \omega_j z_{j:\lambda}(k)$; based on $fit_i(k)$

Calculate the evolution path:

$$p_c(k+1) = (1-c_c).p_c(k) + H_\sigma(k+1)\sqrt{c_c(2-c_c)}.\frac{\sqrt{\mu_{eff}}}{\sigma(k)}\left(\langle z(k+1)\rangle_w - \langle z(k)\rangle_w\right)$$

Adapt the covariance matrix:

$$C(k+1) = (1-c_{\text{cov}}).C(k) + c_{\text{cov}}\frac{1}{\mu_{\text{cov}}}p_c(k+1)(p_c(k+1))^T$$
$$+ c_{\text{cov}}\left(1-\frac{1}{\mu_{\text{cov}}}\right)\sum_{j=1}^{\mu}\frac{\omega_j}{\sigma(k)^2}\left(z_{i:\lambda}(k+1) - \langle z(k)\rangle_w\right)\left(z_{i:\lambda}(k+1) - \langle z(k)\rangle_w\right)^T$$

Calculate the conjugate evolution path:

$$p_\sigma(k+1) = (1-c_\sigma).p_\sigma(k)$$
$$+\sqrt{c_\sigma(2-c_\sigma)}.B(k)D(k)^{-1}B(k)^T\frac{\sqrt{\mu_{eff}}}{\sigma(k)}\left(\langle z(k+1)\rangle_w - \langle z(k)\rangle_w\right)$$

Calculate the global step size:

$$\sigma(k+1) = \sigma(k).\exp\left(\frac{c_\sigma}{d_\sigma}\left(\frac{\|p_\sigma(k+1)\|}{E(\|\aleph(0,I)\|)} - 1\right)\right)$$

Calculate potential candidate solutions:

$$z_i(k+1) = \aleph\left(\langle z(k)\rangle_w, \sigma(k)^2 C(k)\right) = \langle z(k)\rangle_w + \sigma(k)B(k)D(k)\aleph(0,I)$$

END FOR

UNTIL termination criterion is satisfied;

END

2.5 Harmony Search Algorithm (HSA)

The harmony search (HS) is based on natural musical performance processes that occur when a musician searches for a better state of harmony, i.e., during jazz improvisation. This algorithm was introduced by Z. W. Geem in 2001 [6]. Very similar to musical instruments, where a combination of pitches determine the quality of harmony generated, here in HS algorithm, the suitability of a set of values of the decision variables is judged by its objective function. As in musical harmony, if a combination of decision variables can produce good result, then this solution vector is stored in memory, and this helps to produce an improved solution in near future. First, the optimization problem fit($\underline{z}$) is formulated and the candidate solution vector $\underline{z}$, using the decision variables. The free parameters of the HS algorithm, i.e., harmony memory size (*HMS*) or the number of solution vectors in the harmony memory, harmony memory considering rate (*HMCR*), pitch adjusting rate (*PAR*), and the termination criteria or the maximum number of iterations or generations of improvisation ($iter_{\max}$), are also selected at this stage. Then, *HMS* number of harmonies, i.e., randomly generated candidate solution vectors, are generated and stored, to form the initial harmony memory (*HM*) matrix, according to the sorted values of the respective objective function $g(\underline{z})$, given as follows [6, 8, 28–34]:

$$HM = \begin{bmatrix} z_1^1 & z_2^1 & \cdots & z_q^1 \\ z_1^2 & z_2^2 & \cdots & z_q^2 \\ \vdots & \vdots & \ddots & \vdots \\ z_1^{HMS} & z_2^{HMS} & \cdots & z_q^{HMS} \end{bmatrix} \tag{2.11}$$

In the next step, a new harmony vector $\underline{z}' = (z_1', z_2', \ldots, z_q')]] >$ is improvised, either by choosing a value for a decision variable from the past history, stored in *HM* matrix, or by choosing any value in its permissible universe of discourse, with probability $HMCR \in (0, 1)$. Hence, a new decision variable z_j' in $\underline{z}'$ can be determined as follows [10, 30–34]:

$$z_j' \in \left\{ z_j^1, z_j^2, \ldots, z_j^{HMS} \right\} \quad \text{with probability } HMCR$$

or

$$z_j' \in \hat{Z}_j \quad \text{with probability } (1 - HMCR) \tag{2.12}$$

Each decision variable in the new vector $\underline{z}'$, if generated from *HM* matrix, is further examined for a potential pitch-adjustment possibility with a probability

PAR. In pitch adjustment, the decision variable adjusts its value according to the formula [8, 10, 30–34]

$$z_j' = z_j' + bw \times ud(-1, +1) \tag{2.13}$$

where *bw* is an arbitrary distance bandwidth and *ud* is a uniform distribution.

Algorithm 4: HS algorithm

BEGIN

Setfitness function $fit(\underline{z})$, $z_d \in Z_d$, $d = 1,2,\ldots,n$, $\underline{Z}$ is a candidate solution vector comprising the set of decision variable z_d, $\underline{z} = [z_1, z_2, \cdots, z_n]^T \in R^n$, Z_d is the universe of discourse of the *d*th decision variable z_d i.e. $z_{d-\min} \le Z_d \le z_{d-\max}$ and *n* is the number of decision variable.

Create *N* harmonies and make randomized initialization of their *n* dimensional positions *Z*;

Define harmony memory considering rate (*HMCR*);

Define pitch adjusting rate (*PAR*) and other parameters;

FOR$i = 1$ to *N*

  Calculate fitness: $fit_i(k)$;

END FOR

Generate harmony memory (*HM*) with random harmonies, ranked by fitness value;

Initialize iteration number $k = 1$;

  REPEAT:

  FOR $d = 1$ to *n*

    IF (*rand<HMCR*)

      Choose a value from *HM* for the variable *d*.

      IF (*rand<PAR*)

        Adjust the value by adding certain amount based on a *bandwidth* (*bw*) uniformly distributed over [-1, 1].

      END IF

    ELSE

    Choose a random value.

    END IF

  END FOR

  Calculate fitness of newly generated harmony;

  Accept the new harmony (solution) if better;

  Update *HM* with new harmony and ranked by fitness value;

UNTIL termination criterion is satisfied

END

Then, the objective function $g(\underline{z}')$ for this new harmony is evaluated. If $g(\underline{z}') \Rightarrow \min(g(\underline{z}^1), g(\underline{z}^2), \ldots, g(\underline{z}^{HMS}))$, then the solution vector in existing *HM* matrix that produced the worst harmony is replaced with this newly improvised harmony. Then, the *HM* matrix is rearranged by sorting all decision vectors according to their objective function values. Then, the termination criterion is checked. The termination criterion can be set in two ways, either by choosing the maximum number of HS improvisation/iteration ($iter_{\max}$) or by using a minimum error criteria, i.e., continue the HS improvisation until a maximum pre-specified allowable error is attained by the candidate controller. The basic HS algorithm, conceptualized as a minimization problem, is summarized in Algorithm 4 [8, 10, 34].

2.6 Summary

The stochastic optimization techniques described in this chapter are most popularly utilized during the last two decades in different engineering applications. Here we have explained the very basic ideas and structures of these techniques, as developed in their initial stages. Over the years, various stochastic optimization techniques have undergone many modifications and many variants have been proposed, which have been successfully employed in various domains like, controller design [10, 12–14, 36], robotic path planning and navigation [31], image processing [35, 37–39], and other applications [40, 41]. It is widely believed that, as these techniques are stochastic and derivative free in nature, these algorithms generally have higher probability of finding the global optimum, compared to the conventional optimization techniques, especially in cases of complex, high-dimensional engineering problems. Usually, both the mono-objective and multi-objective evaluations of fitness functions can be potentially carried out successfully, using these contemporary optimization algorithms.

References

1. J. Holland, *Adaption in Natural and Artificial Systems* (University of Michigan Press, Ann Arbor, MI, 1975)
2. D.E. Goldberg, *Genetic Algorithms in Search, Optimization and Machine Learning* (Kluwer Academic Publishers, Boston, 1989)
3. R.C. Eberhart, J. Kennedy, A new optimizer using particle swarm theory, in *Proceedings of 6th International Symposium on Micromachine Human Science* (Nagoya, Japan, 1995), pp 39–43
4. J. Kennedy, R.C. Eberhart, Particle swarm optimization. Proc. IEEE Int. Conf. Neural Networks **4**, 1942–1948 (1995)
5. N. Hansen, A. Ostermeier, A. Gawelczyk, On the adaptation of arbitrary normal mutation distributions in evolution strategies: the generating set adaptation, in *Proceedings of 6th International Conference on Genetic Algorithms* (San Francisco, CA, 1995), pp. 57–64
6. Z.W. Geem, J.H. Kim, G.V. Loganathan, A new heuristic optimization algorithm: harmony search. Simulation **76**(2), 60–68 (2001)
7. E. Rashedi, H. Nezamabadi-pour, S. Saryazdi, GSA: a gravitational search algorithm, *Inf. Sci.* **179**(13), 2232–2248 (2009

8. K.D. Sharma, Hybrid methodologies for stable adaptive fuzzy control, Doctoral Dissertation, Jadavpur University, India, 2012
9. P. Engelbrecht, in *Fundamentals of Computational Swarm Intelligence* (Wiley, 2006)
10. K.D. Sharma, in *A Comparison among Multi-Agent Stochastic Optimization Algorithms for State Feedback Regulator Design of a Twin Rotor MIMO System*, Handbook of Research on Natural Computing for Optimization Problems (IGI Global, 2016), pp. 409–448
11. S. Da Ros, G. Colusso, T.A. Weschenfelder, L. de Marsillac Terra, F. de Castilhos, M.L. Corazza, M. Schwaab, A comparison among stochastic optimization algorithms for parameter estimation of biochemical kinetic models. Appl. Soft Comput. **13**, 2205–2214 (2013)
12. P.P. Bonissone, P.S. Khedkar, Y.-T. Chen, Genetic algorithms for automated tuning of fuzzy controllers: a transportation application, in *Proceedings of 5th IEEE International Conference on Fuzzy Systems* (New Orleans, USA, 1996), pp. 674–680
13. R. Maiti, K.D. Sharma, G. Sarkar, PSO based parameter estimation and PID controller tuning for 2-DOF nonlinear twin rotor MIMO system, *Int. J. Automation and Control*, Inderscience, Accepted, 2017
14. P. Biswas, R. Maiti, A. Kolay, K.D. Sharma, G. Sarkar, PSO based PID controller design for twin rotor MIMO system, in *Proceedings of International Conference on Control, Instrumentation, Energy & Communication (CIEC14)*, pp. 106–111, 2014
15. G. Venter, J.S. Sobieski, Particle swarm optimization, in *43rd AIAA Conference* (Colorado, 2002)
16. Y. Shi, R.C. Eberhart, A modified particle swarm optimizer, in *Proceedings of IEEE International Conference on Evolution Computer* (Alaska, May 1998)
17. C. Eberhart, Y. Shi, Comparing inertia weights and constriction factors in particle swarm optimization, in *Proceedings of 2000 Congress Evolution Computer*, 2000, pp. 84–89
18. K.D. Sharma, A. Chatterjee, F. Matsuno, A Lyapunov theory and stochastic optimization based stable adaptive fuzzy control methodology, in *Proceedings of SICE Annual Conference 2008*, Japan, pp. 1839–1844, August 20–22
19. K.D. Sharma, A. Chatterjee, A. Rakshit, A hybrid approach for design of stable adaptive fuzzy controllers employing Lyapunov theory and particle swarm optimization. IEEE Trans. Fuzzy Syst. **17**(2), 329–342 (2009)
20. K.D. Sharma, A. Chatterjee, A. Rakshit, A random spatial *lbest* PSO based hybrid strategy for designing adaptive fuzzy controllers for a class of nonlinear systems. *IEEE Trans. Instr. Meas.*, accepted for publication
21. S. Das A. Abraham, A. Konar, in *Particle Swarm Optimization and Differential Evolution Algorithms: Technical Analysis, Applications and Hybridization Perspectives*, Studies in Computational Intelligence (SCI) vol. 116, pp. 1–38, 2008
22. N. Hansen, A. Ostermeier, Completely derandomized self-adaptation in evolution strategies. *Evol. Comp.* **9**, 159–195 (2001)
23. N. Hansen, S. D. Muller, P. Koumoutsakos, Reducing the time complexity of the derandomized evolution strategy with covariance matrix adaptation (CMA-ES). *Evol. Comp.* **11**, 1–18 (2003)
24. A. Auger, N. Hansen, A restart CMA evolution strategy with increasing population size. *Cong. Evol. Comp.* **2**, 1769–1776 (2005)
25. N. Hansen, S. Kern, Evaluating the CMA evolution strategy on multimodal test functions, in *Parallel Problem Solving from Nature*(PPSN VIII), Lecture Notes in Computer Science, vol. 3242, pp. 282–291, 2004
26. P.N. Suganthan, N. Hansen, J.J. Liang, K. Deb, Y.-P. Chen, A. Auger, S. Tiwari, *Problem Definitions and Evaluation Criteria for the CEC 2005 Special Session on Real-Parameter Optimization* (Technical Report. Nanyang Tech. University, Singapore, 2005)
27. K.D. Sharma, A. Chatterjee, P. Siarry, A. Rakshit, CMA—H^{∞} hybrid design of robust stable adaptive fuzzy controllers for non-linear systems, in *Proceedings of 1st International Conference Frontiers in Optimization: Theory and Applications* (*FOTA*-2016), November 2016, Kolkata, India

28. K.S. Lee, Z.W. Geem, A new structural optimization method based on the harmony search algorithm. Comput. Struct. **82**, 781–798 (2004)
29. K.S. Lee, Z.W. Geem, A new meta-heuristic algorithm for continuous engineering optimization: harmony search theory and practice. Comp. Methods Appl. Mech. Eng. **194** (36–38), 3902–3933 (2005)
30. K.D. Sharma, A. Chatterjee, A. Rakshit, Design of a hybrid stable adaptive fuzzy controller employing Lyapunov theory and harmony search algorithm. IEEE Trans. Contr. Syst. Tech. **18**, 1440–1447 (2010)
31. K. Das Sharma, A. Chatterjee, A. Rakshit, Harmony search-based hybrid stable adaptive fuzzy tracking controllers for vision-based mobile robot navigation. Mach. Vis. Appl. **25**(2), 405–419 (2014)
32. K. Das Sharma, A. Chatterjee, A. Rakshit, Harmony search algorithm and Lyapunov theory based hybrid adaptive fuzzy controller for temperature control of air heater system with transport-delay. Appl. Soft Comput. **25**, 40–50 (2014)
33. M. Mahdavi, M. Fesanghary, E. Damangir, An improved harmony search algorithm for solving optimization problems. Appl. Math. Comput. **188**, 1567–1679 (2007)
34. K.D. Sharma, Stable fuzzy controller design employing group improvisation based harmony search algorithm. Int. J. Control Autom. Syst. **11**(5), 1046–1052 (2013)
35. T. Chakroborti, K.D Sharma, A. Chatterjee, A novel local extrema based gravitational search algorithm and its application in face recognition using one training image per class, *Engineering Applications of Artificial Intelligence*, vol. 34, pp. 13–22, September, 2014
36. A. Roy, K.D. Sharma, Gravitational search algorithm and Lyapunov theory based stable adaptive fuzzy logic controller, in *Proceedings of 1st International Conference Computational Intelligence: Modelling, Techniques and Applications* (*CIMTA*-2013), vol. 10, pp. 581–586, *Procedia Technology*, 2013
37. S.S. Das, K.D. Sharma and J.N. Bera, A simple visual secret sharing scheme employing particle swarm optimization, in *Proceedings of International Conference on Control, Instrumentation, Energy and Communication* (*CIEC*14), January 2014, Kolkata, India
38. A. Chander, A. Chatterjee, P. Siarry, A new social and momentum component adaptive PSO algorithm for image segmentation. Expert Syst. Appl. **38**, 4998–5004 (2011)
39. M. Maitra, A. Chatterjee, A hybrid cooperative–comprehensive learning based PSO algorithm for image segmentation using multilevel thresholding. Expert Syst. Appl. **34**, 1341–1350 (2008)
40. S. Kar, K.D Sharma, M. Maitra, Gene selection from microarray gene expression data for classification of cancer subgroups employing pso and adaptive k-nearest neighborhood technique. *Expert Syst. Appl.* **42**, pp. 612–627, January 2015
41. T.K. Maji, P. Acharjee, Multiple solutions of optimal PMU placement using exponential binary PSO algorithm for smart grid applications. IEEE Trans. Ind. Appl. **53**(3), 2550–2559 (2017)

Chapter 3
Fuzzy Controller Design I: Stochastic Algorithm-Based Approach

3.1 Introduction

In recent years, fuzzy logic controllers (FLCs) have been widely used as a successful practical approach for designing and implementing control systems. The design of fuzzy controllers permitted the designer for simple inclusion of heuristic knowledge or linguistic information about how to control a plant rather than requiring the exact mathematical model. Although the original interest was initiated by the urge to design controllers for those systems which are very difficult to be characterized mathematically, with time, study on FLCs expanded extensively for both model-free and model-based systems. In recent times, more and more attentions are paid to systematic design of FLCs from the point of view of stable behavior and optimization of performance indices. Several such stable and optimized FLCs have been proposed utilizing, e.g., neural network [1], GA [2, 3], PSO [4, 5], ant colony [6].

We shall now describe the formulation of stable fuzzy controllers utilizing supervisory control action and then present a systematic design of these controllers employing stochastic optimization algorithms (SOAs). Particle swarm optimization (PSO), a swarm intelligence (SI) technique based on the natural process of flocking of birds in search of food, and harmony search algorithm (HSA), conceptualized using the musical process of searching for a perfect state of harmony, are the two representative SOAs discussed in this chapter to design the FLCs. Considering the generic behavior of the SOAs, a generalized approach of FLC design has also been presented here. Two benchmark systems are utilized to demonstrate the performances of the designed controllers in simulation case studies.

K. Das Sharma et al., *Intelligent Control*, Cognitive Intelligence and Robotics,
https://doi.org/10.1007/978-981-13-1298-4_3

3.2 Stable Fuzzy Controllers

The fuzzy controller design issues are always concerned about the stability of the designed controllers. In one possible approach, the structure and the free parameters of the fuzzy controller have to be specified to ensure the stability of the closed-loop system with the fuzzy controller. In another possible approach, the fuzzy controller is designed without any stability considerations and then an additional supervisory control law can be incorporated to guarantee the stability requirements. Thus, it is a more liberal approach of designing stable fuzzy controllers, and here, a non-fuzzy supervisory control law has been employed as a second-level safeguard with the fuzzy controller, to stabilize the closed-loop system. The nature of this supervisory control law is such that whenever the main fuzzy control law fails to achieve the control objectives, the supervisory control action will be in active mode; otherwise, it remains idle [4, 7].

Let us consider that the control objective is to design a fuzzy control strategy for an nth order nonlinear plant given as [4, 8, 9]:

$$\left.\begin{aligned} x^{(n)} &= f(\underline{x}) + bu \\ y &= x \end{aligned}\right\} \tag{3.1}$$

where $f(\cdot)$ is an unknown continuous function, $u \in R$ and $y \in R$ are the input and output of the plant, and b is an unknown positive constant. It is assumed that the state vector is given as $\underline{x} = (x_1, x_2, \ldots, x_n)^{\mathrm{T}} = (x, \dot{x}, \ldots, x^{(n-1)})^{\mathrm{T}} \in R^n$.

Assumption 3.1 In order that (3.1) be controllable, we require that $b \neq 0$ and without loss of generality we can assume that $b > 0$.

The system under control may have a reference model, in model-based control scheme, given as

$$\left.\begin{aligned} x_m^{(n)} &= f(\underline{x}_m, w) \\ y_m &= x_m \end{aligned}\right\} \tag{3.2}$$

where $w(t)$ is the excitation signal input to the reference model, and the state vector of the reference model is $\underline{x}_m = (x_m, \dot{x}_m, \ldots, x_m^{(n-1)})^{\mathrm{T}} \in R^n$.

If the control scheme is model free, then the $y_m(t)$ itself represents the reference signal.

The control objective is to force the plant output $y(t)$ to follow a given bounded reference signal $y_m(t)$ under the constraints that all closed-loop variables involved must be bounded to guarantee the closed-loop stability of the system. Thus, the tracking error is $e = y - y_m$.

The objective is to design a stable fuzzy controller for the system described in (3.1). Fuzzy logic controllers (FLCs) have evolved as popular control solutions for those classes of plants whose input–output characteristics are not precisely known. In case of the system in (3.1), this is given by the conditions when $f(\cdot)$ and b are not precisely known, which is widely prevalent in practical situations. Here we need to find the structure of the fuzzy controller as well as a feedback control strategy $u = u_c(\underline{x}|\underline{\theta})$, using fuzzy logic system, along with a supervisory control law, to be appended if required, such that the following conditions are satisfied [4, 5]:

(i) The closed-loop system must be globally stable in the sense that all variables, $\underline{x}(t), \underline{\theta}(t)$ and $u(\underline{x}|\underline{\theta})$ must be uniformly bounded; i.e., $|\underline{x}(t)| \leq M_x < \infty$, $|\underline{\theta}(t)| \leq M_\theta < \infty$ and $|u_c(\underline{x}|\underline{\theta})| \leq M_u < \infty$, where M_x, M_θ and M_u are set by the designer. Here, $\underline{\theta}$ is the parameter vector of the fuzzy controller.
(ii) The tracking error $e(t)$ should be as small as possible under the constraints in (i) and the $\underline{e} - \underline{\theta}$ space should be stable in the large for the system.

In case of a stable fuzzy controller to accomplish these control objectives, let the error vector be $\underline{e} = (e, \dot{e}, \ldots, e^{(n-1)})^{\mathrm{T}}$ and $\underline{k} = (k_1, k_2, \ldots, k_n)^{\mathrm{T}} \in R^n$ be such that all the roots of the Hurwitz polynomial $s^n + k_n s^{n-1} + \cdots + k_2 s + k_1$ are in the left half of s-plane.

Now the ideal control law for the system in (3.1) is given as follows [5, 7, 8]:

$$u^* = \frac{1}{b}\left[-f(\underline{x}) + y_m^{(n)} + \underline{k}^{\mathrm{T}}\underline{e}\right] \tag{3.3}$$

where $y_m^{(n)}$ is the nth derivative of the output of the reference model/the reference signal.

For some specific class of plants, (3.1) and (3.3) lead to the error differential equation as [4, 7]:

$$e^{(n)} = -k_1 e - k_2 \dot{e} - \cdots - k_n e^{(n-1)} \tag{3.4}$$

This definition implies that u^* guarantees perfect tracking, i.e., $y(t) \equiv y_m(t)$ if $Lt_{t \to \infty}\, e(t) = 0$. Here the vector $\underline{k}$ describes the desired closed-loop dynamics for the error. In practical situations, since f and b are not known precisely, the ideal u^* of (3.3) cannot be implemented in real practice. Thus, a suitable solution can be to design a fuzzy logic system to approximate this optimal control law [8].

Let, us assume that the FLC is constructed using a zero-order Takagi-Sugeno (T-S) fuzzy system. Then, $u_c(\underline{x}|\underline{\theta})$ for the FLC is given in the form [4, 5, 7]:

$$u_c(\underline{x}|\underline{\theta}) = \frac{\sum_{l=1}^{N} \theta_l * \alpha_l(\underline{x})}{\sum_{l=1}^{N} \alpha_l(\underline{x})} = \underline{\theta}^{\mathrm{T}} * \underline{\xi}(\underline{x}) \tag{3.5}$$

where $\underline{\theta} = [\theta_1\, \theta_2 \ldots \theta_N]^{\mathrm{T}}$ = the vector of the output singletons, $\alpha_l(\underline{x})$ = the firing degree of rule "l", i.e.,

$$\alpha_l(\underline{x}) = \prod_{i=1}^{r} \mu_i^l(x_i) \tag{3.6}$$

N = the total number of rules, $\mu_i^l(x_i)$ = the membership value of the ith input membership function (MF) in the activated lth rule, $\underline{\xi}(\underline{x})$ = vector containing normalized firing strength of all fuzzy IF-THEN rules $= (\xi_1(\underline{x}), \xi_2(\underline{x}), \ldots, \xi_N(\underline{x}))^{\mathrm{T}}$ and

$$\xi_l(\underline{x}) = \frac{\alpha_l(\underline{x})}{\sum_{l=1}^{N} \alpha_l(\underline{x})} \tag{3.7}$$

Now, utilizing (3.3) and (3.4), the error differential equation will be

$$\begin{aligned} e^{(n)} &= y_m^{(n)} - y^{(n)} = y_m^{(n)} - x^{(n)} \\ &= y_m^{(n)} - f(\underline{x}) - bu_c(\underline{x}|\underline{\theta}) \\ &= -\underline{k}^T \underline{e} + b[u^* - u_c(\underline{x}|\underline{\theta})] \end{aligned} \tag{3.8}$$

In an equivalent form, it can be written as follows [4]:

$$\dot{\underline{e}} = \Lambda_c \underline{e} + \underline{b}_c[u^* - u_c(\underline{x}|\underline{\theta})] \tag{3.9}$$

$$\text{where, } \Lambda_c = \begin{bmatrix} 0 & 1 & 0 & 0 & \cdots & 0 & 0 \\ 0 & 0 & 1 & 0 & \cdots & 0 & 0 \\ \vdots & \vdots & \vdots & \vdots & \ddots & \vdots & \vdots \\ 0 & 0 & 0 & 0 & \cdots & 0 & 1 \\ -k_1 & -k_2 & \cdots & \cdots & \cdots & -k_{n-1} & -k_n \end{bmatrix} \text{ and } \underline{b}_c = \begin{bmatrix} 0 \\ \vdots \\ 0 \\ b \end{bmatrix} \tag{3.10}$$

Now, to ensure stability it is assumed that the control law $u(t)$ is actually the summation of the fuzzy control, $u_c(\underline{x}|\underline{\theta})$, and an additional supervisory control strategy $u_s(\underline{x})$, given as:

$$u(t) = u_c(\underline{x}|\underline{\theta}) + u_s(\underline{x}) \tag{3.11}$$

So, the closed-loop error equation in (3.9) becomes [4]:

$$\dot{\underline{e}} = \Lambda_c \underline{e} + \underline{b}_c[u^* - u_c(\underline{x}|\underline{\theta}) - u_s(\underline{x})] \tag{3.12}$$

Let, the Lyapunov function be defined as follows:

$$V_e = \frac{1}{2} \underline{e}^{\mathrm{T}} P \underline{e} \tag{3.13}$$

where P is a symmetric positive definite matrix satisfying the Lyapunov equation,

$$\Lambda_c^{\mathrm{T}} P + P\Lambda_c = -Q \tag{3.14}$$

Here, Q is a positive definite matrix.

Thus, using (3.12) and (3.14), it can be written as

$$\begin{aligned}\dot{V}_e &= -\frac{1}{2}\underline{e}^{\mathrm{T}} Q\underline{e} + \underline{e}^{\mathrm{T}} P\underline{b}_c[u^* - u_c(\underline{x}|\underline{\theta}) - u_s(\underline{x})] \\ &\leq -\frac{1}{2}\underline{e}^{\mathrm{T}} Q\underline{e} + |\underline{e}^{\mathrm{T}} P\underline{b}_c|(|u^*| + |u_c|) - \underline{e}^{\mathrm{T}} P\underline{b}_c u_s\end{aligned} \tag{3.15}$$

$u_s(\underline{x})$ should be so designed that $\dot{V}_e$ should be negative semi-definite, i.e., $\dot{V}_e \leq 0$.

Assumption 3.2 The functions f^U and b_L can be determined such that $f^U \geq |f(\underline{x})|$ and $0 < b_L \leq b$; that is, it is assumed that the upper bound of $|f(\underline{x})|$ and the lower bound of b are known.

Now, $u_s(\underline{x})$ can be constructed by observing (3.12) and (3.15) and can be presented as [4, 10]:

$$u_s(\underline{x}) = I_1^* \mathrm{sgn}\left(\underline{e}^{\mathrm{T}} P\underline{b}_c\right)\left[|u_c| + \frac{1}{b_L}\left(f^U + |y_m^{(n)}| + |\underline{k}^T \underline{e}|\right)\right] \tag{3.16}$$

where $\begin{cases} I_1^* = 1 \text{ if } |\underline{x}| \geq M_x \\ I_1^* = 0 \text{ if } |\underline{x}| < M_x \end{cases}$.

With this $u_s(\underline{x})$ in (3.16) and considering the case $I_1^* = 1$, it can be written as

$$\begin{aligned}\dot{V}_e &\leq -\frac{1}{2}\underline{e}^{\mathrm{T}} Q\underline{e} + |\underline{e}^{\mathrm{T}} P\underline{b}_c|\left[\frac{1}{b}\left(|f(\underline{x})| + |y_m^{(n)}| + |\underline{k}^{\mathrm{T}}\underline{e}|\right) - \frac{1}{b_L}\left(f^U + |y_m^{(n)}| + |\underline{k}^{\mathrm{T}}\underline{e}|\right)\right] \\ &\leq -\frac{1}{2}\underline{e}^{\mathrm{T}} Q\underline{e} \leq 0\end{aligned} \tag{3.17}$$

Thus, using supervisory control $u_s(\underline{x})$, one can guarantee the boundedness in $\underline{x}$. Hence, the closed-loop stability is guaranteed [5, 7, 10]. Further, it can be seen from (3.16) that the supervisory control law will be in action, whenever $\underline{x}$ touch the boundary, i.e. $|\underline{x}| = M_x$, so a switching chattering due to $u_s(\underline{x})$ will come into reality. This phenomenon of $u_s(\underline{x})$ introduces an oscillation into the system. To overcome this problem, I_1^* can be changed continuously from 0 to 1, rather than utilizing a discrete switching. This can be achieved as [7]

$$I_1^* = \begin{cases} 0 & \text{if } |\underline{x}| < a \\ \frac{|\underline{x}| - a}{M_x - a} & \text{if } a \leq |\underline{x}| < M_x \\ 1 & \text{if } |\underline{x}| \geq M_x \end{cases} \tag{3.18}$$

where $0 < a < M_x$ and the value of a will be specified by the designer.

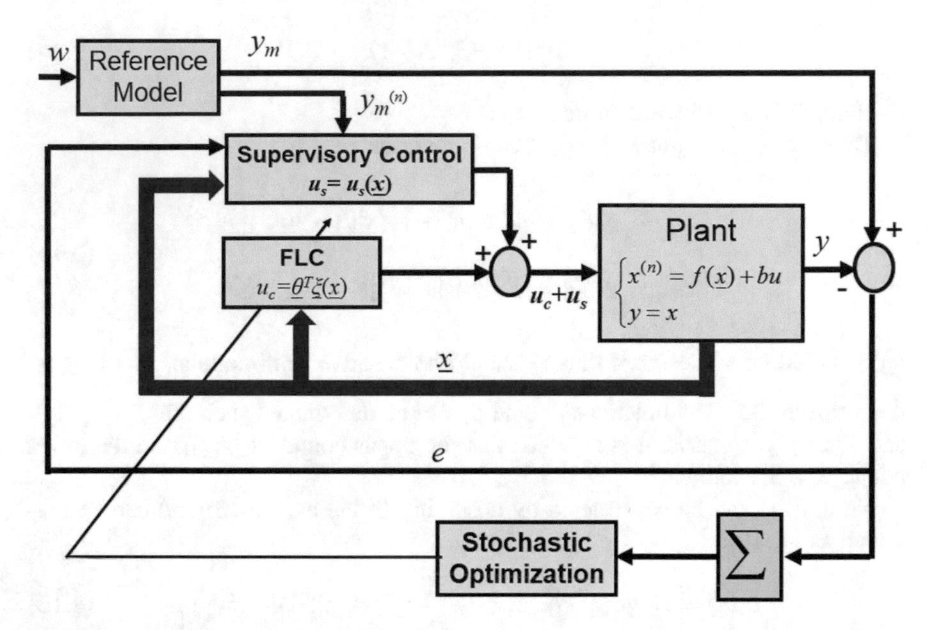

Fig. 3.1 A schematic architecture of fuzzy control action augmented by supervisory control

This modification in supervisory control scheme increases the smoothness in control action and considerably suppresses the oscillations in the plant response, without hampering the stability of the overall system. A schematic diagram of fuzzy control action augmented by supervisory control is shown in Fig. 3.1.

Note It should be noted here that when the reference signal y_m is the output of a reference model excited with a given input signal w and the reference plant has the same dynamics as that of the controlled plant and can be expressed by $\underline{k}$, then the input states include w along with $\underline{x}$ [5].

3.3 PSO-Based Design Methodology

In the PSO computational algorithm, population dynamics simulates bio-inspired behavior, i.e., a "bird flock's" behavior which involves sharing of information and allows particles to take profit from the discoveries and previous experience of all other particles during the search of food [11, 12].

Some early applications of PSO in designing classical PID controllers are found in [13, 14]. Even ANN-based controllers were designed using PSO [15]. One of the earliest applications of PSO in designing fuzzy controllers was demonstrated by the authors [5] in 2009. During the same time, Liao et al. also showed a successful application of PSO in fuzzy controller design for temperature control of a reheating furnace [16]. In recent times, Sahu et al. applied a hybrid version of PSO in designing a fuzzy PID-type controller for automatic generation control of interconnected power systems [17]. Bevrani et al. proposed an online tuning of fuzzy controller by a PSO algorithm in an AC microgrid [18]. This section details a systematic approach of designing zero-order variable structure T-S-type fuzzy controller employing PSO for nonlinear systems.

3.3.1 PSO-Based Fuzzy Controller Design

To design a zero-order T-S type fuzzy controller with stochastically adaptive controller structure and free parameters, the values of the scaling gains, the settings of controller structure (i.e., number of MFs in each variable, number of rules), the supports of input MFs, the positions of output singletons, etc., have to be automatically determined. In the conventional fuzzy controller design methods available in the literature [7, 9], all these parameters are usually hand tuned by trial and error method. Also, these FLCs are usually employed with a fixed structure, with a fixed number of rules and fixed supports for each input MF. The PSO-based design methodology not only stochastically tunes the scaling gains, output singletons and other free parameters properly, but also it can determine the optimum controller structure.

In PSO-based controller design, a particle's position in solution space is a vector containing all required information to construct a fuzzy controller, e.g., (i) information about the number and positions of the MFs, (ii) number of rules, (iii) values of scaling gains, (iv) positions of the output singletons. This particle vector is formed as follows [4, 5, 19]:

$$\begin{aligned}\underline{Z} = [&\text{structural flags for MFs}|\text{center locations of the MFs}|\ldots\\ &|\text{scaling gains}|\text{positions of the output singletons}]\end{aligned} \tag{3.19}$$

Each structural flag bears the information about the existence or non-existence of an MF in its corresponding variable. It can only take binary values where "0" indicates the non-existence and "1" indicates the existence of the corresponding MF. However, PSO is an algorithm where each entry in the particle vector can take continuous values. Hence, for each structural flag in the vector, the universe of discourse is set as [0, 1]. Subsequently, the flag is set to 0, if the continuous value of the variable is <0.5, and the flag is set to 1, if the variable is ≥ 0.5. The total number of 1's in the structural flags keeps changing in each iteration, and hence, the total number of MFs in which an input variable is fuzzified also changes in each iteration. This causes a change in the structure of the FLC in each iteration as it changes the total number of MFs for each input variable, which changes the total number of rules of the fuzzy rule base and hence changes the total number of active output singletons. Each input is fuzzified using triangular MFs, and the center locations of the MFs in the particle vector store the peaks of the MFs. Each input is fuzzified in the range [−1, 1] with two fixed triangular MFs having their peaks fixed at −1 and 1, respectively. All intermediate MFs for that input variable are flexible in nature. They can be either active or inactive during an iteration, and their peaks are also adapted by PSO in each iteration. For each MF, the peaks of the immediate adjacent active MFs on either side of its own peak forms the left and right base support of it. Whether the immediate adjacent MF is an active MF or not is determined by the status of its corresponding structural flag. Hence, some of the center locations of MFs are ignored while evaluating the FLC output, because their corresponding entries in the structural flags are zero. A similar logic holds true for the output singletons in a particle vector. Here, those singletons become inactive whose antecedent parts contain one or more input MFs which become inactive in that iteration [5, 20].

To design an FLC utilizing particle swarm optimization-based approach (PSOBA) algorithm, first the population of particles in the swarm is chosen; then, the algorithm will decide the structure of the candidate controller according to its structural flags. The candidate controller simulation (CCS) algorithm, as shown in Fig. 3.2, is implemented for each on the candidate controller to calculate the fitness function, the integral absolute error (IAE), which can be defined as $\text{IAE} = \sum_{n=0}^{PST} e(n)\Delta t_c$, where PST = plant simulation time and Δt_c = step size or sampling time. Then, according to the value of the fitness function (IAE) of each particle in each iteration, the *pBest* and *gBest* candidate controllers are calculated and the particles' position and velocity in the search space is updated by the global version PSO method [11, 12]. The PSOBA algorithm will stop searching the solution space when the number of

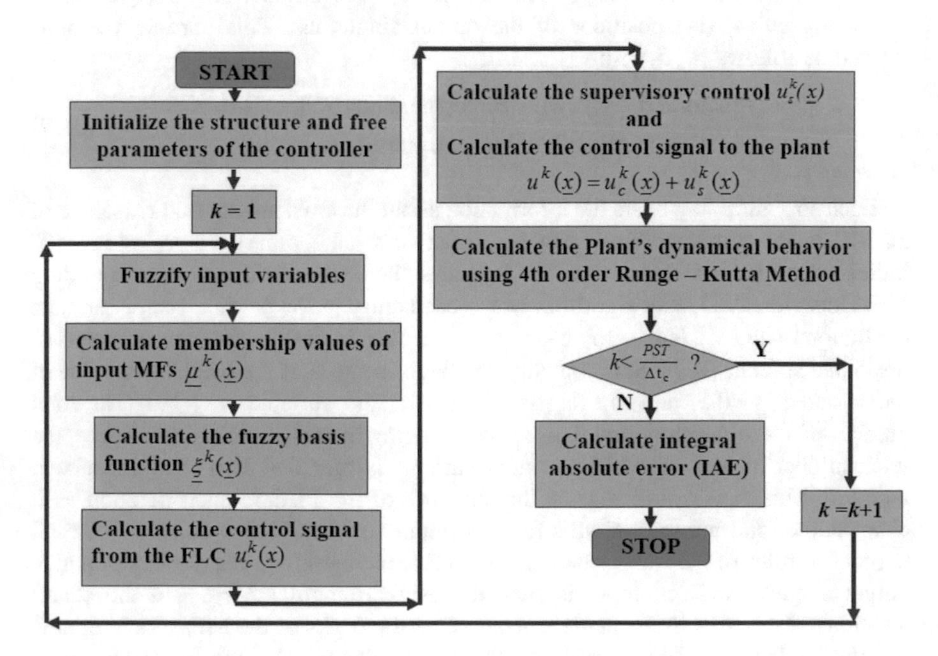

Fig. 3.2 CCS algorithm in flowchart form

iterations specified by the designer is reached or a pre-specified error is attained by the controller.

3.3.2 The Design Algorithm

The outline of the global version PSO-based controller design algorithm is given as follows:

(i) Select the number of particles of the swarm as *NoP*.

(ii) Initialize the *j*th particle's position and velocity randomly throughout the design space (solution space) using the following formulae:

$$\underline{Z}_0^j = \underline{Z}_{\min} + r_1(\underline{Z}_{\max} - \underline{Z}_{\min}) \tag{3.20}$$

$$\underline{V}_0^j = \frac{\underline{Z}_{\min} + r_2(\underline{Z}_{\max} - \underline{Z}_{\min})}{\Delta t_c} \tag{3.21}$$

where r_1, r_2 are the random numbers between 0 and 1, $\underline{Z}_{\min}$ is the vector of lower bounds and $\underline{Z}_{\max}$ is the vector of upper bounds for the controller settings and Δt_c is the time step.

(iii) The stopping criterion can be set in two ways, either by choosing the maximum iteration ($iter_{\max}$) or by using a minimum error criteria, i.e., continue the PSO adaptation until a maximum pre-specified allowable error is attained by the controller [21].

(iv) Determine the structure of the candidate controller based on structural flags as incorporated in the particles position $\underline{Z}$, (see (3.19)).

(v) Calculate the fitness value, e.g., *IAE*, of the controller for each particle that acts as a candidate controller in this work, by using CCS algorithm as shown in flowchart representation in Fig. 3.2.

(vi) Search the fitness value for each particle till now, in all iterations, to obtain its own best fitness value (*pBestFF*) in history and get the corresponding position vector ($\underline{pBestZ}$), i.e., the controller settings corresponding to that fitness value.

(vii) Determine the particle in the design space with the best fitness value (*gBestFF*), till now, among all the particles and denote its position as $\underline{gBestZ}$.

(viii) For each particle, calculate the velocity according to the following formula [22]:

$$\underline{V}_{g+1}^j = w_g \underline{V}_g^j + \frac{c_1 r_1(\underline{pBestZ}_g^j - \underline{Z}_g^j)}{\Delta t_c} + \frac{c_2 r_2(\underline{gBestZ} - \underline{Z}_g^j)}{\Delta t_c} \tag{3.22}$$

where w_g is the inertia weight of the particle at iteration g, as introduced by

Shi and Eberhart [22], and is given by

$$w_g = \frac{(iter_{\max} - g) * (w_{\text{start}} - w_{\text{end}})}{iter_{\max}} + w_{\text{end}} \tag{3.23}$$

Here, $iter_{\max}$ is the maximum number of iterations of PSO. Normally, w_g is reduced linearly from w_{start} to w_{end} throughout the PSO generation. A typical combination is $w_{\text{start}} = 0.9$ and $w_{\text{end}} = 0.4$.
c_1 and c_2 in (3.21) are the "trust" parameters and are usually set to 2. The trust parameters indicate how much confidence the current particle has in itself (given by c_1, depicting the individual behavior) and how much confidence it has in the swarm (given by c_2, depicting the social behavior). Each particle's position vector is updated according to the following equation:

$$\underline{Z}_{g+1}^{j} = \underline{Z}_{g}^{j} + \underline{V}_{g+1}^{j} \Delta t_c \tag{3.24}$$

where $\underline{Z}_{g+1}^{j}$ represents the position of particle "j" at $(g + 1)$th iteration and $\underline{V}_{g+1}^{j}$ represents the corresponding velocity vector. A unit time step (Δt_c) is considered throughout this book.

(ix) Each particle's velocity, in each dimension, is clamped to a maximum velocity $\underline{V}_{\max}$ as

$$\text{if } \underline{V}_g^j > \underline{V}_{\max}, \text{ then } \underline{V}_g^j = \underline{V}_{\max}$$
$$\text{if } \underline{V}_g^j < -\underline{V}_{\max}, \text{ then } \underline{V}_g^j = -\underline{V}_{\max}$$

where $\underline{V}_{\max}$ is specified by the designer according to the nature of the system.

(x) Repeat steps (iv) to (ix), until the termination criterion is fulfilled.

The presentation of PSOBA, in flowchart form, is shown in Fig. 3.3.

3.4 HSA-Based Design Methodology

HS algorithm (HSA) is based on the musical process of searching for perfect state of harmony utilizing the aesthetic and acoustic criteria. The HSA performs in a very similar way to musical instruments, where a combination of pitches determines the quality of harmony generated. Here, the suitability of a set of values of the decision variables is judged by the corresponding value of the objective function. As in musical harmony, if a combination of decision variables can produce good result, then this solution vector is stored in memory and this helps to produce an improved

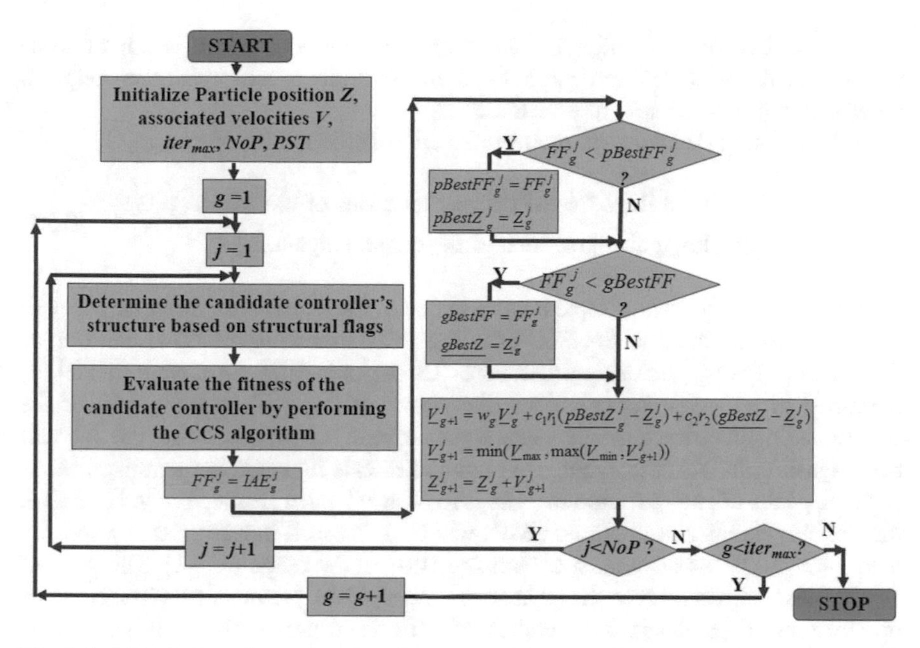

Fig. 3.3 PSOBA algorithm in flowchart form. *j* particle counter, *NoP* number of particles, *g* PSO generation counter

solution in near future. The intention of using HS technique for engineering design optimization is to provide improved results in less iteration.

Till date, there are very few applications of HSA in control systems domain reported, compared to GA or PSO. Wang et al. applied HSA-based fuzzy controllers in power plant control [23]. In Chap. 8, we shall demonstrate a design of fuzzy controllers, tuned by HSA, for temperature control of an air heater system with transport delay [4, 24]. Mukherjee and his group utilized HSA-based fuzzy control schemes in different applications like, in isolated hybrid power system [25], for automatic generation control of multi-unit multi-area deregulated power system [26, 27].

Similar to the previous section, this section describes the HSA-based stochastically optimal/near optimal fuzzy controller design for nonlinear systems.

3.4.1 HSA-Based Fuzzy Controller Design

In a very similar fashion to that of PSO, HSA can be potentially utilized to design a fuzzy controller, where the structure and the other free parameters, e.g., supports of the input MFs, input and output scaling gains of the controller can be meta-heuristically adapted. While designing the HSA-based optimal/near optimal controller, all these parameters are obtained automatically by encoding them as part of the solution vector, which is here termed as harmony. Thus, each of these parameters forms a decision variable in the solution space. This also helps to determine the optimal structure of the FLC.

In HSA-based design, a harmony in solution space is formed as [4, 28]

$$\begin{aligned} Z = & [\text{structural flags for MFs}|\text{center locations of the MFs}|\ldots \\ & |\text{scaling gains}|\text{positions of the output singletons}] \end{aligned} \tag{3.25}$$

The structural flags are implemented to determine whether a particular MF will be present for an input to the FLC or not, exactly following the same concept, as described in PSO-based design (see Sect. 3.3.1). As in HSA also, each variable is permitted to take continuous values, the structural flag will be assigned 1 if the corresponding decision variable value is greater than 0.5 or; otherwise, the flag will be assigned 0. In this case again, each input to FLC is fuzzified in the range $[-1, 1]$ with the peaks of each extreme MFs kept fixed at -1 and 1, respectively. Here also, the possible number of maximum MFs will be 7 for each input variable where we determine the peaks of other 5 MFs using HSA in the range $[-1, 1]$. The peak of any MF also corresponds to the right base support of the previous MF. The contents of structural flags determine number of MFs in which each input variable is fuzzified and accordingly the peaks are also organized.

Figure 3.4 shows the flowchart form of presentation of HSA-based approach (HSABA) algorithm for the FLC with supervisory control action. Here, the initial pool of *HMS* (harmony memory size) number of candidate solution vectors has been created, and, on the basis of simulating the system for each candidate controller by using the candidate controller simulation (CCS) algorithm, as shown in Fig. 3.2, and evaluating the fitness function, the initial pool has been sorted to generate initial *HM* (harmony memory) matrix. In this case also, the fitness function is chosen as the integral absolute error (*IAE*) where $IAE = \sum_{n=0}^{PST} e(n)\Delta t_c$, and PST = plant simulation time, Δt_c = step size or sampling time.

Note As the HS algorithm developed by Geem [29] improvises a single harmony in each iteration, so, the accuracy of the fitness evaluation is not good enough as compared to other population-based peer algorithms like, GA, PSO. A modification has been presented in [28], where the entire harmony matrix (HM) has been improvised in each iteration. Along with this modification, a variable HMCR and PAR has also been utilized in [4, 24, 28] to increase the accuracy of the fitness evaluation. This book uses this modified HS algorithm throughout the book and is referred as HS algorithm. Subsequently, the controller designed with the help of this algorithm is referred as HSABA.

3.4.2 The Design Algorithm

The outline of the HSA-based controller design algorithm is given as follows [28]:

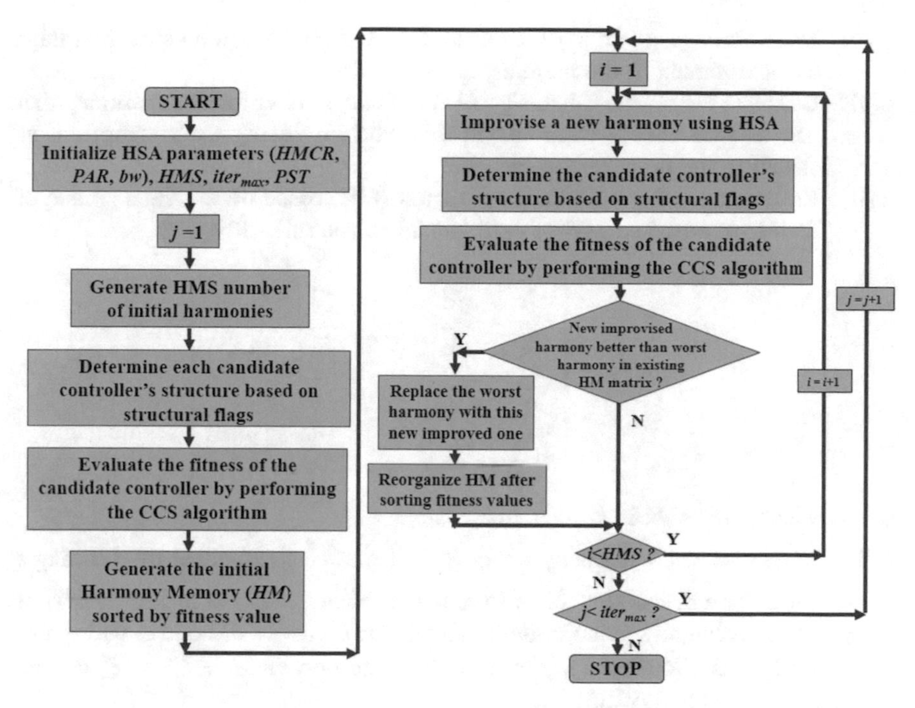

Fig. 3.4 Flowchart representation of HSABA algorithm. *i* harmony counter, *j* harmony improvisation/generation counter

(i) Select the value of harmony memory size (*HMS*), i.e., the number of harmonies in the harmony memory (*HM*).
(ii) Select the parameters of the HAS, i.e., harmony memory considering rate (*HMCR*), pitch adjusting rate (*PAR*) and bandwidth (*bw*).
(iii) The stopping criterion can be set in two ways, either by choosing the maximum iteration ($iter_{\max}$) or by using a minimum error criteria, i.e., continue the HSA adaptation until a maximum pre-specified allowable error is attained by the controller.
(iv) Initialize *HMS* number of randomly generated harmonies and store them in the *HM* matrix [29].

$$HM = \begin{bmatrix} z_1^1 & z_2^1 & \cdots & z_q^1 \\ z_1^2 & z_2^2 & \cdots & z_q^2 \\ \vdots & \vdots & \ddots & \vdots \\ z_1^{HMS} & z_2^{HMS} & \cdots & z_q^{HMS} \end{bmatrix} \quad (3.26)$$

where $z_j^i = z_{j\min}^i + r_1(z_{j\max}^i - z_{j\min}^i)$, $i = 1,2, \ldots, \text{HMS}$, $j = 1,2, \ldots, q$ and q is the number of decision variables in harmony, i.e., the dimension of harmony.

(v) Determine the structure of the candidate controller based on structural flags as incorporated in the harmony $\underline{Z}$, (see (3.25)).

(vi) Calculate the fitness value, e.g., *IAE*, of the controller for each harmony that acts as a candidate controller in this work, by using CCS algorithm, as shown in the flowchart in Fig. 3.2.

(vii) Rearrange the harmony memory matrix (*HM*) based on the fitness value of each harmony, i.e., *IAE* of each candidate controller [29, 30].

$$HM = \begin{bmatrix} z_1^1 & z_2^1 & \cdots & z_q^1 & IAE^1 \\ z_1^2 & z_2^2 & \cdots & z_q^2 & IAE^2 \\ \vdots & \vdots & \ddots & \vdots & \vdots \\ z_1^{HMS} & z_2^{HMS} & \cdots & z_q^{HMS} & IAE^{HMS} \end{bmatrix} \tag{3.27}$$

where $IAE^1 < IAE^2 < \cdots < IAE^{HMS}$.

(viii) Improvise a new harmony vector $\underline{Z}^{i'} = \left(z_1^{i'}, z_2^{i'}, \ldots, z_q^{i'}\right)$ either by choosing a value for a decision variable from the past history, stored in *HM* matrix, or by choosing any value in its permissible universe of discourse, with probability $HMCR \in (0, 1)$. Hence, a new decision variable $z_j^{i'}$ in $\underline{Z}'$ can be determined as [28–30]:

$$z_j^{i'} \in \left\{z_j^1, z_j^2, \ldots, z_j^{\text{HMS}}\right\} \quad \text{with probability HMCR} \tag{3.28A}$$

or

$$z_j^{i'} \in \hat{Z}_j \, \text{with probability } (1 - \text{HMCR}) \tag{3.28B}$$

Each decision variable of the new vector $\underline{Z}'$, if generated from *HM* matrix, is further examined for a potential pitch-adjustment possibility with a probability *PAR*. In pitch-adjustment, the decision variable adjusts its value according to the formula

$$z_j^{i'} = z_j^{i'} + bw \times ud(-1, +1) \tag{3.29}$$

where *bw* is an arbitrary distance bandwidth and *ud* is a uniform distribution.

(ix) Determine the structure of the candidate controller based on the new harmony $\underline{Z}'$ and then evaluate the fitness value IAE'. If $IAE' \Rightarrow \min(IAE^1, IAE^2, \ldots, IAE^{HMS})$, then replace the harmony in existing *HM* matrix that produced the worst harmony, with this new harmony. Then, re-arrange the *HM* matrix by sorting all decision vectors according to their fitness values.

(x) Repeat steps (v) to (ix) until the termination criterion is fulfilled.

The flowchart for the HSABA is shown in Fig. 3.4.

3.5 Stochastic Algorithm-Based Design Methodology: *A Generalized Approach*

After scrutinizing the architecture of the stochastic optimization algorithms, as described in Chap. 2, it is evident that the general structure for all these algorithms follows a similar philosophy. All these search algorithms are essentially multi-agent, population based and introduce stochasticity during their operations. In conceptual realm, the communication procedures among the search agents are different in different algorithms, as they are inspired from different physical, biological, or other phenomena. During implementations, only the updating procedure of the candidate solution vectors (CSVs) is different from each other. So, this section introduces a generic design approach of fuzzy controllers by adapting the structure, free parameters, scaling gains and output singletons.

3.5.1 *The Generic Design Algorithm*

The outline of the SOA-based controller design algorithm is given as follows [4, 5, 28]:

(i) Select the value of the number of candidate solution vectors (N_{CSV})
(ii) Select the free parameters of the stochastic optimization algorithm (SOA).
(iii) The stopping criterion can be set in two ways, either by choosing the maximum iteration ($iter_{max}$) or by using a minimum error criteria, i.e., continue the SOA adaptation until a maximum pre-specified allowable error is attained by the controller.
(iv) Initialize N_{CSV} number of randomly generated candidate solution vectors (CSV).
(v) Determine the structure of the candidate controller based on structural flags as incorporated in the candidate solution vector $\underline{Z}$, (see (3.19) or (3.25)).
(vi) Calculate the fitness value, e.g., *IAE*, of the controller for each CSV that acts as a candidate controller considered here, by using CCS algorithm as shown in the flowchart in Fig. 3.2.
(vii) Improvise a new CSV, based on the updating rules of the SOA.
(viii) Repeat steps (v) to (vii) until the termination criterion is fulfilled.

The SOABA, in flowchart form, is shown in Fig. 3.5.

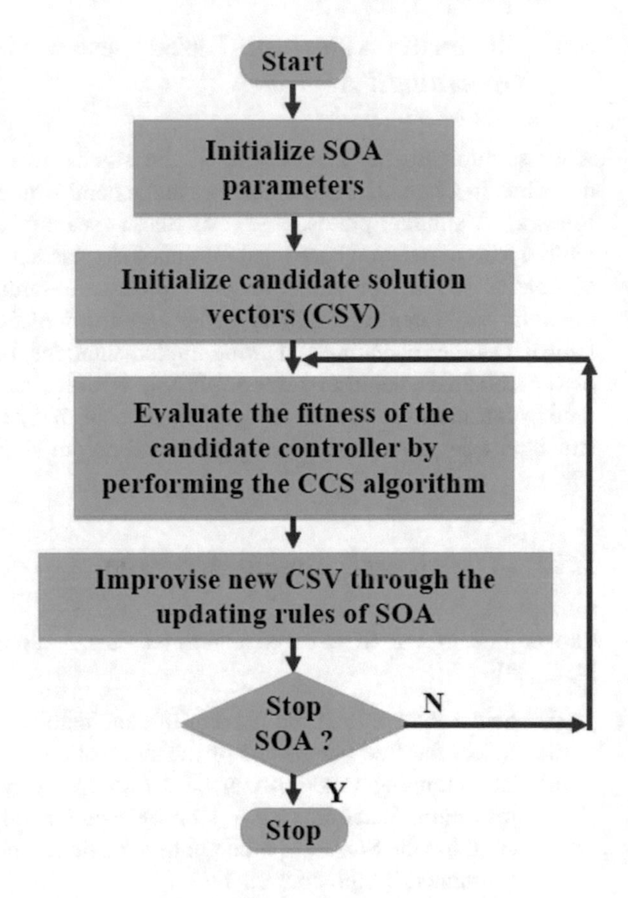

Fig. 3.5 SOABA algorithm, in flowchart form

3.6 A Comparative Case Study of Different Algorithms

To evaluate the effectiveness of the proposed schemes, two benchmark case studies are considered, previously utilized in several research works [4, 5, 8, 10, 19]. To study the performances in terms of *IAE*, nonlinear systems with fixed structure configurations and a variable structure configuration, with minimum 2 and maximum 7 numbers of MFs for each input, are simulated, respectively. The process models are simulated each using a fixed step fourth order Runge–Kutta method with sampling time $\Delta t_c = 0.01$ s. In all simulation case studies, the plant is evaluated for 10 s, after the training is completed by different optimization strategies, to compare the results.

3.6.1 Case Study I: DC Motor Containing Nonlinear Friction Characteristics

The controlled plant under consideration is a second-order DC motor containing nonlinear friction characteristics described by the following model [5, 8]:

$$\left.\begin{array}{l} \dot{x}_1 = x_2 \\ \dot{x}_2 = -\frac{f(x_2)}{J} + \frac{C_T}{J}u \\ y = x_1 \end{array}\right\} \tag{3.30}$$

where $y = x_1$ is the angular position of the rotor (in rad), x_2 is the angular speed (in rad/s) and u is the current fed to the motor (in A). The plant parameters are $C_T = 10$ Nm/A, $J = 0.1$ kgm^2 and the nonlinear friction torque is defined as:

$$f(x_2) = 5\tan^{-1}(5x_2)\,\text{Nm} \tag{3.31}$$

The control objective is to make the angular position y follow a reference signal given by

$$\ddot{y}_m = 400w(t) - 400y_m - 40\dot{y}_m \tag{3.32}$$

(i.e., the reference model has a double pole at $s_{1,2} = -20$) where $w(t)$ is a square wave of random amplitude and y_m is the output of $w(t)$ filtered by the second-order linear filter (given by (3.32)).

The control signal obtained as the output from the fuzzy controller, in response to input signals, is given below as follows:

$$u_c = u_c(\underline{x}, \underline{Z}) \tag{3.33}$$

In this simulation, input MFs are distributed evenly throughout the corresponding universe of discourse. The other parameters are chosen as $P = \begin{bmatrix} p_0 & p_1 \\ p_1 & p_2 \end{bmatrix} = \begin{bmatrix} 5 & 0.01 \\ 0.01 & 0.001 \end{bmatrix}$ such that $p_1k_1 > 0$ and $p_2 > \frac{p_1}{k_2}$, where $k_i = a_{im}$, where a_{im} can be obtained from the reference model, described in general for an nth order system as follows:

$$y_m^{(n)} = b_{1m}w - a_{1m}y_m - a_{2m}\dot{y}_m - \cdots - a_{nm}y_m^{(n-1)} \tag{3.34}$$

Here y_m is the output of a linear reference model, having a numerator of degree zero. The other parameters are chosen as follows: $b = 100$, $b_L = 95$ and $f^U = 1.5$.

The controller is trained by applying a signal $w(t)$, which changes its value randomly in the interval (−1.5, 1.5) in every 0.5 s. as shown in Fig. 3.6a, to the reference model, which gives an output $y_m(t)$ as shown in Fig. 3.6b. This reference

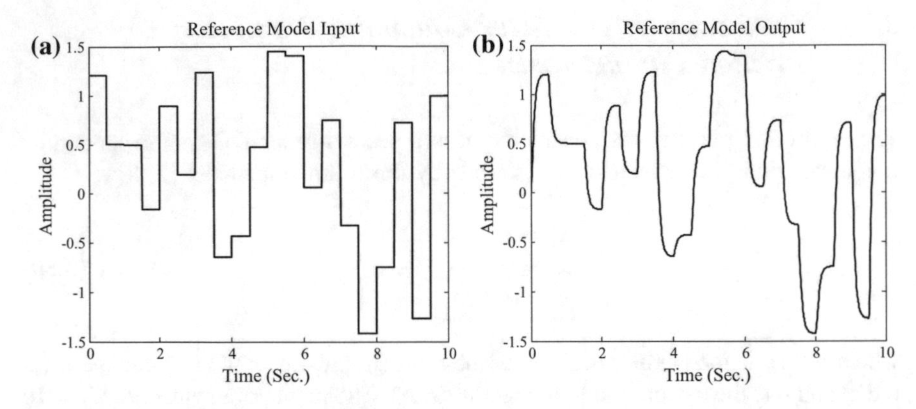

Fig. 3.6 Plot of **a** reference model input signal, $w(t)$ and **b** reference model output signal $y_m(t)$

signal $y_m(t)$ is chosen as it contains infinitely many frequencies and leads to good exploration of controller input space [5].

3.6.1.1 Performance Analysis of PSOBA

To perform this simulation, population size of ten particles is taken. For each simulation, 200 iterations of PSOBA are carried out. In fixed structure controller design, a fixed number of evenly distributed input MFs are used for the initial population of PSOBA. PSOBA algorithm further tunes the free parameters and the positions of the output singletons to minimize the tracking error. In variable structure controller design, the structural flags are utilized to automatically vary the structure of the candidate controller, in each iteration. The input MFs and the corresponding positions of the output singletons are set according to the active structural flags as stated in Sect. 3.3. *IAE* between the reference signal y_m and the motor output y is considered as the fitness function for optimization. To maintain the stochastic behavior of the optimization process, twenty number of simulations are performed and average *IAE* is calculated. The results of the simulation studies are tabulated in Table 3.1.

Table 3.1 Comparison of simulation results for case study I (Nonlinear DC motor system): PSO-based design strategies

Control strategy	No. of MFs	No. of rules	Avg. IAE	Best IAE	Std. Dev.
PSOBA	$3 \times 3 \times 3$	27	0.6737	0.6682	0.0050
PSOBA	$5 \times 5 \times 5$	125	0.3505	0.3018	0.0411
PSOBA	$7 \times 7 \times 7$	343	0.3091	0.2763	0.0428
PSOBA-V	$4 \times 6 \times 5$	120	–	0.4028	–

The simulations are carried out for three fixed structures of $3 \times 3 \times 3$, $5 \times 5 \times 5$ and $7 \times 7 \times 7$ input MFs and a variable structure with a potentially biggest possible structure of $7 \times 7 \times 7$ MFs. Figures 3.7, 3.8, 3.9, and 3.10 show temporal variations of system responses for the fixed structure controllers with $3 \times 3 \times 3$,

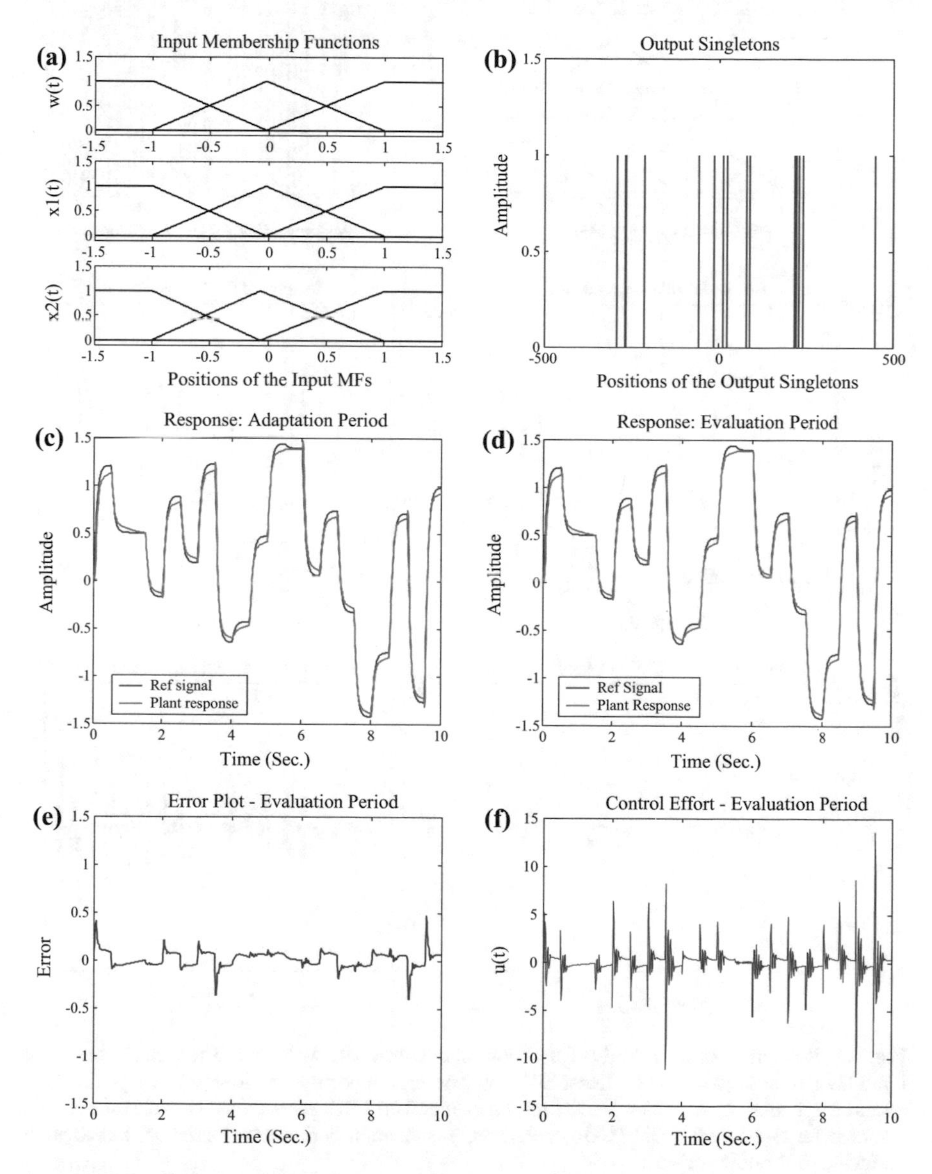

Fig. 3.7 Responses of the adaptive fuzzy controller with PSOBA, for $3 \times 3 \times 3$ configuration of input MFs, of case study I (Nonlinear DC motor system): **a** positions of input MFs, **b** positions of output singletons, **c** adaptation period response after 100 PSO generations, **d** evaluation period response for 0–10 s after 200 PSO generations, **e** evaluation period error variation, **f** evaluation period control effort variation

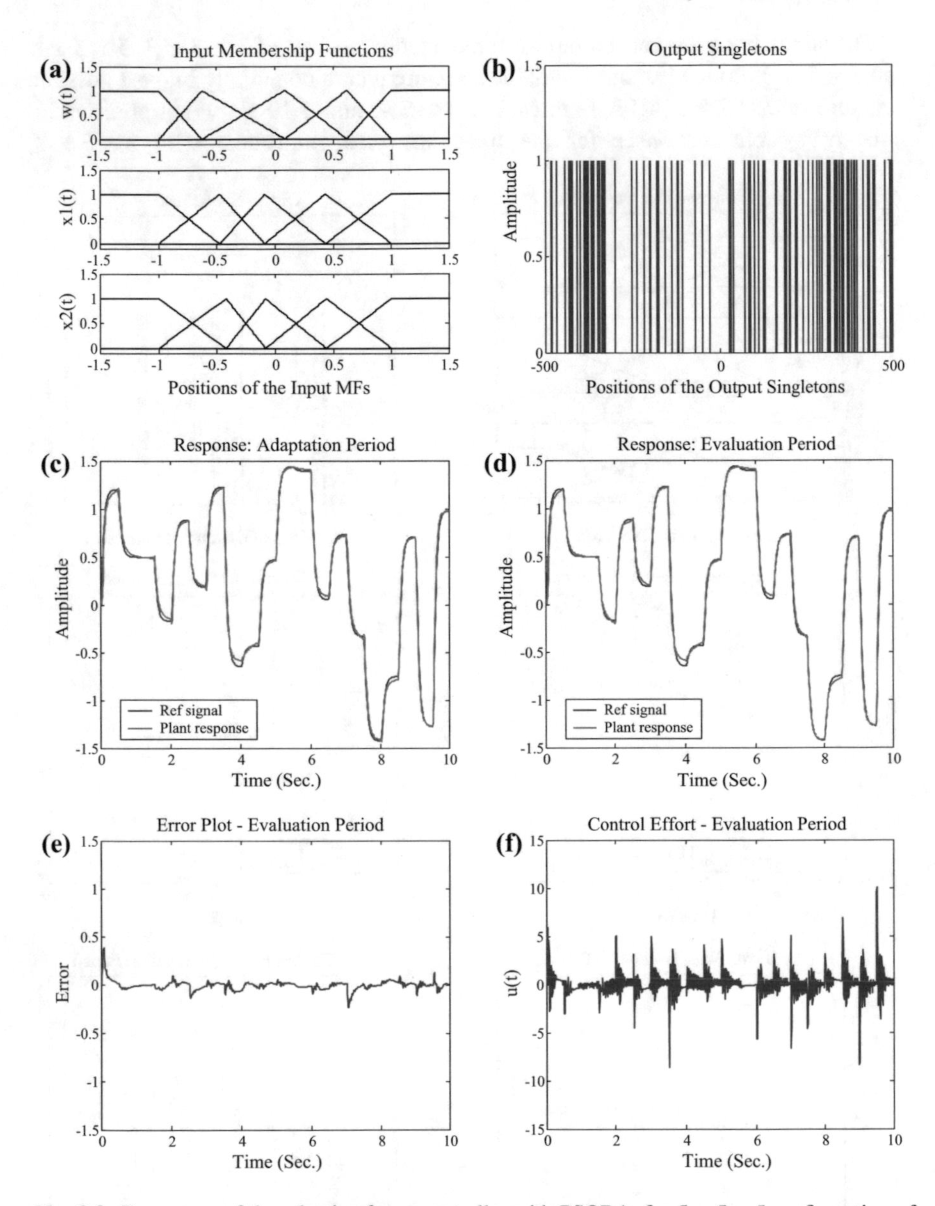

Fig. 3.8 Responses of the adaptive fuzzy controller with PSOBA, for $5 \times 5 \times 5$ configuration of input MFs, of case study I (Non-linear DC motor system): **a** positions of input MFs, **b** positions of output singletons, **c** adaptation period response after 100 PSO generations, **d** evaluation period response for 0–10 s after 200 PSO generations, **e** evaluation period error variation, **f** evaluation period control effort variation

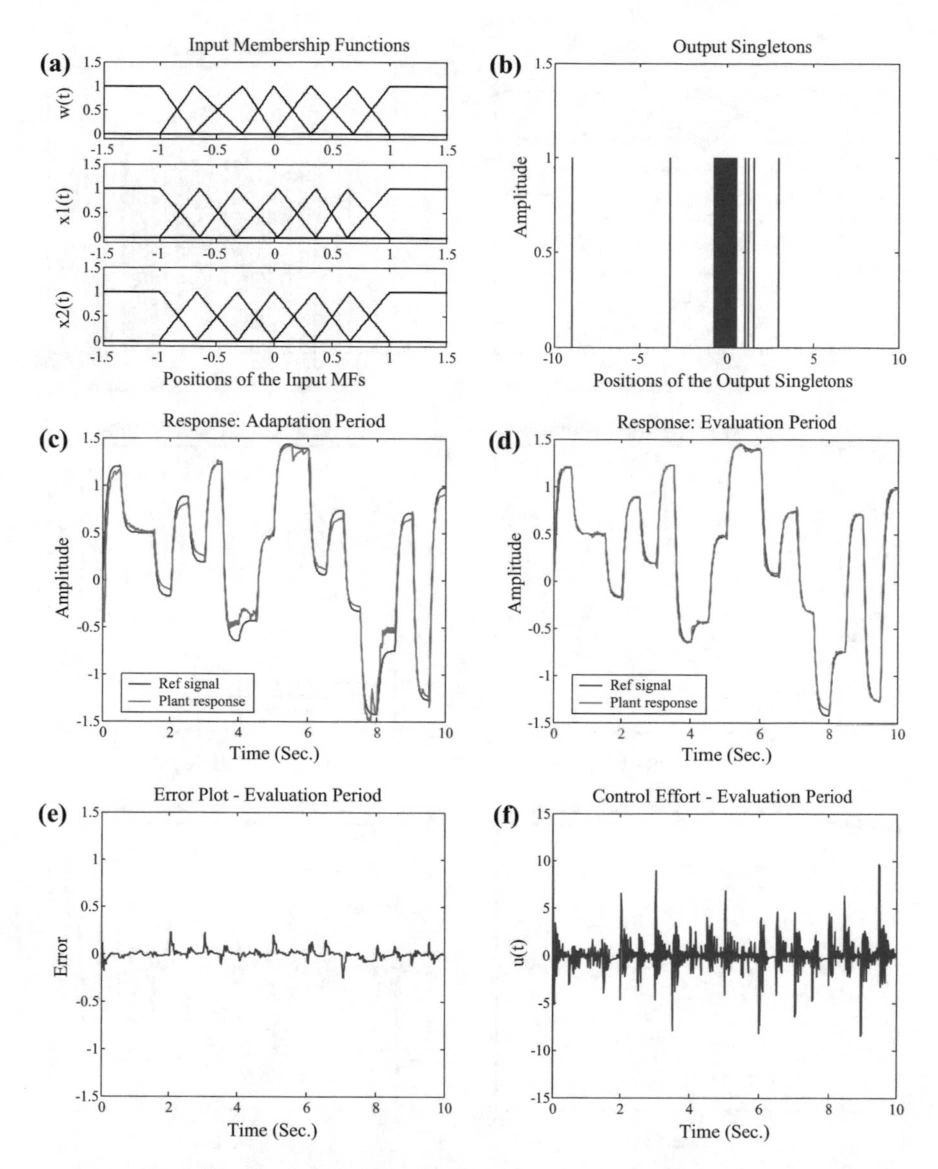

Fig. 3.9 Responses of the adaptive fuzzy controller with PSOBA, for 7 × 7 × 7 configuration of input MFs, of case study I (Nonlinear DC motor system): **a** positions of input MFs, **b** positions of output singletons, **c** adaptation period response after 100 PSO generations, **d** evaluation period response for 0–10 s after 200 PSO generations, **e** evaluation period error variation, **f** evaluation period control effort variation

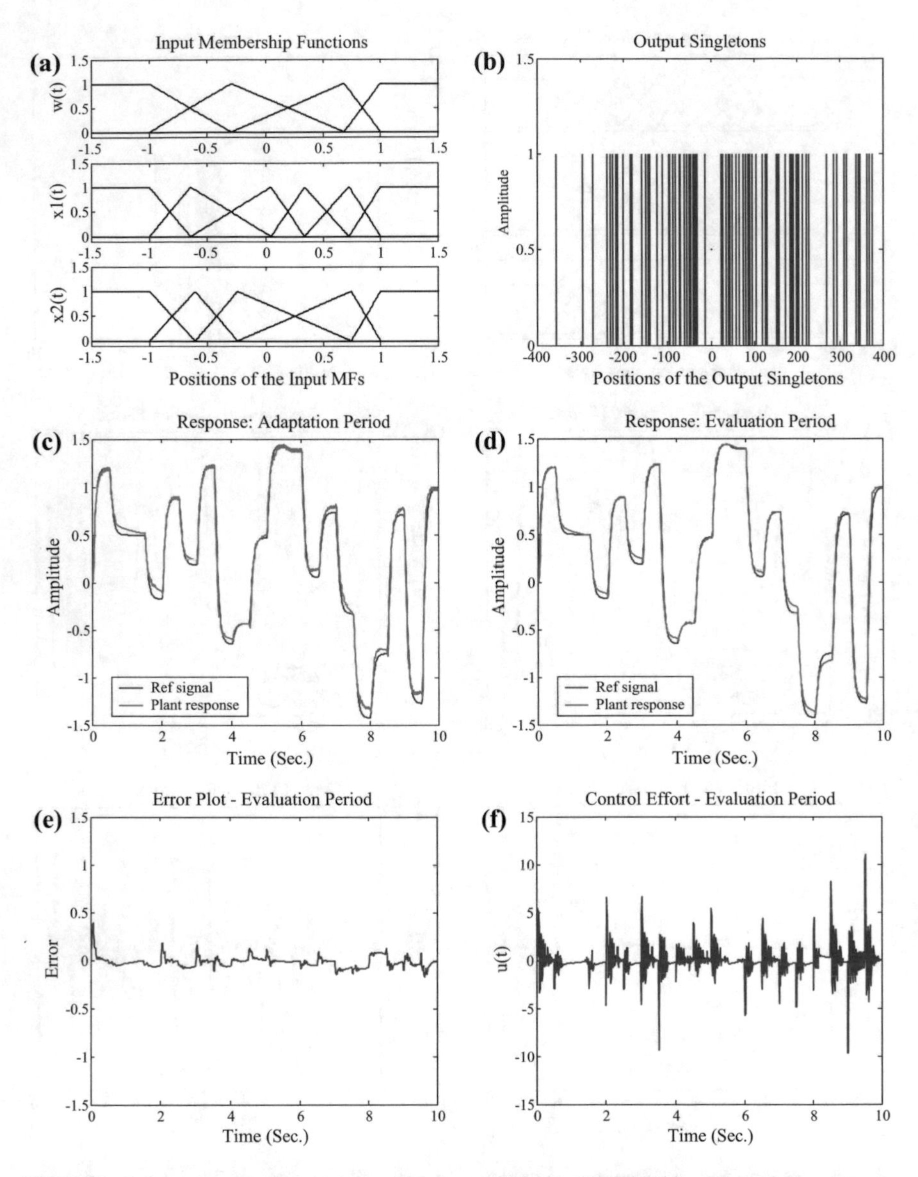

Fig. 3.10 Responses of the adaptive fuzzy controller with PSOBA, for variable structure configuration of input MFs, of case study I (Nonlinear DC motor system): **a** positions of input MFs, **b** positions of output singletons, **c** adaptation period response after 100 PSO generations, **d** evaluation period response for 0–10 s after 200 PSO generations, **e** evaluation period error variation, **f** evaluation period control effort variation

$5 \times 5 \times 5$, $7 \times 7 \times 7$ and variable structure input MFs, respectively, for PSOBA control strategy. In (a) and (b) of Figs. 3.7, 3.8, 3.9, and 3.10, the input MFs and positions of the output singletons, respectively, after the completion of the adaptation process are shown. Figures 3.7c, 3.8c, 3.9c, and 3.10c show the adaptation period responses after 100 PSO iterations. In (d)–(f) of Figs. 3.7, 3.8, 3.9, and 3.10 the evaluation period responses (of a duration of 10 s), error variations and control effort variations are shown, respectively.

For fixed structure controllers, it can be seen that the performances of the controllers, in terms of *IAE*, improve with increase in the number of rules from $3 \times 3 \times 3$ to $7 \times 7 \times 7$ configuration. Although the *IAE* value is higher for the variable structure case, it can be seen that the best result, in terms of *IAE*, is obtained with only 120 rules.

It can be observed from Figs. 3.7, 3.8, 3.9, and 3.10 that the center locations of all input MFs are so optimized and the algorithms for different control strategies are developed that the input MFs are distributed over the universe of discourse ensuring the crosspoint level is always 0.5 and the crosspoint ratio is 1, as these values provide good performances for FLC design [31].

3.6.1.2 Performance Analysis of HSABA

To perform this case study, a harmony memory (*HM*) of ten harmonies is considered. For each simulation, 200 generations of *HM* improvisation are carried out. The other design parameters are same as those chosen in Sect. 3.6.1.1. Fixed structure and variable structure controllers are designed in similar fashion as in PSOBA, and the *IAE* values are compared in Table 3.2.

It can be seen from Table 4.2 that the performance of the designed FLC does not improve with the mere increase in number of rules. Here the performance of HASBA design scheme with $5 \times 5 \times 5$ input MF configuration is better than other configurations. The performance of variable structure controller, in terms of *IAE* value, is poor, but the total number rules get reduced to only 60. This is a significant reduction in the total number of rules and such a simplified structure of an FLC will be very useful from the point of view of real implementation.

Figures 3.11 and 3.12 show the temporal variations of system responses for the sample case of fixed structure controllers with $5 \times 5 \times 5$ input MFs and variable structure configurations for HSABA scheme, respectively. Each of these figures shows the following features of the case study I as: (a) positions of input

Table 3.2 Comparison of simulation results for case study I (Nonlinear DC motor system): HSABA-based design strategies

Control strategy	No. of MFs	No. of rules	Avg. IAE	Best IAE	Std. Dev.
HSABA	$3 \times 3 \times 3$	27	0.6872	0.6631	0.0150
HSABA	$5 \times 5 \times 5$	125	0.2354	0.2265	0.0085
HSABA	$7 \times 7 \times 7$	343	0.3016	0.2817	0.0148
HSABA-V	$4 \times 3 \times 5$	60	–	0.6669	–

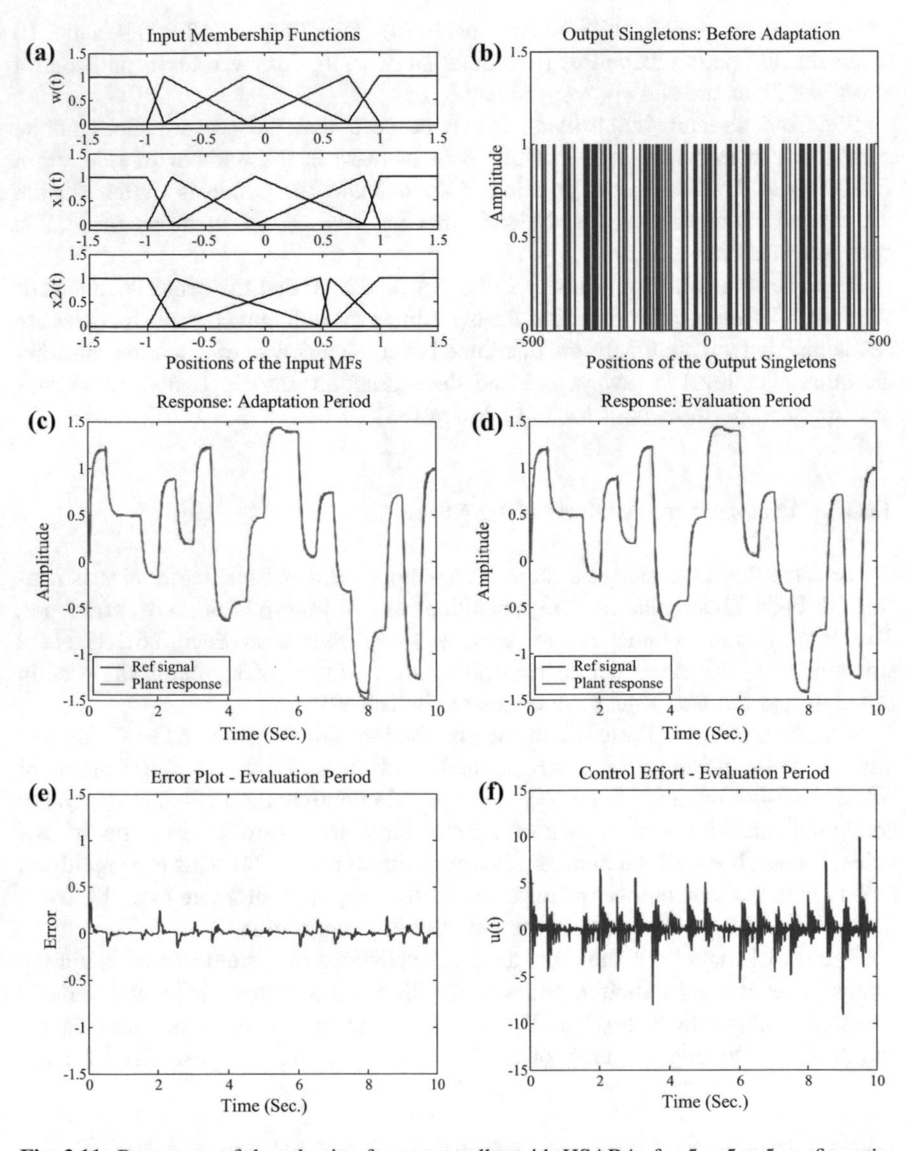

Fig. 3.11 Responses of the adaptive fuzzy controller with HSABA, for 5 × 5 × 5 configuration of input MFs, of case study I (Nonlinear DC motor system): **a** positions of input MFs, **b** positions of output singletons, **c** adaptation period response after 100 HS generations, **d** evaluation period response for 0–10 s after 200 HS generations, **e** evaluation period error variation, **f** evaluation period control effort variation

membership functions, (b) positions of output singletons, (c) adaptation period response after 100 HS generations, (d) evaluation period response for 0–10 s after 200 HS generations, (e) evaluation period error variation, (f) evaluation period control effort variation.

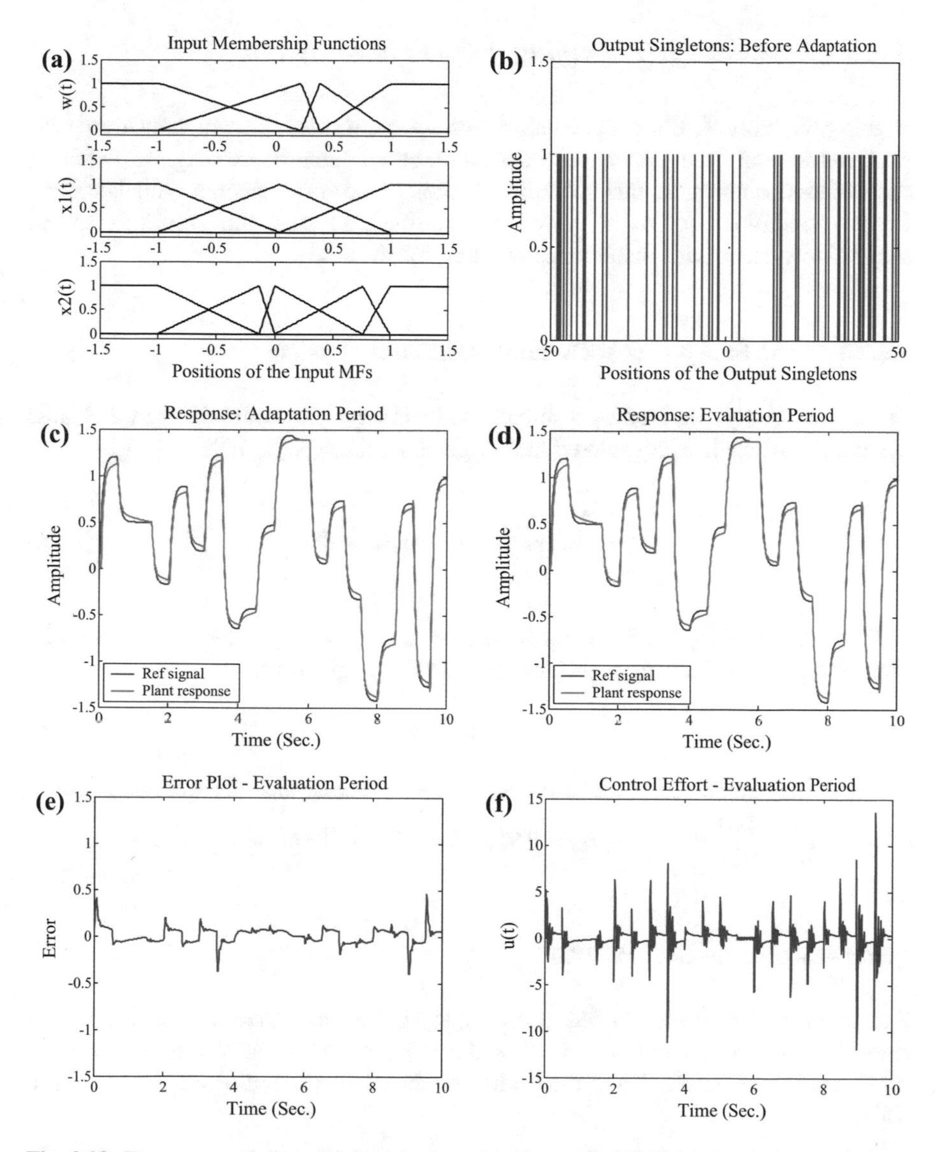

Fig. 3.12 Responses of the adaptive fuzzy controller with HSABA, for variable structure configuration of input MFs, of case study I (Non-linear DC motor system): **a** positions of input MFs, **b** positions of output singletons, **c** adaptation period response after 100 HS generations, **d** evaluation period response for 0–10 s after 200 HS generations, **e** evaluation period error variation, **f** evaluation period control effort variation

3.6.2 *Case Study II: Duffing's Oscillatory System*

In this case study, Duffing's oscillatory system is considered for demonstrating the performance of the stochastic algorithm-based control strategies. After a model-based control, in this section, a model-free control scheme with Duffing's forced oscillation system is considered at first and then the Duffing's forced oscillation system with disturbance is taken for the study.

3.6.2.1 Case Study II (A): Duffing's Oscillatory System

In this case study, the Duffing's forced oscillation system, which is itself a chaotic system if unforced, is considered and is given as follows [5, 10]:

$$\left.\begin{aligned} \dot{x}_1 &= x_2 \\ \dot{x}_2 &= -0.1x_2 - x_1^3 + 12\cos(t) + u(t) \\ y &= x_1 \end{aligned}\right\} \quad (3.35)$$

The control objective is to track the reference signal y_m, where $y_m = \sin(t)$. The control signal obtained from the fuzzy controller is given as:

$$u_c = u_c(\underline{x}|\underline{Z}) \quad (3.36)$$

Here the parameters for this simulation process are chosen as follows: $P = \begin{bmatrix} 1 & 0.01 \\ 0.01 & 0.001 \end{bmatrix}$, $b = 1$, $b_L = 0.95$, $f^U = 12 + |x_1^3|$ and $K = [2\ 1]$.

Performance Analysis of PSOBA

Similar to the approach adopted in case study I, the simulations are carried out for three fixed structures of 3 × 3, 5 × 5 and 7 × 7 input MFs and a variable structure configuration of the FLC and the performances of all simulations are tabulated in Table 3.3.

Table 3.3 Comparison of simulation results for case study II(A) (Duffing's system): PSOBA-based design strategies

Control strategy	No. of MFs	No. of rules	Avg. IAE	Best IAE	Std. Dev.
PSOBA	3 × 3	27	0.2378	0.2109	0.0242
PSOBA	5 × 5	25	0.2729	0.1948	0.0775
PSOBA	7 × 7	49	0.2076	0.1995	0.0119
PSOBA-V	4 × 7	28	–	0.2214	–

In this case study, PSOBA for variable structure configuration evolved with 28 number of rules and also does not contribute a minimum *IAE*. Figures 3.13 and 3.14 show temporal variations of system responses for the fixed structure controller with 5 × 5 input MFs and variable structure configuration, respectively, chosen as

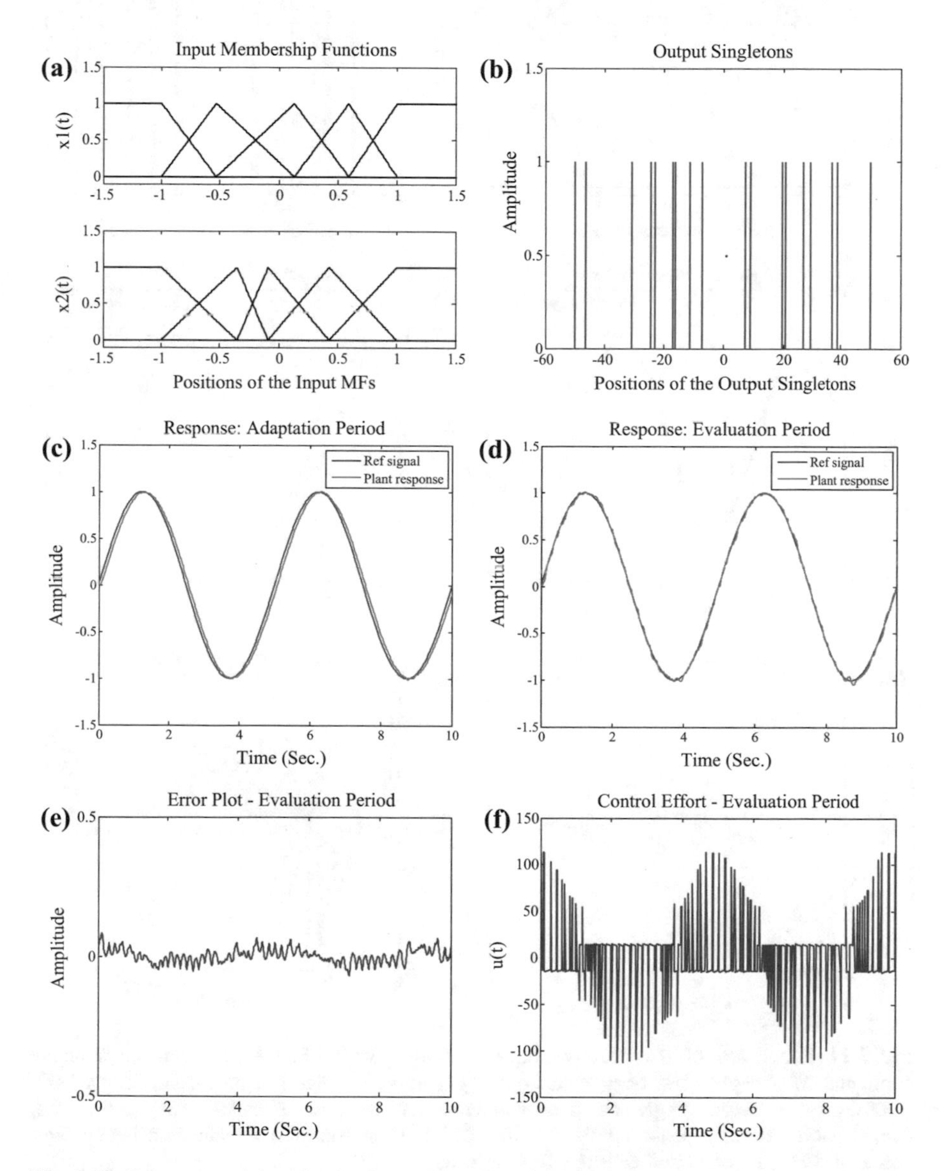

Fig. 3.13 Responses of the adaptive fuzzy controller with PSOBA model, for 5 × 5 configuration of input MFs, of case study II (A) (Duffing's system): **a** positions of input MFs, **b** positions of output singletons, **c** adaptation period response after 100 PSO generations, **d** evaluation period response for 0–10 s after 200 PSO generations, **e** evaluation period error variation, **f** evaluation period control effort variation

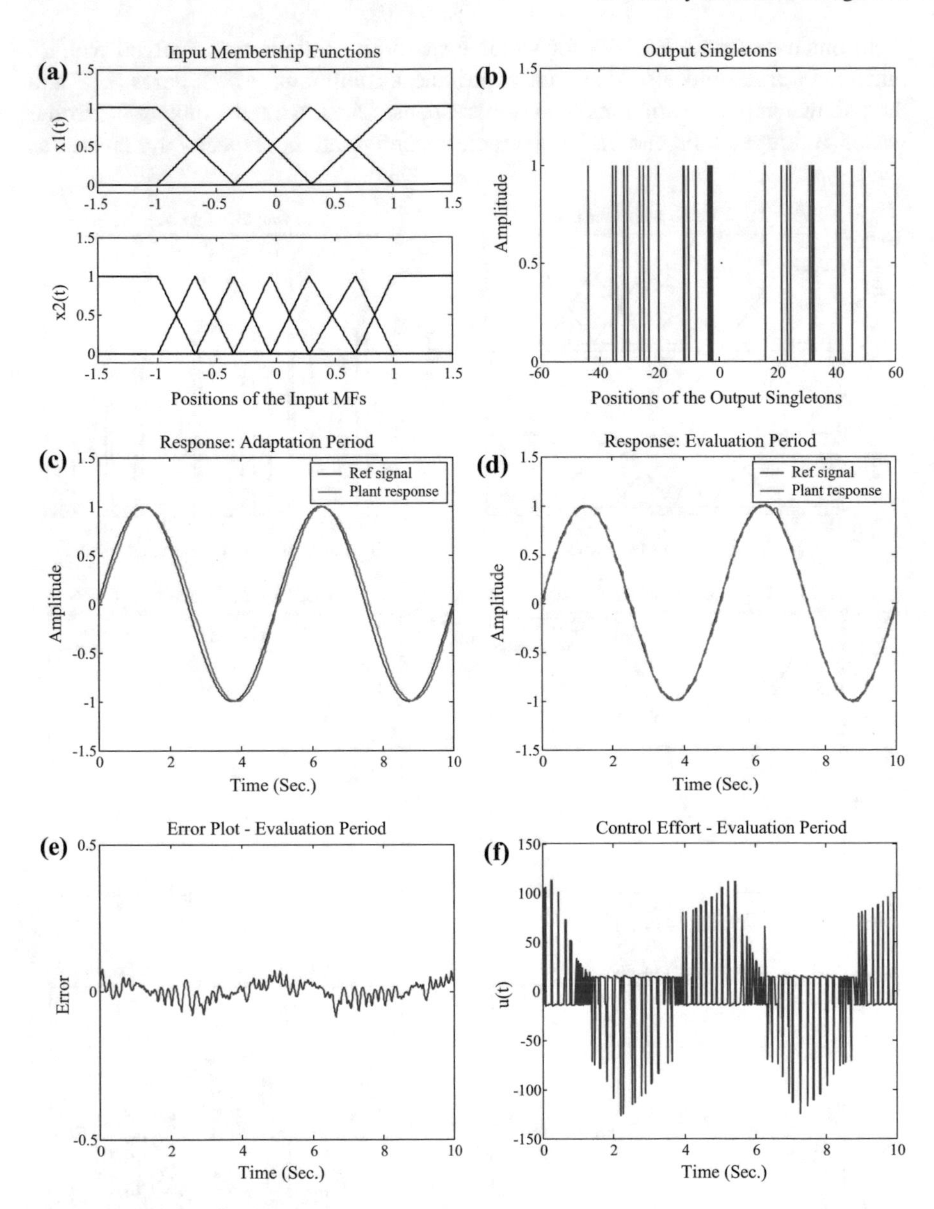

Fig. 3.14 Responses of the adaptive fuzzy controller with PSOBA, for variable structure configuration of input MFs, of case study II (A) (Duffing's system): **a** positions of input MFs, **b** positions of output singletons, **c** adaptation period response after 100 PSO generations, **d** evaluation period response for 0–10 s after 200 PSO generations, **e** evaluation period error variation, **f** evaluation period control effort variation

representative samples. In (a) and (b) of Figs. 3.13 and 3.14, the input MFs and the positions of the output singletons are shown, respectively, after the completion of the adaptation process. Figures 3.13c and 3.14c show the adaptation period responses after 100 PSO iterations, and in (d)–(f) of Figs. 3.13 and 3.14, the evaluation period responses, error variations, and control effort variations are shown, respectively. In this case study also, the values of the crosspoint level and the crosspoint ratio are maintained to be 0.5 and 1, respectively, to achieve better performances from the FLCs designed [26].

Performance Analysis of HSABA

Very similar to the approach considered in case study I, harmony memory of ten harmonies is considered here. For each simulation, 200 iterations of HSA are performed. The fixed structure and variable structure design procedures are carried out following the same philosophy as was adopted in case study I. The performances of different HSABA configurations are tabulated in Table 3.4.

In this case study also, the simulations are carried out for three fixed structures of 3×3, 5×5, and 7×7 input MFs and a variable structure with a potentially biggest possible structure of 7×7 MFs. For fixed structure controllers, it can be seen that the performances of the controllers (demonstrated by the performance index of average *IAE*) improve with increase in the number of rules from 3×3 to 7×7 configuration. However, the best *IAE* value obtained does not support this trend. Therefore, it can be concluded again that, merely increasing number of rules in fuzzy controllers does not necessarily improve their performances in all fronts [19], as the operation of a fuzzy controller depends on many complex relationships between the variables associated with the operation. The reduction obtained in the number of rules in case of variable structure configuration is about 50% (24 out of 49). But it does not provide any extra edge as is evident from the results in Table 3.4. Figures 3.15 and 3.16 show the evaluation phase responses of the controller designed according to HSABA algorithm, with a fixed structure of 5×5 MFs and variable structure MFs, respectively, chosen as representative samples.

Table 3.4 Comparison of simulation results for case study II (A) (Duffing's system): HSABA-based design strategies

Control strategy	No. of MFs	No. of rules	Avg. IAE	Best IAE	Std. Dev.
HSABA	3×3	9	0.2554	0.2417	0.0125
HSABA	5×5	25	0.2220	0.1936	0.0200
HSABA	7×7	49	0.2137	0.2071	0.0093
HSABA–V	4×6	24	–	0.2428	–

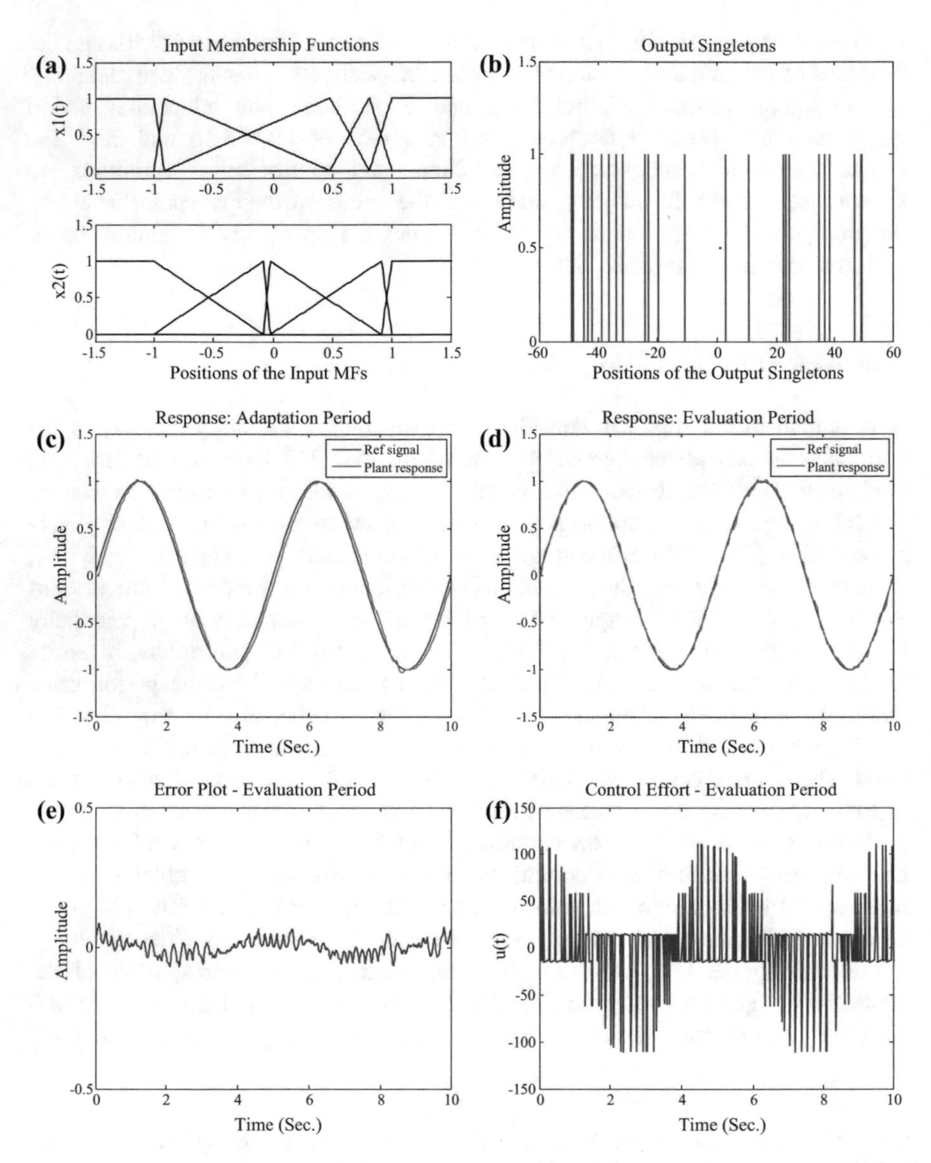

Fig. 3.15 Responses of the adaptive fuzzy controller with HSABA, for 5×5 configuration of input MFs, of case study II (A) (Duffing's system): **a** positions of input MFs, **b** positions of output singletons, **c** adaptation period response after 100 HS generations, **d** evaluation period response for 0–10 s after 200 HS generations, **e** evaluation period error variation, **f** evaluation period control effort variation

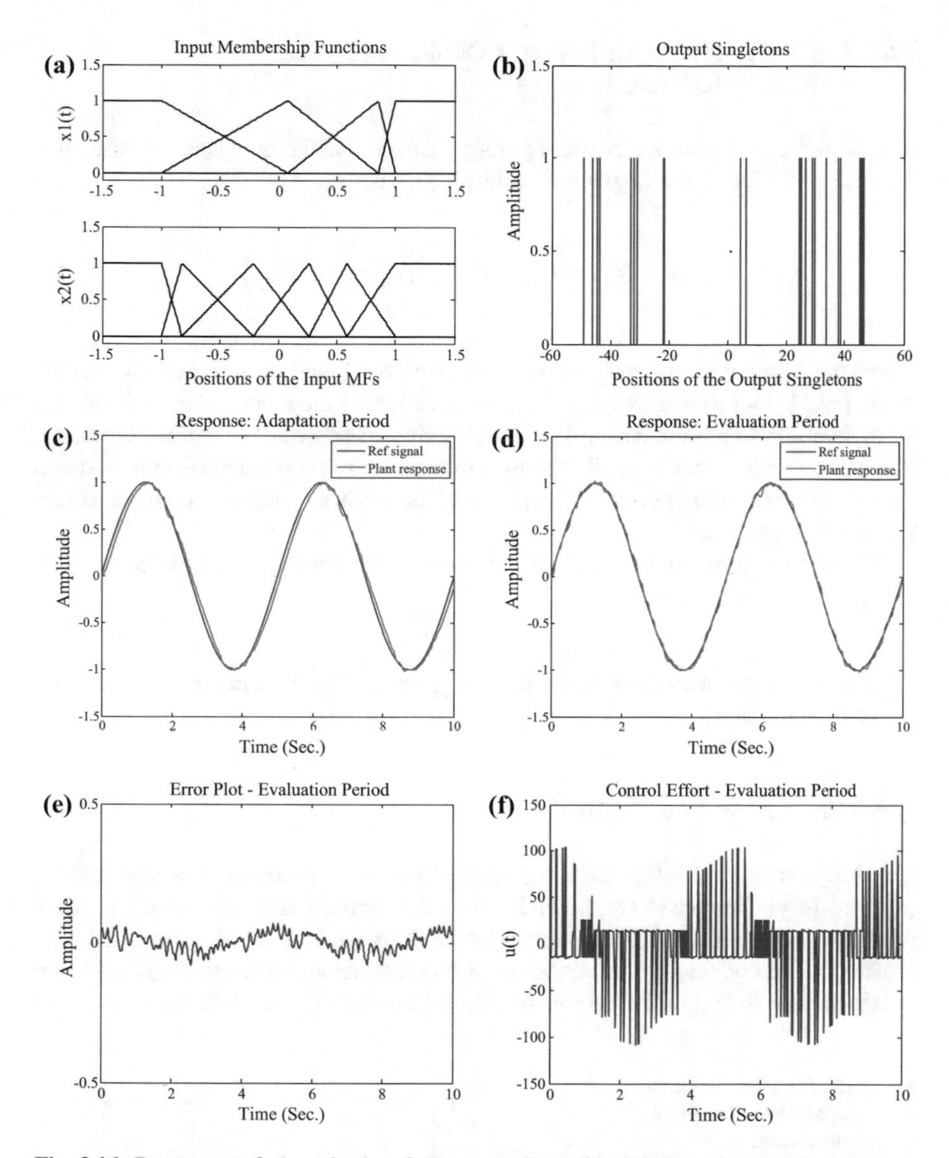

Fig. 3.16 Responses of the adaptive fuzzy controller with HSABA, for variable structure configuration of input MFs, of case study II (A) (Duffing's system): **a** positions of input MFs, **b** positions of output singletons, **c** adaptation period response after 100 HS generations, **d** evaluation period response for 0 to 10 s after 200 HS generations, **e** evaluation period error variation, **f** evaluation period control effort variation

3.6.2.2 Case Study II (B): Duffing's Oscillatory System with Disturbance

In this part of the case study, the Duffing's forced oscillation system with disturbance is considered and is given as follows [5, 10]:

$$\left.\begin{aligned} \dot{x}_1 &= x_2 \\ \dot{x}_2 &= -0.1x_2 - x_1^3 + 12\cos(t) + u(t) + d(t) \\ y &= x_1 \end{aligned}\right\} \tag{3.37}$$

where the external disturbance $d(t)$ is a square wave of random amplitude within the range [−1, 1] and a period of 0.5 s. Figure 3.17 shows the nature of disturbance for 10 s. duration, to present a sample, representative demonstration. In this case study, this type of disturbance is applied to the system for the entire adaptation period and also for the evaluation period. The control objective is to track the reference signal y_m, where $y_m = \sin(t)$.

The control signal obtained from the fuzzy controller is given as follows:

$$u_c = u_c(\underline{x}|\underline{Z}) \tag{3.38}$$

The design parameters for this case study are exactly the same as those chosen for case study II(A).

Performance Analysis of PSOBA

The design strategies of PSOBA are simulated in a similar fashion as in case study II (A), and the performances are shown in Table 3.5. In this case study, the disturbance rejection capabilities of FLC designed by PSOBA strategy are demonstrated successfully. The tracking performances of different structural configurations are more or less similar to that of the case study II(A). Figures 3.18 and 3.19 show temporal

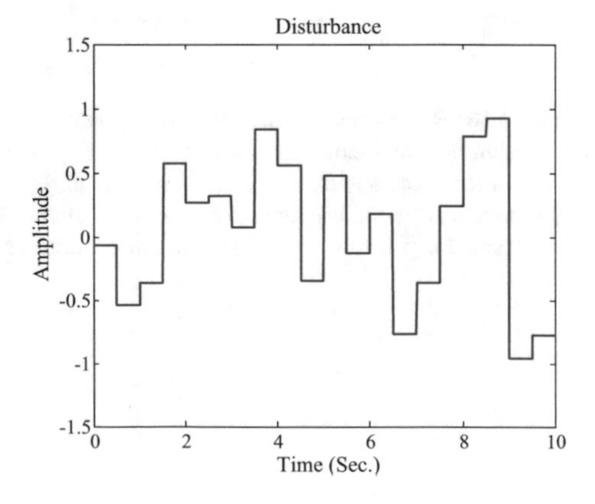

Fig. 3.17 A sample nature of the disturbance applied to the system in case study II (B) (Duffing's system with disturbance)

Table 3.5 Comparison of simulation results for case study II (B) (Duffing's system with disturbance): PSOBA-based design strategies

Control strategy	No. of MFs	No. of rules	Avg. IAE	Best IAE	Std. Dev.
PSOBA	3 × 3	9	0.2268	0.2047	0.0310
PSOBA	5 × 5	25	0.2143	0.2012	0.0185
PSOBA	7 × 7	49	0.2578	0.2311	0.0250
PSOBA-V	4 × 3	12	–	0.2190	–

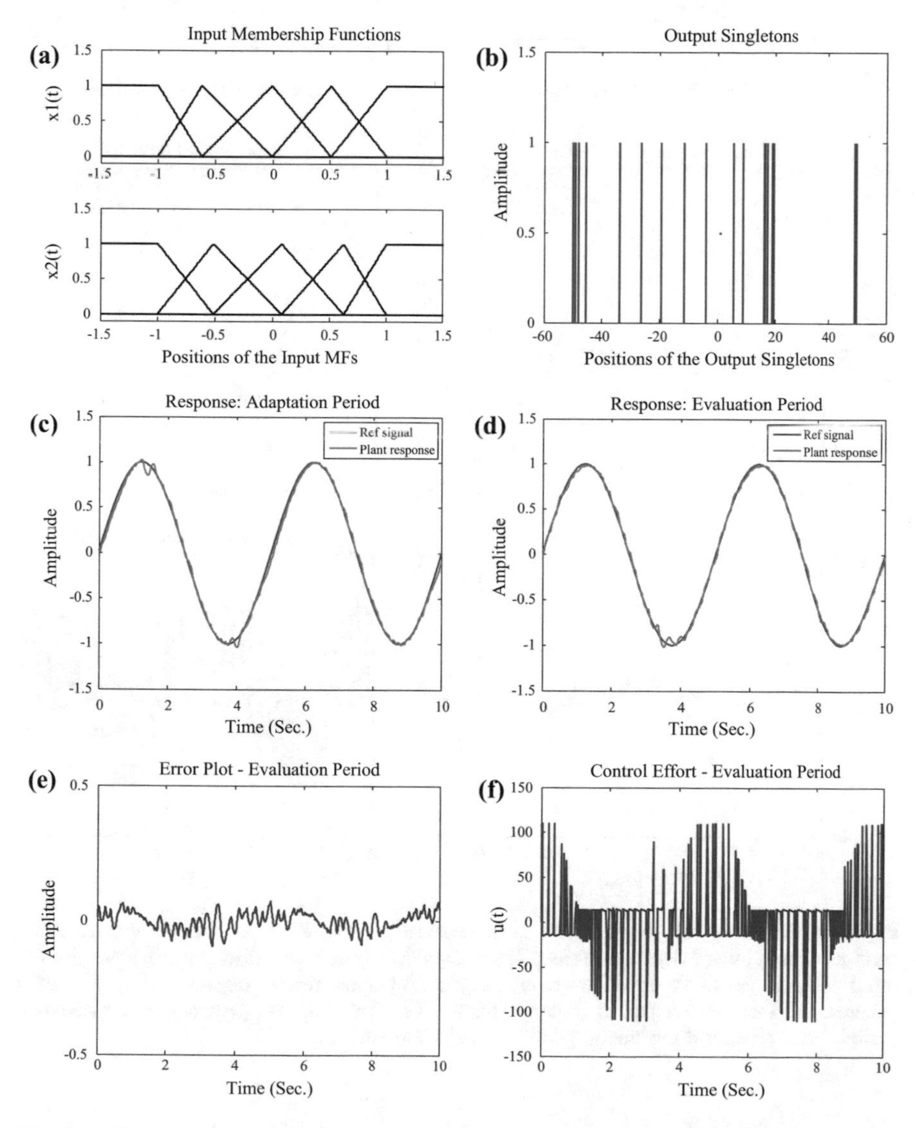

Fig. 3.18 Responses of the adaptive fuzzy controller with PSOBA model, for 5 × 5 configuration of input MFs, of case study II (B) (Duffing's system with disturbance): **a** positions of input MFs, **b** positions of output singletons, **c** adaptation period response after 100 PSO generations, **d** evaluation period response for 0–10 s after 200 PSO generations, **e** evaluation period error variation, **f** evaluation period control effort variation

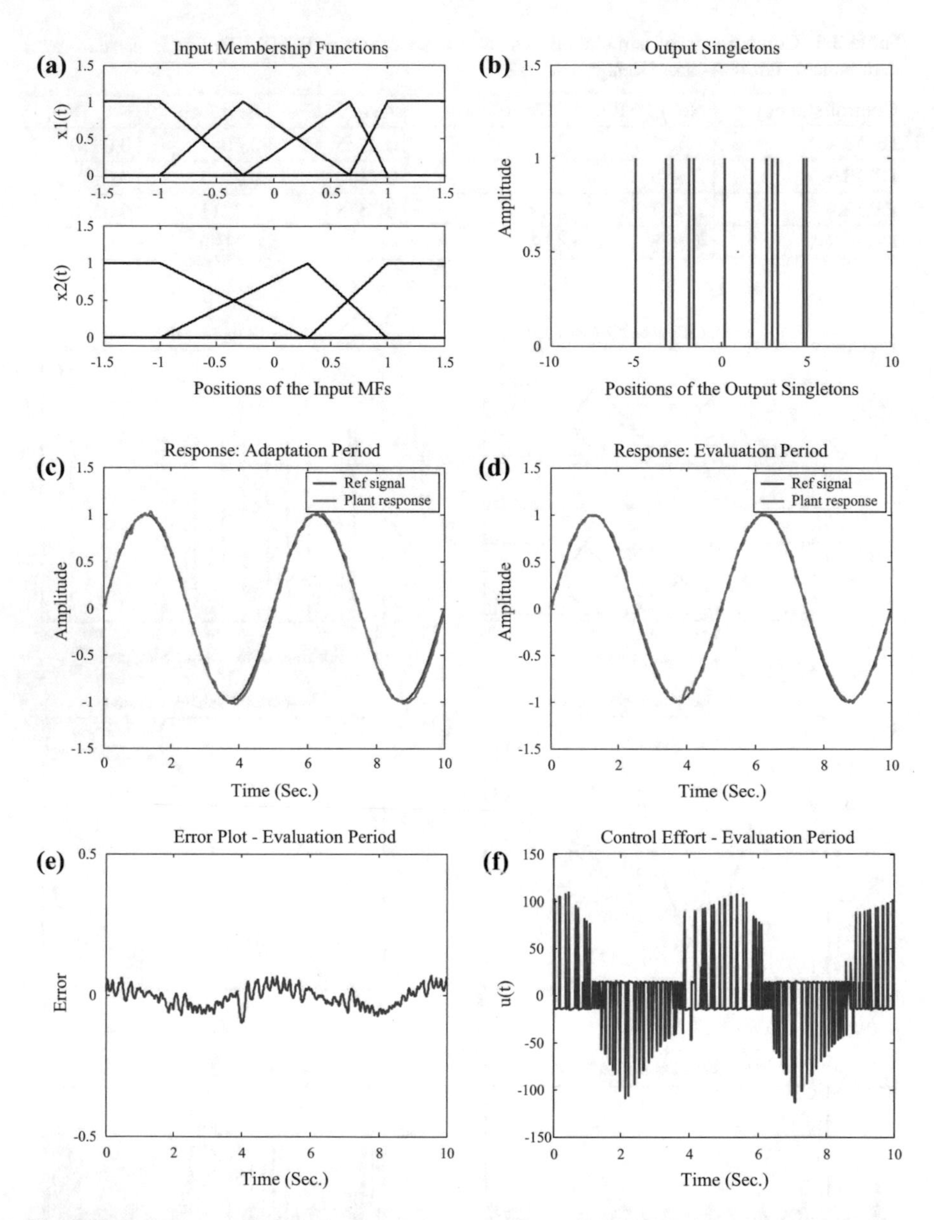

Fig. 3.19 Responses of the adaptive fuzzy controller with PSOBA, for variable structure configuration of input MFs, of case study II (B) (Duffing's system with disturbance): **a** positions of input MFs, **b** positions of output singletons, **c** adaptation period response after 100 PSO generations, **d** evaluation period response for 0–10 s after 200 PSO generations, **e** evaluation period error variation, **f** evaluation period control effort variation

Table 3.6 Comparison of simulation results for case study II (B) (Duffing's system with disturbance): HSABA-based design strategies

Control strategy	No. of MFs	No. of rules	Avg. IAE	Best IAE	Std. Dev.
HSABA	3 × 3	9	0.2609	0.2532	0.0075
HSABA	5 × 5	25	0.2216	0.2201	0.0021
HSABA	7 × 7	49	0.2432	0.2314	0.0104
HSABA-V	3 × 5	15	–	0.2718	–
	3 × 7	21	–	0.2353	–

variations of system responses for the representative fixed structure controller with 5 × 5 input MFs and variable structure configuration, respectively, for PSOBA control strategy, in a very similar fashion like the approach adopted in case study II(A).

Performance Analysis of HSABA

This part of case study also depicts the performance of HSABA control strategy for simultaneous tracking and disturbance rejection. Different configurations of FLCs under consideration are very similar to those chosen in case study II(A), and their results are tabulated in Table 3.6. Corresponding graphical results are presented in Figs. 3.20 and 3.21, following the same approach utilized previously.

3.7 Summary

This chapter has discussed the T-S-type fuzzy control scheme under consideration in detail and the application of supervisory control to ensure the stability is also discussed. A stochastic algorithm-based approach for systematic design of zero-order T-S-type FLC is described. PSO and HSA are the two contemporary stochastic optimization approaches considered in this chapter to design the FLC, and then, a generic stochastic optimization-based philosophy has been introduced. Two benchmark nonlinear systems are utilized to demonstrate the performances of the designed controllers. After analyzing the performances of these controllers, it can be concluded that a mere increase in the number of fuzzy rules does not improve the transient performances of the system in terms of *IAE*. The performance depends on very complex relationships among the number of rules, scaling gains, positions of the output singletons, etc. Overall, the variable structure configuration can significantly reduce the number of rules, keeping tracking performances reasonably satisfactory. In subsequent chapters, different types of adaptive schemes will be elaborated to enhance the performances of the FLCs further.

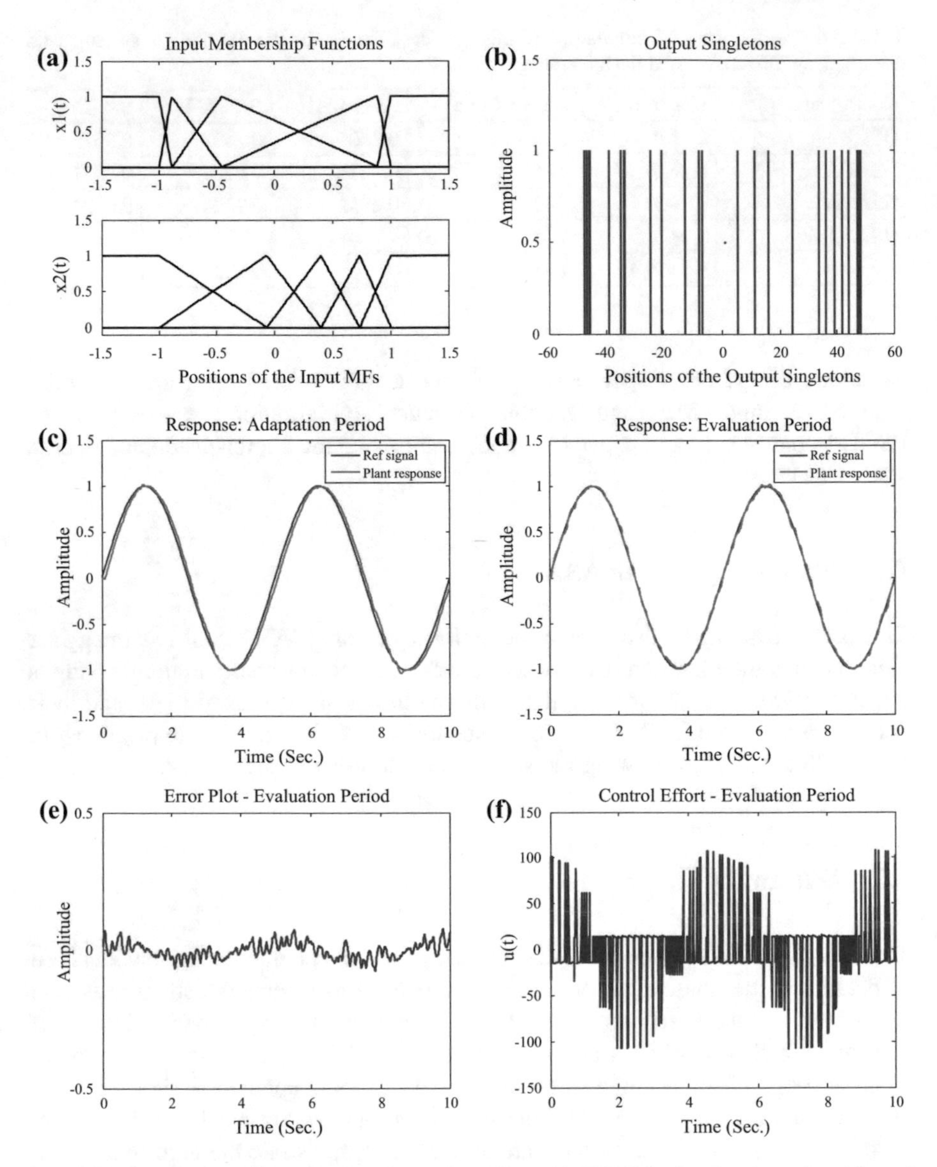

Fig. 3.20 Responses of the adaptive fuzzy controller with HSABA, for 5×5 configuration of input MFs, of case study II (B) (Duffing's system with disturbance): **a** positions of input MFs, **b** positions of output singletons, **c** adaptation period response after 100 HS generations, **d** evaluation period response for 0–10 s after 200 HS generations, **e** evaluation period error variation, **f** evaluation period control effort variation

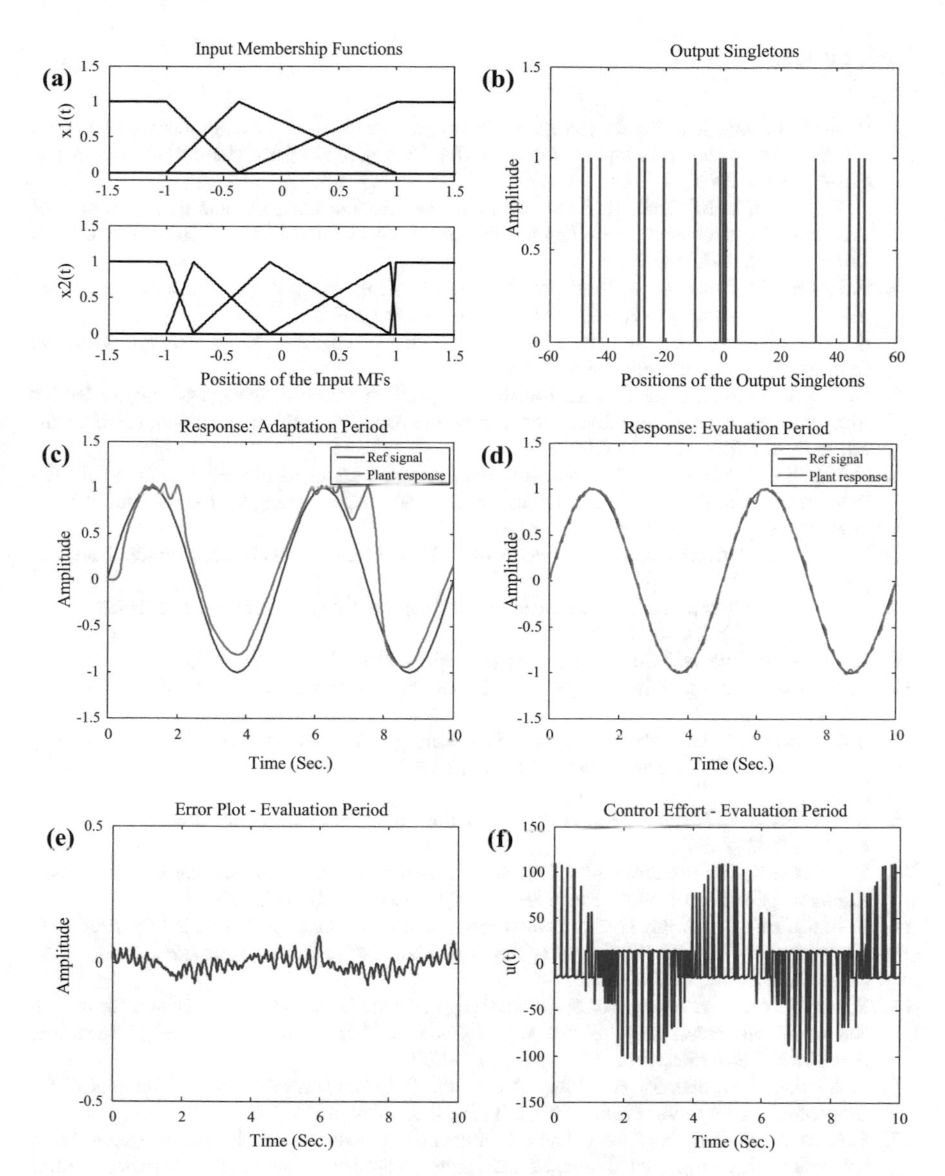

Fig. 3.21 Responses of the adaptive fuzzy controller with HSABA, for variable structure configuration of input MFs, of case study II (B) (Duffing's system with disturbance): **a** positions of input MFs, **b** positions of output singletons, **c** adaptation period response after 100 HS generations, **d** evaluation period response for 0–10 s after 200 HS generations, **e** evaluation period error variation, **f** evaluation period control effort variation

References

1. P. He, S. Jagannathan, Reinforcement learning neural network based controller for non-linear discrete time systems with input constraints. IEEE Trans. Syst. Man Cybern. B Cybern. **37**(2), 425–436 (Apr 2007)
2. F. Cupertino, E. Mininno, D. Naso, B. Turchiano, L. Salvatore, On-line genetic design of anti-windup unstructured controllers for electric drives with variable load. IEEE Trans. Evol. Comput. **8**(4), 347–364 (2004)
3. F. Herrera, M. Lozano, J.L. Verdegay, Tuning fuzzy logic controllers by genetic algorithms. Int. J. Approx. Reason. **12**(3–4), 299–315 (April–May 1995)
4. K. Das Sharma, Hybrid methodologies for stable adaptive fuzzy control, Doctoral Dissertation, Jadavpur University, India, 2012
5. K. Das Sharma, A. Chatterjee, A. Rakshit, A hybrid approach for design of stable adaptive fuzzy controllers employing Lyapunov theory and particle swarm optimization. IEEE Trans. Fuzzy Syst. **17**(2), 329–342 (2009)
6. O. Castillo, R. Martinez-Marroquin, J. Soria, Parameter tuning of membership functions of a fuzzy logic controller for an autonomous wheeled mobile robot using ant colony optimization, in *Proceedings of SMC* (2009), pp. 4770–4775
7. L.-X. Wang, *A Course in Fuzzy Systems and Control* (Englewood Cliffs, Prentice Hall, NJ, 1997)
8. K. Fischle, D. Schroder, An improved stable adaptive fuzzy control method. IEEE Trans. Fuzzy Syst. **7**(1), 27–40 (1999)
9. S.H. Zak, *Systems and Control* (Oxford University Press, New York, 2003)
10. L.X. Wang, Stable adaptive fuzzy control of nonlinear system. IEEE Trans. Fuzzy Syst. **1**(2), 146–155 (May 1993)
11. R.C. Eberhart, J. Kennedy, A new optimizer using particle swarm theory, in *Proceedings of the 6th International Symposium Micromachine Human Science,* Nagoya, Japan (1995), pp 39–43
12. J. Kennedy, R.C. Eberhart, Particle swarm optimization. Proc. IEEE Int. Conf. Neural Netw. **4**, 1942–1948 (1995)
13. S.P. Ghoshal, Optimizations of PID gains by particle swarm optimizations in fuzzy based automatic generation control. Electr. Power Syst. Res. **72**, 203–212 (2004)
14. Y. Liu, J. Bang, S. Wang, Optimization design based on PSO algorithm for PID controller, in *Proceedings of 5th World Congress Intelligent Control and Automation*, Hangzhou, P.R. China, June 2004, pp. 2419–2422
15. Y. Du, N. Wang, J. Zhang, An optimum design method based on PSO algorithm for neuron controllers, in *Proceedings of 5th World Congress Intelligent Control and Automation*, Hangzhou, P.R. China, June 2004, pp. 2617–2621
16. Y.-X. Liao, J.-H. She, M. Wu, Integrated hybrid-PSO and fuzzy-NN decoupling control for temperature of reheating furnace. IEEE Trans. Ind. Elect. **56**(7), 2704–2714 (2009)
17. B.K. Sahu, S. Pati, S. Panda, Hybrid differential evolution particle swarm optimization optimised fuzzy proportional–integral-derivative controller for automatic generation control of interconnected power system. IET Gener. Transm. Distrib. **8**(11), 1789–1800 (2014)
18. H. Bevrani, F. Habibi, P. Babahajyani, M. Watanabe, Y. Mitani, Intelligent frequency control in an AC microgrid: online PSO-based fuzzy tuning approach. *IEEE Trans. Smart Grid* (2012)
19. V. Giordano, D. Naso, B. Turchiano, Combining genetic algorithm and Lyapunov-based adaptation for online design of fuzzy controllers. IEEE Trans. Syst. Man Cybern. B Cybern. **36**(5), 1118–1127 (Oct 2006)
20. K. Das Sharma, A. Chatterjee, and F. Matsuno, A Lyapunov theory and stochastic optimization based stable adaptive fuzzy control methodology, in *Proceedings of SICE Annual Conference 2008*, Japan, Aug 20–22, pp. 1839–1844

21. G. Venter, J.S. Sobieski, Particle swarm optimization, in *43rd AIAA Conference*, Colorado (2002)
22. Y. Shi and R. C. Eberhart, A modified particle swarm optimizer, in *Proceedings of IEEE International Conference Evolution Computation*, Alaska, May 1998
23. L. Wang, R. Yang, P.M. Pardalos, L. Qian, M. Fei, An adaptive fuzzy controller based on harmony search and its application to power plant control. Electr. Power Energy Syst. **53**, 272–278 (2013)
24. K. DasSharma, A. Chatterjee, A. Rakshit, Harmony search algorithm and Lyapunov theory based hybrid adaptive fuzzy controller for temperature control of air heater system with transport-delay. Appl. Soft Comput. **25**, 40–50 (2014)
25. T. Mahto, V. Mukhetjee, Power and frequency stabilization of an isolated hybrid power system incorporating scaling factor based fuzzy logic controller, in *3rd International Conference on Recent Advances in Information Technology, RAIT,* 2016
26. C.K. Shiva, V. Mukherjee, Automatic generation control of multi-unit multi-area deregulated power system using a novel quasi-oppositional harmony search algorithm. IET Gener. Transm. Distrib., pp. 1–11 (2015)
27. C.K. Shiva, V. Mukherjee, A novel quasi-oppositional harmony search algorithm for automatic generation control of power system. Appl. Soft Comput. (2015)
28. K. Das Sharma, A. Chatterjee, A. Rakshit, Design of a hybrid stable adaptive fuzzy controller employing Lyapunov theory and harmony search algorithm. IEEE Trans. Contr. Syst. Tech. **18**(6), 1440–1447 (Nov 2010)
29. Z.W. Geem, J.H. Kim, G.V. Loganathan, A new heuristic optimization algorithm: harmony search. Simulation **76**(2), 60–68 (2001)
30. K.S. Lee, Z.W. Geem, A new structural optimization method based on the harmony search algorithm. Comput. Struct. **82**, 781–798 (2004)
31. D. Driankov, H. Hellendoorn, M.M. Reinfrank, *An Introduction to Fuzzy Control* (Springer, Heidelberg, Germany, 1993)

Part II
Lyapunov Strategy Based Design Methodologies

Chapter 4
Fuzzy Controller Design II: Lyapunov Strategy-Based Adaptive Approach

4.1 Introduction

Fuzzy logic control schemes are widely employed for the ill-defined systems with parametric uncertainty and disturbances. The controller structure and its free parameters are determined on the basis of heuristic knowledge of the plant. However, when heuristics do not provide enough information to specify all the parameters of a fuzzy controller, adaptive schemes based on input–output data are employed to automatically learn these parameters [1–3]. In [4, 5], authors presented two earliest, most notable methodologies for designing direct stable adaptive fuzzy logic controllers.

In [4], a direct adaptive fuzzy controller was developed which was capable of incorporating fuzzy control rules directly into the controllers by means of Lyapunov theory-based adaptation scheme, and guaranteed the global stability of the resulting closed-loop system in the sense that all variables involved are uniformly bounded. It was also shown how the supervisory control could force the state (states) of the controlled plant to be within the bound(s), set by the designer. On the other hand, in [5], the problem of temporarily varying reference input was tackled using a modified version of Lyapunov theory-based adaptation of controller parameters.

In this chapter, the Lyapunov theory-based stable adaptive fuzzy controllers have been utilized to achieve the control objectives and some case studies are also presented to evaluate the performance of the designed controller [2, 4, 5].

4.2 Stable Adaptive Fuzzy Controllers

It is well defined in fuzzy control literature that although the basic principle and mathematical tools for designing both conventional adaptive control law and adaptive fuzzy control law are more or less the same, the incorporation of human

K. Das Sharma et al., *Intelligent Control*, Cognitive Intelligence and Robotics,
https://doi.org/10.1007/978-981-13-1298-4_4

expert knowledge into the fuzzy counterpart makes it more acceptable over a wide range of systems. If the fuzzy control law evolves from a number of fuzzy subsystems to introduce the plant knowledge, then the controller is termed as an indirect adaptive fuzzy controller. On the other hand, when the fuzzy control law constructed out of a single fuzzy system with the knowledge of control actions, the controller is known as direct adaptive fuzzy controller [6]. In this section, first the control objectives are formulated, and then, in a constructive manner, a direct adaptive fuzzy controller has been developed to achieve these control objectives.

Let us consider that the objective is to design an adaptive strategy for an nth-order SISO nonlinear plant given as [4, 7, 8]:

$$\left.\begin{aligned} x^{(n)} &= f(\underline{x}) + bu(t) \\ y &= x \end{aligned}\right\} \tag{4.1}$$

where $f(\cdot)$ is an unknown continuous function, $u \in R$ and $y \in R$ are the input and output of the plant, and b is an unknown positive constant. Let us assume that the state vector is given as $\underline{x} = (x_1, x_2, \ldots, x_n)^{\mathrm{T}} = (x, \dot{x}, \ldots, x^{(n-1)})^{\mathrm{T}} \in R^n$. The system under control has a reference model given as:

$$\left.\begin{aligned} x_m^{(n)} &= f(\underline{x}_m, w) \\ y_m &= x_m \end{aligned}\right\} \tag{4.2}$$

where $w(t)$ is the excitation signal input to the reference model, and the state vector of the reference model is $\underline{x}_m = (x_m, \dot{x}_m, \ldots, x_m^{(n-1)})^{\mathrm{T}} \in R^n$.

The control objective is to force the plant output $y(t)$ to follow a given bounded reference signal $y_m(t)$ under the constraints that all closed-loop variables involved must be bounded to guarantee the closed-loop stability of the system. Thus, the tracking error is $e = y_m - y$. The objective is to design a stable adaptive fuzzy controller for the system described in (4.1). AFLCs have evolved as popular control solutions for those classes of plants whose input–output characteristics are not precisely known. In case of the system in (4.1), this is given by the condition when $f(\cdot)$ and b are not precisely known, which is widely prevalent in practical situations. It is needed to find a feedback control strategy $u = u(\underline{x}|\underline{\theta})$, using fuzzy logic system and an adaptive law for adjusting the parameter vector $\underline{\theta}$ such that the following conditions are satisfied:

(i) The closed-loop system must be globally stable in the sense that all variables, $\underline{x}(t)$, $\underline{\theta}(t)$ and $u(\underline{x}|\underline{\theta})$ must be uniformly bounded, i.e., $|\underline{x}(t)| \le M_x < \infty$, $|\underline{\theta}(t)| \le M_\theta < \infty$ and $|u(\underline{x}|\underline{\theta})| \le M_u < \infty$, where M_x, M_θ, and M_u are set by the designer [2, 4, 8].

(ii) The tracking error $e(t)$ should be as small as possible under the constraints in (i), and the $\underline{e} - \underline{\theta}$ space should be stable in the large for the system [2, 5].

Now the ideal control law for the system in (4.1) and (4.2) is given as [2, 6, 8]:

$$u^* = \frac{1}{b}\left[-f(\underline{x}) + y_m^{(n)} + \underline{k}^{\mathrm{T}}\underline{e}\right] \tag{4.3}$$

where $y_m^{(n)}$ is the nth derivative of the output of the reference model, $\underline{e} = (e, \dot{e}, \ldots, e^{(n-1)})^{\mathrm{T}}$ is the error vector and $\underline{k} = (k_1, k_2, \ldots, k_n)^{\mathrm{T}} \in R^n$ is the vector describing the desired closed-loop dynamics for the error.

This definition implies that u^* guarantees perfect tracking, i.e., $y(t) \equiv y_m(t)$ if $Lt_{t\to\infty}\, e(t) = 0$. But in practical situations, since f and b are not known precisely, the ideal u^* of (4.3) cannot be implemented in real practice. Thus, a suitable solution can be to design a fuzzy logic system to approximate this optimal control law [8–11].

Similar to the discussions presented in Chap. 3, it has been assumed that the AFLC is constructed using a zero-order Takagi–Sugeno (T–S) fuzzy system. Then $u_c(\underline{x}|\underline{\theta})$ for the AFLC is given in the form [2, 6, 12]:

$$u_c(\underline{x}|\underline{\theta}) = \sum_{l=1}^{N} \theta_l * \alpha_l(\underline{x}) \Bigg/ \left(\sum_{l=1}^{N} \alpha_l(\underline{x})\right) = \underline{\theta}^{\mathrm{T}} * \underline{\xi}(\underline{x}) \tag{4.4}$$

where $\underline{\theta} = [\theta_1\, \theta_2 \ldots \theta_N]^{\mathrm{T}}$ = the vector of the output singletons, $\alpha_l(\underline{x})$ = = the firing degree of rule "l", N = the total number of rules, $\mu_i^l(x_i)$ = the membership value of the ith input membership function (MF) in the activated lth rule, $\underline{\xi}(\underline{x})$ = vector containing normalized firing strength of all fuzzy IF-THEN rules = $(\xi_1(\underline{x}), \xi_2(\underline{x}), \ldots, \xi_N(\underline{x}))^{\mathrm{T}}$ and

$$\xi_l(\underline{x}) = \alpha_l(\underline{x}) \Bigg/ \left(\sum_{l=1}^{N} \alpha_l(\underline{x})\right) \tag{4.5}$$

For more details, see Sect. 3.2 in Chap. 3.

Design of the Adaptation Law

The error differential equation can be written as [6, 8, 10, 13]:

$$\dot{\underline{e}} = \Lambda_c \underline{e} + \underline{b}_c[u^* - u_c(\underline{x}|\underline{\theta})] \tag{4.6}$$

where

$$\Lambda_c = \begin{bmatrix} 0 & 1 & 0 & 0 & \cdots & 0 & 0 \\ 0 & 0 & 1 & 0 & \cdots & 0 & 0 \\ \vdots & \vdots & \vdots & \vdots & \ddots & \vdots & \vdots \\ 0 & 0 & 0 & 0 & \cdots & 0 & 1 \\ -k_1 & -k_2 & \cdots & \cdots & \cdots & -k_{n-1} & -k_n \end{bmatrix} \text{ and } \underline{b}_c = \begin{bmatrix} 0 \\ \vdots \\ 0 \\ b \end{bmatrix} \tag{4.7}$$

Let us define the optimal parameters as:

$$\underline{\theta}^* = \arg \min_{\underline{\theta} \in \Re^{\prod_{i=1}^n m_i}} \left[\sup_{\underline{x} \in \Re^n} |u_c(\underline{x}|\underline{\theta}) - u^*| \right] \tag{4.8}$$

and the minimum inherent approximation error as:

$$\varepsilon = u_c(\underline{x}|\underline{\theta}) - u^* \tag{4.9}$$

Using (4.4) and (4.9), the error equation (4.6) becomes:

$$\dot{\underline{e}} = \Lambda_c \underline{e} + \underline{b}_c (\underline{\theta}^* - \underline{\theta})^{\mathrm{T}} \underline{\xi}(\underline{x}) - \underline{b}_c \varepsilon \tag{4.10}$$

Let us define a quadratic form of Lyapunov function as [6]:

$$V_e = \frac{1}{2} \underline{e}^{\mathrm{T}} P \underline{e} + \frac{b}{2v} (\underline{\theta}^* - \underline{\theta})^{\mathrm{T}} (\underline{\theta}^* - \underline{\theta}) \tag{4.11}$$

where v is a positive constant and P is a symmetric positive definite matrix satisfying the Lyapunov equation,

$$\Lambda_c^{\mathrm{T}} P + P \Lambda_c = -Q \tag{4.12}$$

Here Q is a positive definite matrix and V_e in (4.11) is positive, as it has been assumed that $b > 0$.

Using (4.10) and (4.12), we have

$$\dot{V}_e = -\frac{1}{2} \underline{e}^{\mathrm{T}} Q \underline{e} + \underline{e}^{\mathrm{T}} P \underline{b}_c \left[(\underline{\theta}^* - \underline{\theta})^{\mathrm{T}} \underline{\xi}(\underline{x}) - \varepsilon \right] - \frac{b}{v} (\underline{\theta}^* - \underline{\theta})^{\mathrm{T}} \dot{\underline{\theta}} \tag{4.13}$$

Let p_n be the last column of P, then from $\underline{b}_c = [0, 0, \ldots, b]^{\mathrm{T}}$ we have $\underline{e}^{\mathrm{T}} P \underline{b}_c = \underline{e}^{\mathrm{T}} p_n b$.

Thus, (4.13) can be rewritten as:

$$\dot{V}_e = -\frac{1}{2}\underline{e}^{\mathrm{T}}Q\underline{e} + \frac{b}{v}(\underline{\theta}^* - \underline{\theta})^{\mathrm{T}}\left[v\underline{e}^{\mathrm{T}}p_n\underline{\xi}(\underline{x}) - \underline{\dot{\theta}}\right] - \underline{e}^{\mathrm{T}}p_n b v \tag{4.14}$$

If the adaptation law is chosen as:

$$\underline{\dot{\theta}} = v\underline{e}^{\mathrm{T}}p_n\underline{\xi}(\underline{x}) \tag{4.15}$$

The positive constant v is termed as adaptation gain/learning rate.
Thus,

$$\dot{V}_e = -\frac{1}{2}\underline{e}^{\mathrm{T}}Q\underline{e} - \underline{e}^{\mathrm{T}}p_n b v \tag{4.16}$$

Since $Q > 0$ and ε is the minimum inherent approximation error, it can be hoped that by designing the fuzzy control law $u = u(\underline{x}|\underline{\theta})$ with adequate number of rules, the ε will be small enough such that $\frac{1}{2}\underline{e}^{\mathrm{T}}Q\underline{e} > |\underline{e}^{\mathrm{T}}p_n b v|$, which results in $\dot{V} < 0$.

Stability Issues

Now, to ensure stability, it is assumed that the control law $u(t)$ is actually the summation of the fuzzy control, $u_c(\underline{x}|\underline{\theta})$, and an additional non-fuzzy supervisory control strategy, $u_s(\underline{x})$, given as [6, 8]:

$$u(t) = u_c(\underline{x}|\underline{\theta}) + u_s(\underline{x}) \tag{4.17}$$

where $u_s(\underline{x})$, the supervisory control, is used in addition to the AFLC scheme so that the system states can be kept bounded.

Now the closed-loop error equation in (4.6) becomes

$$\underline{\dot{e}} = \Lambda_c\underline{e} + \underline{b}_c[u^* - u_c(\underline{x}|\underline{\theta}) - u_s(\underline{x})] \tag{4.18}$$

where Λ_c and $\underline{b}_c$ are the same as in (4.7).

Again, defining the Lyapunov function as:

$$V_e = \frac{1}{2}\underline{e}^{\mathrm{T}}P\underline{e} \tag{4.19}$$

and using (4.12) and (4.18), it can be written as:

$$\begin{aligned}\dot{V}_e &= -\frac{1}{2}\underline{e}^{\mathrm{T}}Q\underline{e} + \underline{e}^{\mathrm{T}}P\underline{b}_c[u^* - u_c(\underline{x}|\underline{\theta}) - u_s(\underline{x})] \\ &\leq -\frac{1}{2}\underline{e}^{\mathrm{T}}Q\underline{e} + |\underline{e}^{\mathrm{T}}P\underline{b}_c|(|u^*| + |u_c|) - \underline{e}^{\mathrm{T}}P\underline{b}_c u_s(\underline{x})\end{aligned} \tag{4.20}$$

$u_s(\underline{x})$ should be so designed that $\dot{V}_e$ should be negative semi-definite, i.e., $\dot{V}_e \leq 0$.

Assumption 4.1 The functions f^U and b_L can be determined, such that $f^U \geq |f(\underline{x})|$ and $0 < b_L \leq b$ [6, 8, 13].

Now, observing (4.3) and (4.20), we can infer that the supervisory control $u_s(\underline{x})$ is as follows [6–8]:

$$u_s(\underline{x}) = I_1^* \mathrm{sgn}\left(\underline{e}^{\mathrm{T}} P \underline{b}_c\right)\left[|u_c| + \frac{1}{b_L}\left(f^U + |y_m^{(n)}| + |\underline{k}^{\mathrm{T}}\underline{e}|\right)\right] \tag{4.21}$$

where $\begin{cases} I_1^* = 1 \text{ if } V_e > \overline{V} \\ I_1^* = 0 \text{ if } V_e \leq \overline{V} \end{cases}$, and $\overline{V}$ is a constant specified by the designer.

With this $u_s(\underline{x})$ in (4.21) and considering $I_1^* = 1$ case, it can be written that

$$\begin{aligned} \dot{V}_e \leq & -\frac{1}{2}\underline{e}^T Q \underline{e} + \ldots |\underline{e}^T P \underline{b}_c| [\frac{1}{b}\left(|f(\underline{x})| + |y_m^{(n)}| + |\underline{k}^T \underline{e}|\right) \\ & - \frac{1}{b_L}\left(f^U + |y_m^{(n)}| + |\underline{k}^T \underline{e}|\right)] \leq -\frac{1}{2}\underline{e}^T Q \underline{e} \leq 0 \end{aligned} \tag{4.22}$$

Thus, using supervisory control $u_s(\underline{x})$, one can guarantee $V_e \leq \overline{V}$, which implies the boundedness in $\underline{x}$. Hence, the closed-loop stability is guaranteed [6, 8, 14–16]. A schematic diagram of adaptive fuzzy control action augmented by supervisory control is shown in Fig. 4.1.

Note For adaptive fuzzy control, also it should be noted that when the reference signal y_m is the output of a reference model, excited with a given input signal w and

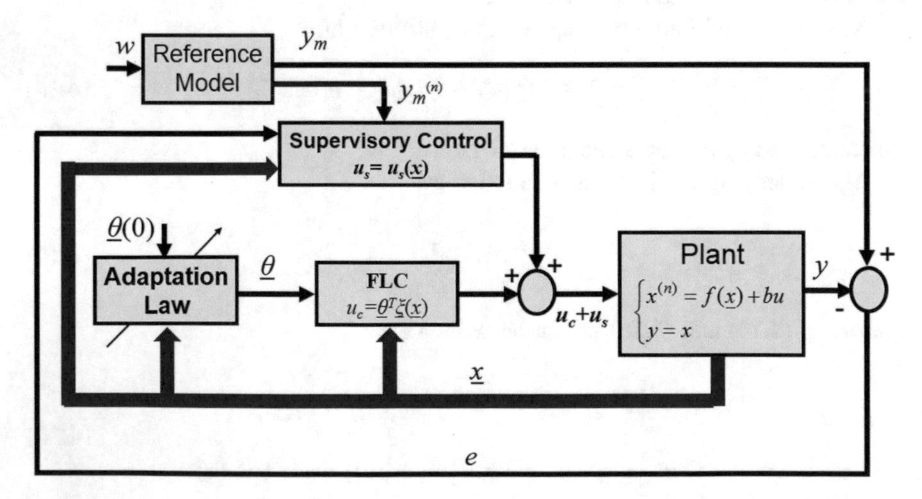

Fig. 4.1 A schematic architecture of adaptive fuzzy control action augmented by supervisory control

the reference plant has the same dynamics as that of the controlled plant and can be expressed by $\underline{k}$, then the input states include w along with $\underline{x}$ [8].

4.3 Lyapunov Strategy-Based Approach (LSBA)

Design methodology for stable direct adaptive fuzzy controllers, based on Lyapunov method, should satisfy the following mathematical conditions [2, 4, 5, 8]:

- The type of the plant must be like (4.1).
- The order of the plant and its relative degree must be equal, i.e., the plant must be an all-pole system.
- The nonlinearity of the plant must be differentiable in an adequate number of times.
- There should exist the first $(n - 1)$ derivatives of output of the plant for the learning algorithm.
- The fuzzy basis function $\underline{\xi}(\underline{x})$ must be exciting for the convergence of the parameter vector $\underline{\theta}$.
- The inherent approximation error must be bounded, and the boundary should be known to the designer.
- The control algorithm and adaptation law are required to be implemented in continuous-time mode, otherwise, if it is in discrete-time mode, then the sampling rate must be high enough so that the plant can be modeled as quasi-continuous.

4.3.1 The Design Algorithm

The Lyapunov strategy-based approach (LSBA) for the said controller design algorithm can be divided into two parts—(a) parameter calculations and (b) controller adaptation.

(a) ***Parameter Calculations***

(i) Specify $k_1, k_2, \ldots, k_n$ such that $s^n + k_n s^{n-1} + \cdots + k_2 s + k_1 = 0$ be a Hurwitz polynomial. The $\underline{k}$ vector can be chosen such that $k_i = a_{im}$, where a_{im} can be obtained from:

$$y_m^{(n)} = b_{1m}w - a_{1m}y_m - a_{2m}\dot{y}_m - \cdots - a_{nm}y_m^{(n-1)} \tag{4.23}$$

Here y_m is the output of a linear reference model having a numerator of degree zero as shown in (4.2) [5, 8, 17, 18].

(ii) Specify a positive definite $n \times n$ matrix Q and solve (4.12) to obtain $P > 0$.
(iii) Specify the design parameters M_x, M_θ, and M_u, based on the nature of the plant and control objectives.
(iv) Define m_i number of fuzzy sets or MFs, F_i^l for fuzzification of input variable x_i, which uniformly cover universe of discourse U_i, where $l = 1, 2, \ldots, m_i$ and $i = 1, 2, \ldots, n$. The $U = U_1 \times U_2 \times \cdots \times U_n$ is usually chosen to be $\{\underline{x} \in R^n : |\underline{x}| \leq M_x\}$.
(v) Construct the fuzzy rule base for fuzzy logic system $u_c(\underline{x}|\theta)$ as:

$$\begin{aligned} R_c^{(l_1,\ldots,l_n)} : &\ \text{IF}\, x_1\, \text{is}\, F_1^{l_1}\, \&\ldots\&\, \text{IF}\, x_n\, \text{is}\, F_n^{l_n} \\ &\ \text{THEN}\, u_c\, \text{is}\, G^{(l_1,\ldots,l_n)} \end{aligned} \tag{4.24}$$

where $G^{(l_1,\ldots l_n)}$ are constrained as $\{\underline{\theta} : |\underline{\theta}| \leq M_\theta\}$ and for zero-order T–S method these are constants of arbitrary values for initial settings. The consequent parts of the rules are utilized to create the singleton vector $\underline{\theta} = [\theta_1\, \theta_2 \ldots \theta_N]^{\mathrm{T}}$, N = total number of rules.

(vi) Define the outputs of fuzzy basis functions

$$\xi_l(\underline{x}) = \frac{\prod_{i=1}^{r} \mu_i^l(x_i)}{\sum_{l=1}^{N} \left(\prod_{i=1}^{r} \mu_i^l(x_i)\right)} \tag{4.25}$$

and create the vector of these basis functions $\underline{\xi}(\underline{x}) = [\xi_1(\underline{x}), \xi_2(\underline{x}), \ldots, \xi_N(\underline{x})]^{\mathrm{T}}$. The $u_c(\underline{x}|\theta)$ is thus constructed as

$$u_c(\underline{x}|\underline{\theta}) = \underline{\theta}^{\mathrm{T}} \underline{\xi}(\underline{x}) \tag{4.26}$$

(b) ***Controller Adaptation***

(i) The feedback control u as in (4.17) is applied to the plant (4.1), where u_c is calculated from (4.4) and u_s from (4.21).
(ii) The following adaptation law is applied to adjust the parameter vector $\underline{\theta}$:

$$\dot{\underline{\theta}} = \begin{cases} v\underline{e}^{\mathrm{T}} \underline{p}_n \underline{\xi}(\underline{x}) & \text{if } (|\underline{\theta}| < M_\theta) \text{ or} \\ & \left(|\underline{\theta}| = M_\theta \text{ and } \underline{e}^{\mathrm{T}} \underline{p}_n \underline{\theta}^{\mathrm{T}} \underline{\xi}(\underline{x}) \leq 0\right) \\ P\left\{v\underline{e}^{\mathrm{T}} \underline{p}_n \underline{\xi}(\underline{x})\right\} & \text{if } \left(|\underline{\theta}| = M_\theta \text{ and } \underline{e}^{\mathrm{T}} \underline{p}_n \underline{\theta}^{\mathrm{T}} \underline{\xi}(\underline{x}) > 0\right) \end{cases} \tag{4.27}$$

where $P\{*\}$ is the projection operator and is given by [4, 5, 8]:

$$P\left\{v\underline{e}^{\mathrm{T}} \underline{p}_n \underline{\xi}(\underline{x})\right\} = v\underline{e}^{\mathrm{T}} \underline{p}_n \underline{\xi}(\underline{x}) - v\underline{e}^{\mathrm{T}} \underline{p}_n \frac{\underline{\theta}\underline{\theta}^{\mathrm{T}} \underline{\xi}(\underline{x})}{|\underline{\theta}|^2} \tag{4.28}$$

The flowchart form of presentation of LSBA algorithm is given in Fig. 4.2.

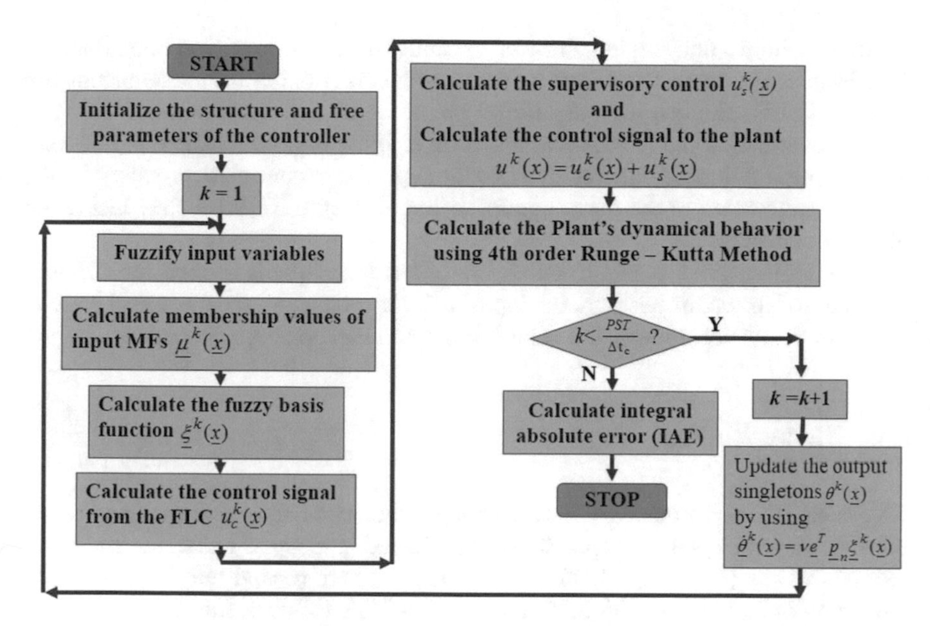

Fig. 4.2 LSBA algorithm in flowchart form. k plant simulation counter, PST plant simulation time, Δt_c controller simulation sampling time

Some Issues to be Noted

The magnitude of the inherent approximation error is mainly dependent on the design of the fuzzification process, i.e., the number of MFs used and the choice of the logical AND operator [5]. In turn, the inherent approximation error determines how good the controller is able to approximate the ideal control law. These design issues are of particular importance as these directly influence the control performance of the design strategy. However, it is quite difficult to present a concrete strategy, because prior knowledge about the ideal control law is absent. Theoretically, the inherent approximation error must be bounded in order to guarantee stability of the system [5]. So, in a general way, it can be stated that, by increasing the number of MFs used for fuzzification, and thus the number of rules used to generate an inference, the approximation capability of the fuzzy system can be improved. It has also been reported in the literature [3, 6, 8, 17, 18] that the product operator to represent the logical AND yields a smoother approximation than the minimum operator. It can also be said that a fuzzy controller with a finite number of rules can only provide a good approximation of an unknown function over a finite region of the input space [4, 11]. Thus, it can be inferred that if the states of the system are bounded, then only the stability of the overall system can be proved. In practice, due to physical constraints, the system state is limited within a finite region, therefore, stability of such systems can be guaranteed in principle. However, there may be some problems if the system state attains the physical limits. The two most significant problems are: (i) components of the plant may be overloaded and (ii) the conditions for stability may be violated. In this situation, some additional control action, like the supervisory control, as proposed in [4], must be augmented to the fuzzy control action to keep the system state inside the allowable boundaries.

Keeping the above points in mind, the LSBA technique is studied for different structures of the controller, e.g., the input MFs for each input variable are chosen as 3 or 5 or 7 and accordingly the resulting output singletons are generated.

4.4 Case Studies

To evaluate the performance of the LSBA algorithm, two benchmark case studies are considered, which were also previously considered in Chap. 3 and in several other research works [6, 8, 12, 14, 17]. The simulations are carried out for three fixed structure configurations with 3, 5, and 7 input MFs for each input variable. The performances of each controller are measured by calculating the integral absolute error (*IAE*), which can be defined as $IAE = \sum_{n=0}^{PST} e(n)\Delta t_c$, where PST = plant simulation time, Δt_c = step size or sampling time, for the reference signal and the actual output of the plant. The process models are simulated each using a fixed-step fourth-order Runge–Kutta method with sampling time $\Delta t_c = 0.01$ s. In all simulation case studies, the plant is simulated for 210 s where the adaptation starts at $t = 0$ s and continues up to $t = 200$ s when the adaptation is switched off. Then the plant is evaluated for 10 s with this reference signal. During this period, adaptation is turned off (i.e., $v = 0$) and the controller operates with already adapted parameters [8].

4.4.1 Case Study I: DC Motor Containing Nonlinear Friction Characteristics

The controlled plant under consideration is, once more, a second-order DC motor containing nonlinear friction characteristics described by the following model [5, 6, 8]:

$$\left.\begin{aligned} \dot{x}_1 &= x_2 \\ \dot{x}_2 &= -\frac{f(x_2)}{J} + \frac{C_T}{J}u \\ y &= x_1 \end{aligned}\right\} \tag{4.29}$$

where $y = x_1$ is the angular position of the rotor (in rad), x_2 is the angular speed (in rad/s), and u is the current fed to the motor (in A). The plant parameters are C_T = 10 Nm/A, J = 0.1 kgm^2, and the nonlinear friction torque is defined as:

$$f(x_2) = 5\tan^{-1}(5x_2)\ \text{Nm} \tag{4.30}$$

The control objective is to make the angular position y follow a reference signal given by

$$\ddot{y}_m = 400w(t) - 400y_m - 40\dot{y}_m \tag{4.31}$$

The detailed discussions about the plant and the reference model are presented before in Sect. 3.6.1 of Chap. 3.

The control signal obtained as the output from the fuzzy controller, in response to input signals, is given below:

$$u_c = u_c(\underline{x}, w|\underline{\theta}) \tag{4.32}$$

In this LSBA design methodology, only the positions of the output singletons (θ) are optimized and the positions of the input MFs and scaling gains for the input and output variables of the fuzzy controller are manually set *a priori*. The value of the adaptation gain is chosen as v = 5, giving a good trade-off between adaptation rate and small amplitudes of parameter oscillations which are due to inherent approximation error. The input and output scaling gains of the AFLC are chosen so as to map the ranges of the corresponding variables into the interval [−1, 1]. The other design parameters of the controllers are kept same, as chosen before in Sect. 3.6.1 of Chap. 3.

The controller is trained by applying a signal *w*(*t*), which changes its values randomly in the interval (−1.5, 1.5) in every 0.5 s as shown in Fig. 4.3a, to the reference model, which gives an output $y_m(t)$ as shown in Fig. 4.3b. This reference

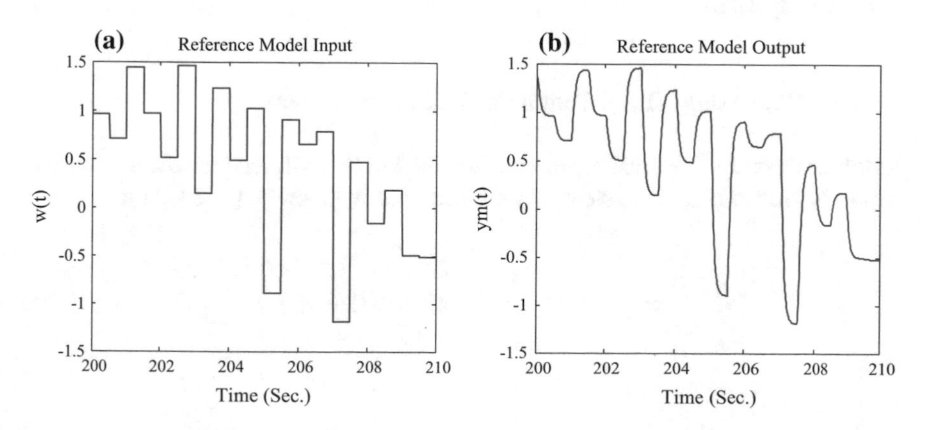

Fig. 4.3 Plot of **a** reference model input signal, *w*(*t*) and **b** reference model output signal $y_m(t)$

signal $y_m(t)$ is chosen as it contains infinitely many frequencies and leads to good exploration of controller input space [8]. The simulation is carried out for 210 s (training + evaluation time), but Fig. 4.3 shows the reference model excitation and response for the last 10 s duration, i.e., from 200 to 210 s, as a sample representation.

Figures 4.4, 4.5, and 4.6 show temporal variations of system responses for the fixed structure controllers with $3 \times 3 \times 3$, $5 \times 5 \times 5$, and $7 \times 7 \times 7$ input MFs, respectively. Figures 4.4a, 4.5a, and 4.6a show the positions of input MFs, and Figs. 4.4b, 4.5b, and 4.6b show the positions of the output singletons of the respective FLC's. In Figs. 4.4c, 4.5c, and 4.6c, the adaptation period responses for intermediate 100–110 s are shown. In Figs. 4.4d, 4.5d, and 4.6d the evaluation period responses for 200–210 s are shown. In Figs. 4.4e, 4.5e, and 4.6e, the variations in component vector θ for 0–200 s are shown. The evaluation period error and control effort variations are shown in subplots (f) and (g) of Figs. 4.4, 4.5, and 4.6, for $3 \times 3 \times 3$, $5 \times 5 \times 5$, and $7 \times 7 \times 7$ input MF configurations, respectively. Table 4.1 compares the simulation results for case study I for $3 \times 3 \times 3$, $5 \times 5 \times 5$, and $7 \times 7 \times 7$ input MF controller structure configurations. It can be seen from the table that the performances of the controllers (demonstrated by the performance index of *IAE*) improve with increase in the number of rules from $3 \times 3 \times 3$ to $5 \times 5 \times 5$ configuration. However, for $7 \times 7 \times 7$ configuration, results obtained are worse than those obtained for $5 \times 5 \times 5$ configuration. Hence, again, we can conclude that merely increasing the number of rules in fuzzy controllers does not necessarily improve their performances [5, 8].

4.4.2 *Case Study II: Duffing's Oscillatory System*

In this case study, LSBA algorithm is implemented for the Duffing's system in case study II(A) and the same algorithm is applied for Duffing's system with disturbance in case study II(B).

4.4.2.1 Case Study II(A): Duffing's Oscillatory System

In this case study, we once again consider the Duffing's forced oscillation system, which is itself a chaotic system, if unforced, and is given as [4, 14, 18]:

$$\left.\begin{aligned} \dot{x}_1 &= x_2 \\ \dot{x}_2 &= -0.1x_2 - x_1^3 + 12\cos(t) + u(t) \\ y &= x_1 \end{aligned}\right\} \quad (4.33)$$

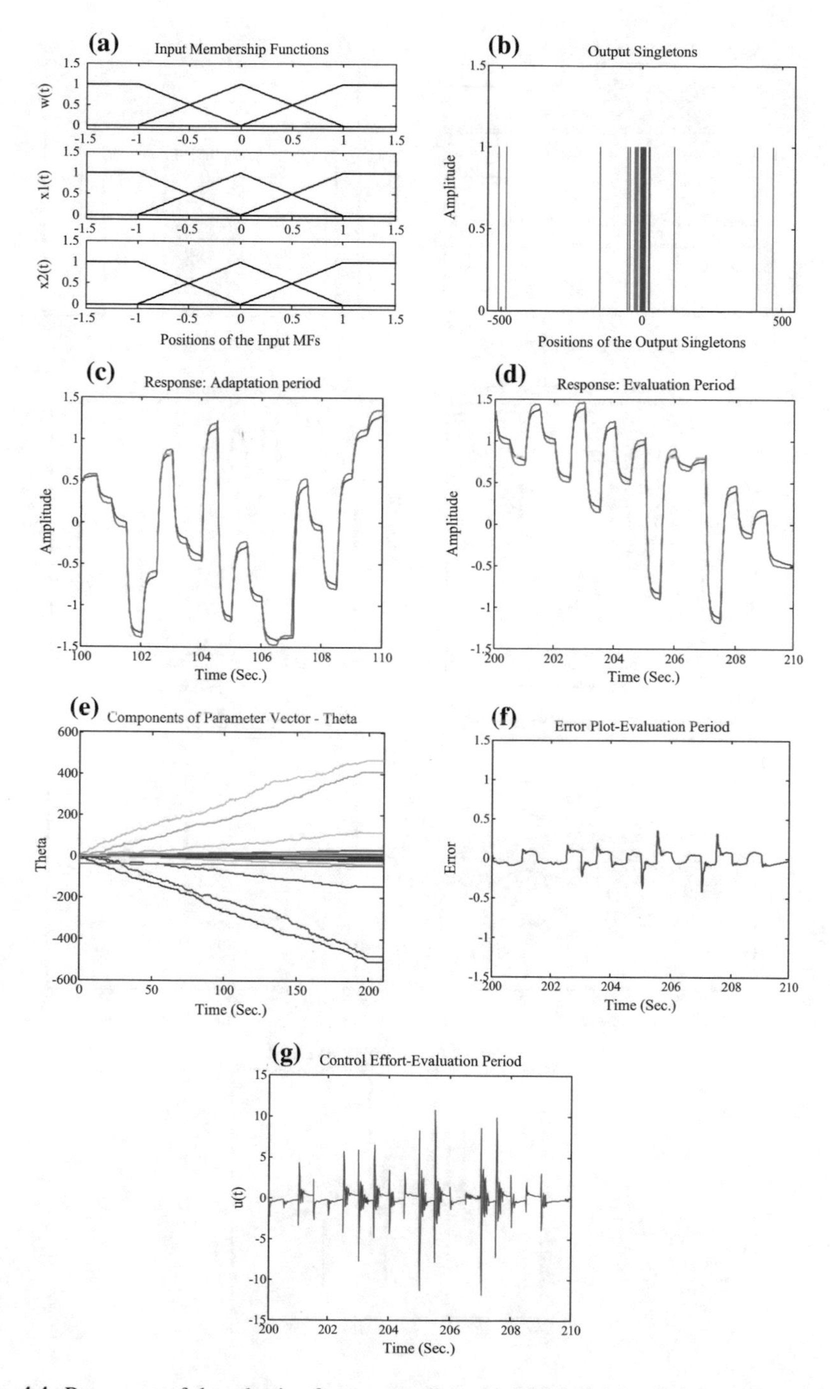

Fig. 4.4 Responses of the adaptive fuzzy controller with LSBA, for $3 \times 3 \times 3$ configuration of input MFs, of case study I (nonlinear DC motor system): **a** positions of input MFs, **b** positions of output singletons, **c** adaptation period response for 100–110 s, **d** evaluation period response for 200–210 s, **e** variation of component vector θ, **f** evaluation period error variation, and **g** evaluation period control effort variation

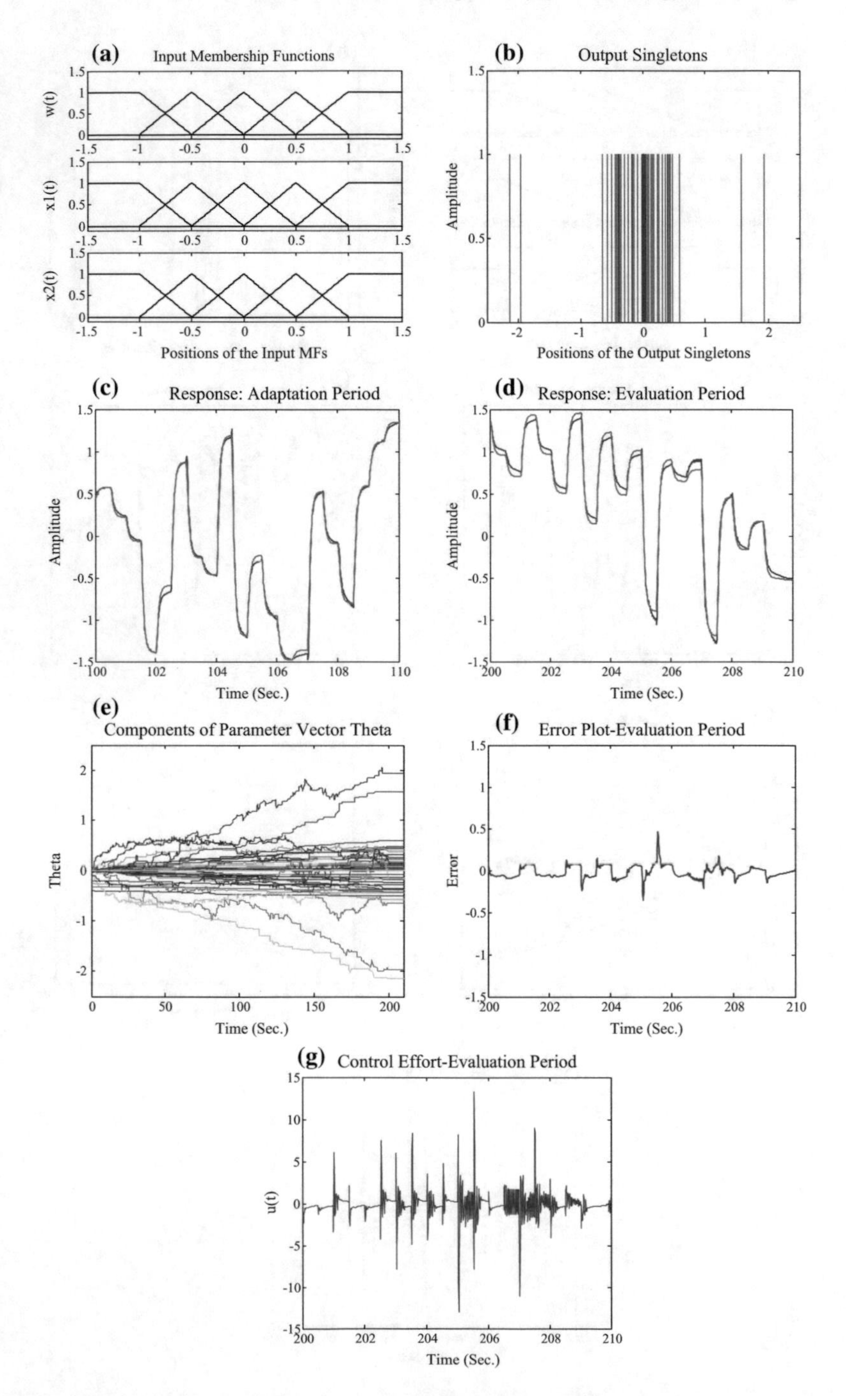

Fig. 4.5 Responses of the adaptive fuzzy controller with LSBA, for $5 \times 5 \times 5$ configuration of input MFs, of case study I (nonlinear DC motor system): **a** positions of input MFs, **b** positions of output singletons, **c** adaptation period response for 100–110 s, **d** evaluation period response for 200–210 s, **e** variation of component vector θ, **f** evaluation period error variation, and **g** evaluation period control effort variation

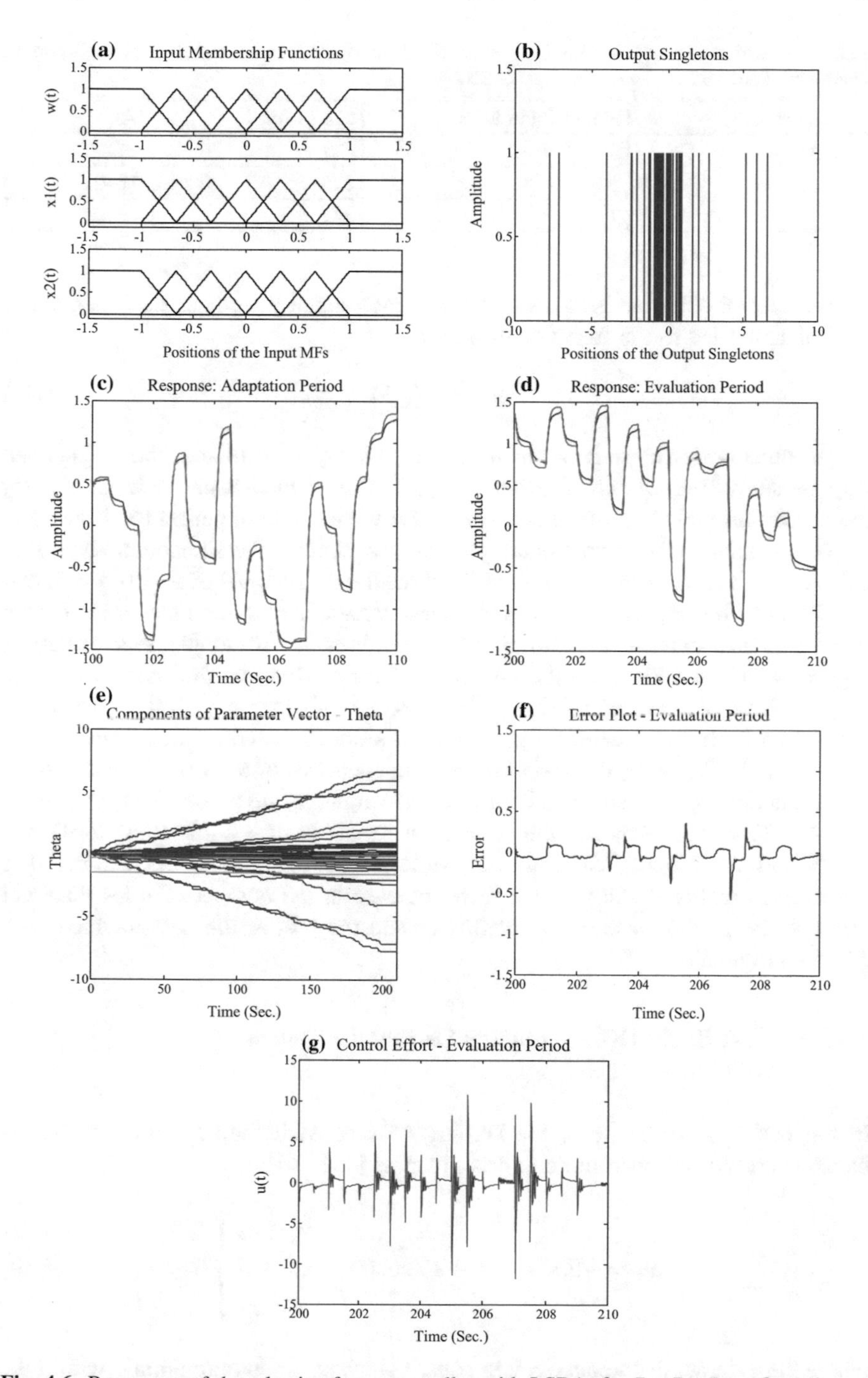

Fig. 4.6 Responses of the adaptive fuzzy controller with LSBA, for $7 \times 7 \times 7$ configuration of input MFs, of case study I (nonlinear DC motor system): **a** positions of input MFs, **b** positions of output singletons, **c** adaptation period response for 100–110 s, **d** evaluation period response for 200–210 s, **e** variation of component vector θ, **f** evaluation period error variation, and **g** evaluation period control effort variation

Table 4.1 Comparison of simulation results for case study I (nonlinear DC motor system): Lyapunov theory-based design strategy (LSBA)

Control strategy	No. of input MFs	No. of rules	IAE
LSBA	3 × 3 × 3	27	0.7472
	5 × 5 × 5	125	0.6402
	7 × 7 × 7	343	0.7400

The control objective is to track the reference signal y_m, where $y_m = \sin(t)$. The control signal fed to the fuzzy controller is

$$u_c = u_c(\underline{x}|\underline{\theta}) \tag{4.34}$$

The other design considerations for this case study are similar as those discussed for case study II(A) in Sect. 3.6.2 of Chap. 3. The adaptation gain v is set to 2 for the adaptation period and then set to zero for the evaluation period [4, 7, 8, 14].

Very similar to the approach adopted in case study I, the simulations based on LSBA design strategy are carried out for three fixed structures of 3 × 3, 5 × 5, and 7 × 7 input MFs. Figure 4.7 shows temporal variations of system responses for the fixed structure controllers with 5 × 5 input MFs, as a sample representative. Figures 4.7a and 4.7b show the positions of input MFs and the positions of the output singletons, respectively. In Fig. 4.7c, the adaptation period response for 100–110 s is shown, whereas Fig. 4.7d shows the evaluation period response for 200–210 s. In Fig. 4.7e, the variation of component vector θ for 0–200 s is shown. The evaluation period error and control effort variations are shown in (f) and (g) of Fig. 4.7. Table 4.2 compares the simulation results of case study II(A) for 3 × 3, 5 × 5, and 7 × 7 input MF controller structure configurations. Once more, it is observed in this case study that a mere increase in the number of rules does not improve the performance of the controller and the best results are obtained with 5 × 5 configurations.

4.4.2.2 Case Study II(B): Duffing's Oscillatory System with Disturbance

In this part of the case study, the Duffing's forced oscillation system with disturbance is considered once more and is given as [4, 8, 14]:

$$\left.\begin{aligned} \dot{x}_1 &= x_2 \\ \dot{x}_2 &= -0.1x_2 - x_1^3 + 12\cos(t) + u(t) + d \\ y &= x_1 \end{aligned}\right\} \tag{4.35}$$

where the external disturbance d is a square wave of random amplitude within the range [−1, 1] and a period of 0.5 s. Figure 4.8 shows a representative variation of disturbance for 0–10 s.

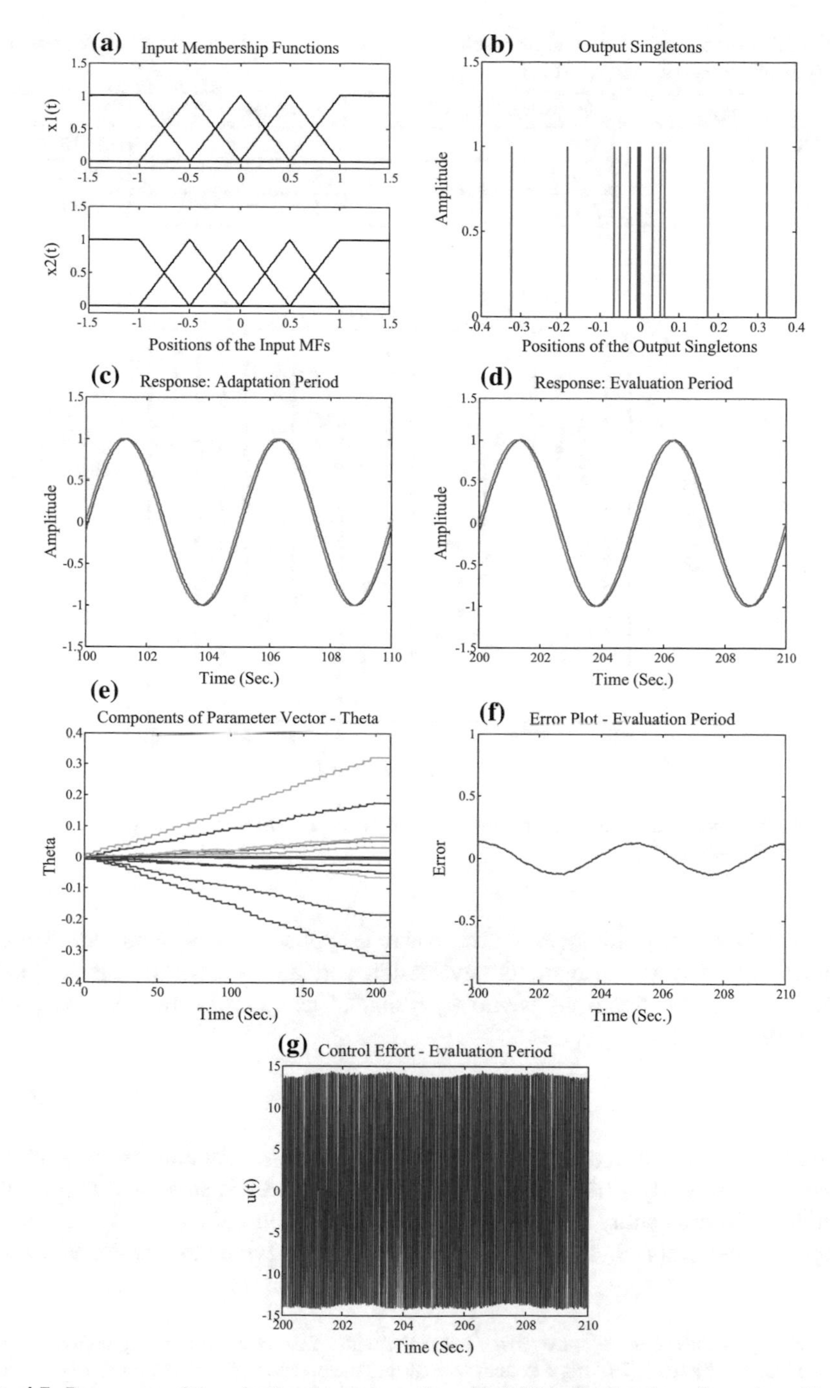

Fig. 4.7 Responses of the adaptive fuzzy controller with LSBA, for 5 × 5 configuration of input MFs, of case study II(A) (Duffing's system): **a** positions of input MFs, **b** positions of output singletons, **c** adaptation period response for 100–110 s, **d** evaluation period response for 200–210 s, **e** variation of component vector θ, **f** evaluation period error variation, and **g** evaluation period control effort variation

Table 4.2 Comparison of simulation results for case study II(A) (Duffing's system): Lyapunov theory-based design strategy (LSBA)

Control strategy	No. of input MFs	No. of rules	IAE
LSBA	3 × 3	9	0.7916
	5 × 5	25	0.7874
	7 × 7	49	0.7931

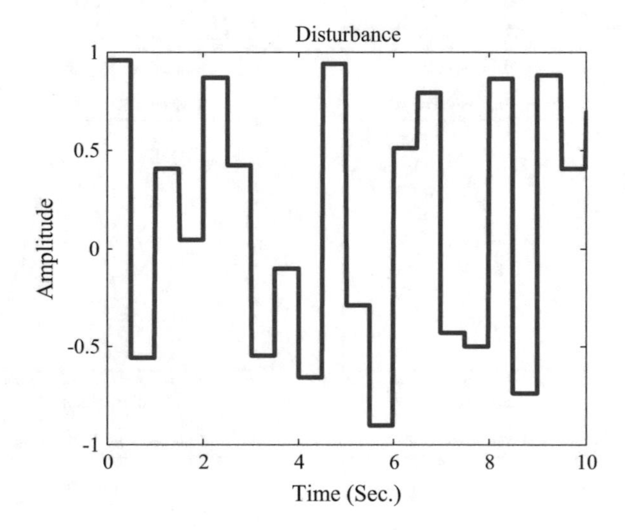

Fig. 4.8 A sample variation of disturbance applied to the system in case study II(B) (Duffing's system with disturbance)

In this case study, this type of disturbance is applied to the system for the entire adaptation period and also for the evaluation period. The control objective is to track the reference signal y_m, where $y_m = \sin(t)$. The control signal obtained from the fuzzy controller is given as:

$$u_c = u_c(\underline{x}|\underline{\theta}) \tag{4.36}$$

Here also, the same parameters are chosen for this simulation process as was chosen in case study II(A). The design strategy of LSBA is simulated in a similar fashion as in case study II(A). Here also the adaptation gain v is set to 2 for the adaptation period [4, 8, 14]. In a similar style as in previous two case studies, the

Fig. 4.9 Responses of the adaptive fuzzy controller with LSBA, for 5 × 5 configuration of input MFs, of case study II(B) (Duffing's system with disturbance): **a** positions of input MFs, **b** positions of output singletons, **c** adaptation period response for 100–110 s, **d** evaluation period response for 200–210 s, **e** variation of component vector θ, **f** evaluation period error variation, and **g** evaluation period control effort variation ▶

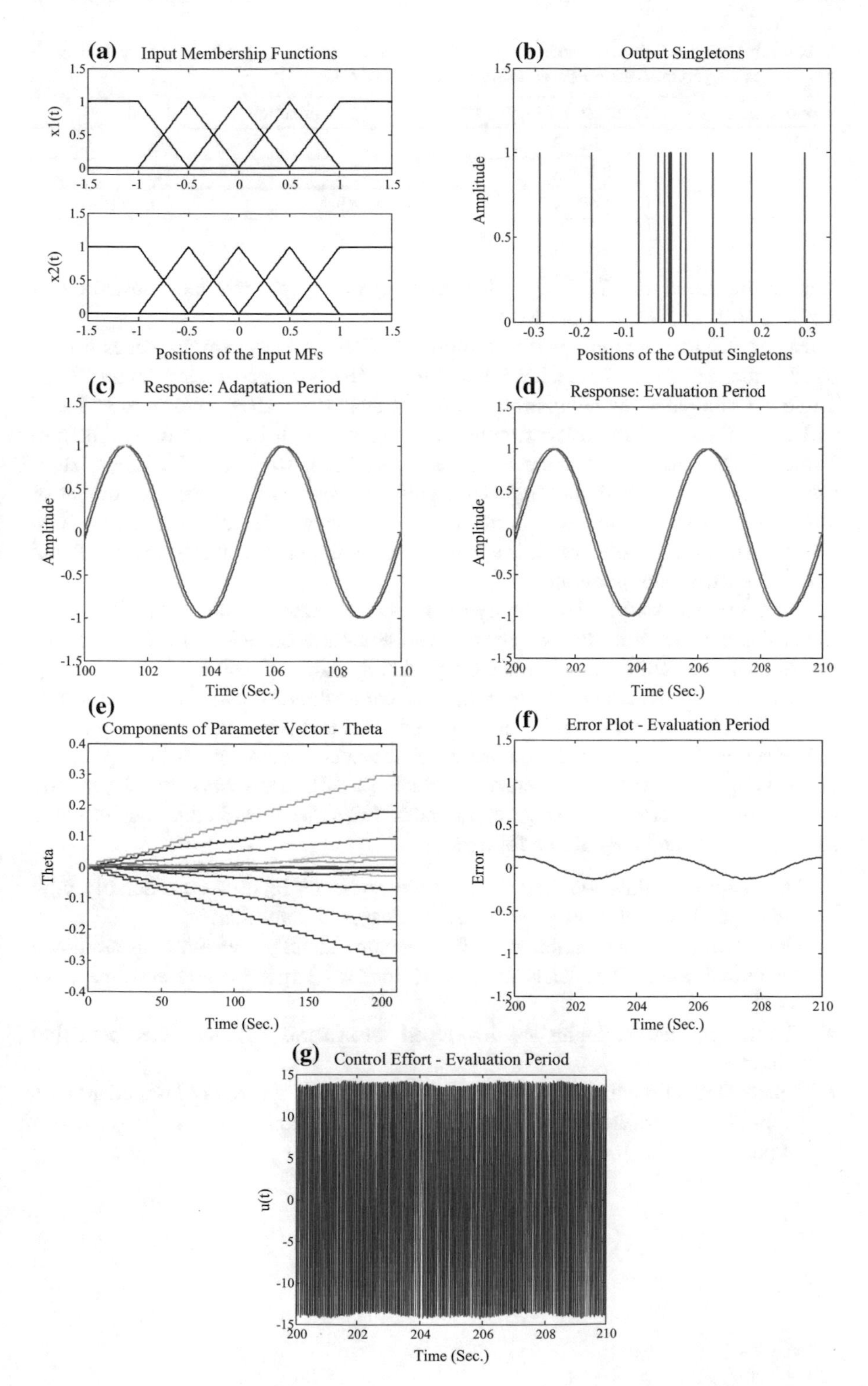

(a) Input Membership Functions
x1(t)
x2(t)
Positions of the Input MFs
(b) Output Singletons
Amplitude
Positions of the Output Singletons
(c) Response: Adaptation Period
Amplitude
Time (Sec.)
(d) Response: Evaluation Period
Amplitude
Time (Sec.)
(e) Components of Parameter Vector - Theta
Theta
Time (Sec.)
(f) Error Plot - Evaluation Period
Error
Time (Sec.)
(g) Control Effort - Evaluation Period
u(t)
Time (Sec.)

Table 4.3 Comparison of simulation results for case study II(B) (Duffing's system with disturbance): Lyapunov theory-based design strategy (LSBA)

Control strategy	No. of input MFs	No. of rules	IAE
LSBA	3×3	9	0.7967
	5×5	25	0.8006
	7×7	49	0.7940

simulations based on LSBA algorithm are worked out for three fixed structures of 3×3, 5×5, and 7×7 input MFs.

Figure 4.9 shows representative temporal variations of system responses for the fixed structure controllers with 5×5 input MFs, similar to case study II(A). Table 4.3 compares the simulation results for case study II(B), with 3×3, 5×5, and 7×7 input MF controller structure configurations. It has been observed from Table 4.3 that the performances of AFLC's, designed following the LSBA algorithm, in terms of *IAE*, do not improve significantly with the increase in number of input MFs. Although the best performance is obtained, in this case study, with 7×7 input MF configuration, it is a very marginal improvement over the 3×3 and 5×5 input MF configurations.

Therefore, the usefulness of the Lyapunov-based strategies, utilized to design the direct stable AFLC's in this chapter that can guarantee closed-loop stability of the system, is aptly demonstrated. In the original adaptation scheme in [4], the learning procedures need the adaptation of controller parameters for every variation in reference signal. In comparison, for the design procedure discussed in this chapter, the use of stable AFLC allows the controller parameters to converge even in presence of a temporally varying reference signal [2, 5]. However, the Lyapunov strategy-based approach of designing direct AFLC's discussed in this chapter poses some problems and they are as follows:

- The adaptation time required to track the reference input is considerably high and depends on the configuration of the designed controller.
- The inherent approximation error that remains in fully achieving an unknown theoretical control law leads the system towards larger integral absolute error (*IAE*).
- The design methodologies are developed for systems with unlimited actuation authority.
- The designs of controllers have been proposed based on control laws adapting a parameter vector containing the positions of the fuzzy sets of the controller outputs only.

4.5 Summary

This chapter has formulated the control problem for the design of stable adaptive fuzzy controllers. The conventional Lyapunov-based design strategy has been implemented for two benchmark nonlinear systems for different controller configurations and the drawbacks of this design method are summarized.

The next chapter will describe the interesting, contemporary concept of hybridization of Lyapunov theory-based local adaptation and stochastic optimization-based global optimization, for designing direct stable adaptive fuzzy controllers.

References

1. K.J. Astrom, B. Wittenmark, *Adaptive Control, Reading* (Addison-Wesley Pub. Co., Massachusetts, 1989)
2. K. Das Sharma, *Hybrid Methodologies for Stable Adaptive Fuzzy Control*, Doctoral Dissertation, Jadavpur University, India, 2012
3. K.M. Passino, S. Yurkovich, Fuzzy control, in *Handbook on Control*, ed. by W. Levine (CRC, Boca Raton, FL, 1996), pp. 1001–1017
4. L.X. Wang, Stable adaptive fuzzy control of nonlinear system. IEEE Trans. Fuzzy Syst. **1**(2), 146–155, May (1993)
5. K. Fischle, D. Schroder, An improved stable adaptive fuzzy control method. IEEE Trans. Fuzzy Syst. **7**(1), 27–40 (1999)
6. L.-X. Wang, *A Course in Fuzzy Systems and Control* (Englewood Cliffs, NJ, Prentice Hall, 1997)
7. K. Das Sharma, A. Chatterjee, F. Matsuno, A Lyapunov theory and stochastic optimization based stable adaptive fuzzy control methodology, in *Proceedings of SICE Annual Conference 2008*, Japan, 20–22 Aug, pp. 1839–1844
8. K. Das Sharma, A. Chatterjee, A. Rakshit, A hybrid approach for design of stable adaptive fuzzy controllers employing Lyapunov theory and particle swarm optimization. IEEE Trans. Fuzzy Syst. **17**(2), 329–342 (2009)
9. Y.P. Glorennec, Adaptive fuzzy control, in *Proceedings of 4th IFSA Congress*, Brussels, 1991, vol. Eng, pp. 33–36
10. C.-Y. Su, Y. Stepanenko, Adaptive control of a class of nonlinear systems with fuzzy logic. IEEE Trans. Fuzzy Syst. **2**, 285–294 (1994)
11. V. Giordano, D. Naso, B. Turchiano, Combining genetic algorithm and Lyapunov-based adaptation for online design of fuzzy controllers. IEEE Trans. Syst. Man Cybern. B Cybern. **36**(5), 1118–1127, Oct (2006)
12. S.H. Zak, *Systems and Control* (Oxford University Press, New York, 2003)
13. D.L. Tsay, H.Y. Chung, C.J. Lee, The adaptive control of nonlinear systems using the Sugeno-type of fuzzy logic. IEEE Trans. Fuzzy Syst. **7**(2), 225–229 (1999)
14. K. Das Sharma, A. Chatterjee, A. Rakshit, Design of a hybrid stable adaptive fuzzy controller employing Lyapunov theory and harmony search algorithm. IEEE Trans. Contr. Syst. Technol. **18**(6), 1440–1447, Nov (2010)
15. S.J. Huang, W.C. Lin, Adaptive fuzzy controller with sliding surface for vehicle suspension control. IEEE Trans. Fuzzy Syst. **11**(4), 550–559 (2003)
16. T. Das, I.N. Kar, Design and implementation of an adaptive fuzzy logic-based controller for wheeled mobile robots. IEEE Trans. Contr. Syst. Technol. **14**(3), 501–510 (2006)

17. S. Isaka, A. Sebald, A. Karimi, N. Smith, M. Quinn, On the design and performance evaluation of adaptive fuzzy controllers, in *Proceedings of IEEE Conference on Decision and Control*, Austin, TX, Dec 1988, pp. 1068–1069
18. T. Wang, S. Tong, Adaptive fuzzy output feedback control for SISO nonlinear systems. Int. J. Innov. Comput. Inf. Contr. **2**(1), 51–60 (2006)

Chapter 5
Fuzzy Controller Design III: Hybrid Adaptive Approaches

5.1 Introduction

In this chapter, an interesting, contemporary approach for designing stable adaptive fuzzy controllers is described, which employs a hybridization of conventional Lyapunov theory-based approach (LSBA) [1, 2] and a stochastic optimization algorithm (SOA)-based approach (SOABA). The objective is to design a self-adaptive fuzzy controller, optimizing both its structures and free parameters, such that the designed controller can guarantee desired stability and simultaneously it can provide satisfactory transient performance. The design methodology for the controller simultaneously utilizes the good features of SOA (capable of providing good global search capability and high degree of automation) and Lyapunov theory-based tuning (providing fast adaptation utilizing local search method). The SOA-based global optimization method is employed to determine the most suitable complete optimized vector containing flags for determining controller structure and the free parameters of the adaptive fuzzy logic controller (AFLC) e.g., positions of input membership functions (MFs), scaling gains, and positions of output singletons. For each such global candidate solution of controller design, local adaptation by LSBA is carried out to further adapt the output singletons only. The LSBA helps to guarantee closed-loop stability with satisfactory speed of convergence. On the other hand, the SOABA helps to achieve desired automation for the design methodology of the AFLC by getting rid of many manually tuned parameters in LSBA design procedure [2–4].

Within the purview of the present book, three types of hybridization of locally operative LSBA and globally optimal SOABA [5] have been described. These approaches are called hybrid (abbreviated as Hybd) adaptation strategy-based approaches. Cascade and concurrent combinations of LSBA and SOABA are the first two hybrid approaches described in this chapter. To get rid of the tedious manual tuning part of the conventional LSBA design procedure, in hybrid cascade model, the structure of the controller and the values of the free parameters for

K. Das Sharma et al., *Intelligent Control*, Cognitive Intelligence and Robotics,
https://doi.org/10.1007/978-981-13-1298-4_5

LSBA are determined by the SOA-based approach. But the adaptation of the controller parameters for tracking the temporally varying reference signal still requires a considerable amount of time [6, 7]. So, the drawbacks of conventional LSBA cannot be fully overcome in this approach. In hybrid concurrent approach, the LSBA performs its local tuning operation within the supervision of SOABA, with the objective of keeping the inherent approximation error limited and the tracking of the input reference signal, may be temporal in nature, is expected to be satisfactory enough [2, 4, 8]. However, it has been realized during the extensive experimentations that SOABA alone performs better than the hybrid concurrent approach, for some classes of nonlinear systems [2]. Hence, the third hybrid strategy for designing AFLCs has been suggested and is termed as hybrid preferential approach [2, 4, 8]. In this case, the movements of the candidate solution vector (CSV) in the multidimensional search space are sometimes governed by the hybrid concurrent method and sometimes by the SOA method, depending on the performance of the corresponding fitness evaluations. Therefore, the best candidate solution in this hybrid preferential approach evolves from a second level of hybridization process [2]. In this chapter, the two contemporary SOAs, namely nature-inspired particle swarm optimization (PSO) [9–12] and music-inspired harmony search algorithm (HSA) [13–15], are employed for hybridization purpose. The hybrid approaches have all been implemented for the same benchmark case studies which were considered in Chaps. 3 and 4. The obtained results illustrate, on the whole, the superiority of the hybrid preferential approach over the other approaches [2].

5.2 Hybrid Design Methodology Using Lyapunov Strategy

The control objectives and fuzzy controller formulation in hybrid design methodology are quite similar to that of the Lyapunov-based strategy. For brevity of discussion, the basics are put forth in this chapter. The details are available in Chap. 4.

Control Objectives

Let us once more consider that the objective is to design an adaptive strategy for an nth-order SISO nonlinear plant given as [6, 16]:

$$\left.\begin{aligned} x^{(n)} &= f(\underline{x}) + bu \\ y &= x \end{aligned}\right\} \tag{5.1}$$

where $f(\cdot)$ is an unknown continuous function, $u \in R$ and $y \in R$ are the input and output of the plant, and b is an unknown positive constant. Let us assume that the state vector is given as $\underline{x} = (x_1, x_2, \ldots, x_n)^{\mathrm{T}} = (x, \dot{x}, \ldots x^{(n-1)})^{\mathrm{T}} \in R^n$. The system under control has a reference model given as:

$$\left.\begin{array}{l} x_m^{(n)} = f(\underline{x}_m, w) \\ y_m = x_m \end{array}\right\} \tag{5.2}$$

where $w(t)$ is the excitation signal input to the reference model, and the state vector of the reference model is $\underline{x}_m = (x_m, \dot{x}_m, \ldots x_m^{(n-1)})^{\mathrm{T}} \in R^n$.

The control objective is to force the plant output $y(t)$ to follow a given bounded reference signal $y_m(t)$ under the constraints that all closed-loop variables involved must be bounded to guarantee the closed-loop stability of the system [1, 2]. So objective is to design a stable adaptive fuzzy controller for the system described in (5.1) where, the tracking error is $e = y_m - y$.

Now the ideal control law for the system in (5.1) and (5.2) is given as [2, 4, 8]:

$$u^* = \frac{1}{b}\left[-f(\underline{x}) + y_m^{(n)} + \underline{k}^{\mathrm{T}}\underline{e}\right] \tag{5.3}$$

where $y_m^{(n)}$ is the nth derivative of the output of the reference model, $\underline{e} = (e, \dot{e}, \ldots, e^{(n-1)})^{\mathrm{T}}$ is the error vector and $\underline{k} = (k_1, k_2, \ldots, k_n)^{\mathrm{T}} \in R^n$ is the vector describing the desired closed-loop dynamics for the error.

The ideal control law cannot be realized in practice, as f and b are not known precisely, the fuzzy logic controller is a suitable solution to approximate this optimal control law. Now, to ensure the stability, it has been assumed that the control law $u(t)$ is given by the summation of a fuzzy control, $u_c(\underline{x}|\underline{\theta})$, and an additional supervisory control $u_s(\underline{x})$ [1, 16]. Thus, $u(t)$ is given as:

$$u(t) = u_c(\underline{x}|\underline{\theta}) + u_s(\underline{x}) \tag{5.4}$$

It has been assumed that the AFLC is constructed using a zero-order Takagi–Sugeno (T–S) fuzzy system. Then $u_c(\underline{x}|\underline{\theta})$ for the AFLC is given in the form [1, 4]:

$$u_c(\underline{x}|\underline{\theta}) = \sum_{l=1}^{N} \theta_l * \alpha_l(\underline{x}) \Bigg/ \left(\sum_{l=1}^{N} \alpha_l(\underline{x})\right) = \underline{\theta}^{\mathrm{T}} * \underline{\xi}(\underline{x}) \tag{5.5}$$

where $\underline{\theta} = [\theta_1\, \theta_2 \ldots \theta_N]^{\mathrm{T}}$ = the vector of the output singletons, $\alpha_l(\underline{x}) = \prod_{i=1}^{r} \mu_i^l(x_i)$ = the firing degree of rule "l", N = the total number of rules, $\mu_i^l(x_i)$ = the membership value of the ith input membership function (MF) in the activated lth rule, $\underline{\xi}(\underline{x})$ = vector containing normalized firing strength of all fuzzy IF-THEN rules $= (\xi_1(\underline{x}), \xi_2(\underline{x}), \ldots, \xi_N(\underline{x}))^{\mathrm{T}}$ and

$$\xi_l(\underline{x}) = \alpha_l(\underline{x}) \Bigg/ \left(\sum_{l=1}^{N} \alpha_l(\underline{x})\right) \tag{5.6}$$

The supervisory control can be formed as in Chap. 3 and is given as [1, 2]:

$$u_s(\underline{x}) = I_1^* \operatorname{sgn}\left(\underline{e}^{\mathrm{T}} P \underline{b}_c\right)\left[|u_c| + \frac{1}{b_L}\left(f^U + |y_m^{(n)}| + |\underline{k}^{\mathrm{T}}\underline{e}|\right)\right] \tag{5.7}$$

Let us define a quadratic form of tracking error as $V_e = \frac{1}{2}\underline{e}^{\mathrm{T}}P\underline{e}$ where P is a symmetric positive definite matrix satisfying the Lyapunov equation [17]. With the use of $u_s(\underline{x})$, it can be shown that $\dot{V}_e \leq -\frac{1}{2}\underline{e}^{\mathrm{T}}Q\underline{e} \leq 0$, where Q is a positive definite matrix [1, 16, 17]. Thus, as $P > 0$, boundedness of V_e implies the boundedness in $\underline{x}$. Hence, the closed-loop stability is guaranteed [2].

The zero-order T–S-type fuzzy control $u_c(\underline{x}|\underline{\theta})$ can be so constructed that it will produce a linear weighted combination of adapted parameter vector $\underline{\theta}$. Thus, a simple singleton-based adaptation law, as proposed in [2, 4, 8], can be given as:

$$\dot{\underline{\theta}} = v\underline{e}^{\mathrm{T}}\underline{p}_n\,\underline{\xi}(\underline{x}) \tag{5.8}$$

where $v > 0$ is the adaptation gain or learning rate and $\underline{p}_n$ is the last column of P.

Stochastic Optimization Algorithm-Based Design of Fuzzy Controller

A systematic design strategy for fuzzy controllers, employing SOA, has already been detailed in Chap. 3 of this book. PSO, HSA, etc., can be potentially utilized to design the fuzzy controller, where the structure and the other free parameters, e.g., supports of the input MFs, input and output scaling gains etc., of the controller can be stochastically adapted. While designing the SOA-based optimal/near-optimal controller, all these parameters are obtained automatically by encoding them as part of the solution vector, here it is termed as a candidate solution vector (CSV). Thus, each of these parameters forms a decision variable in the solution space. This also helps to determine the optimal structure of the FLC [2, 5].

In SOA-based generic design, a CSV in solution space is formed as [4, 5]:

$$\begin{aligned}\underline{Z} = [&\text{structural flags for MFs} \mid \text{center locations of the MFs} \mid \ldots \\ &\mid \text{scaling gains} \mid \text{positions of the output singletons}]\end{aligned} \tag{5.9}$$

The structural flags are implemented to determine whether a particular MF will be present for an input to the FLC or not, exactly with the same concept as described in PSO-based design (see Art. 3.3.1). As in SOA, each variable is permitted to take continuous values, the structural flag will be set to 1 if the corresponding decision variable value is greater than 0.5, otherwise the flag will be set to 0. In this case again, each input to FLC is fuzzified in the range [−1, 1] with the peaks of each extreme MF kept fixed at −1 and 1, respectively. Here also the possible number of maximum MFs will be 7 for each input variable, where, we determine the peaks of other 5 MFs using SOA in the range [−1, 1]. The peak of any MF also corresponds to the right base support of the previous MF. The contents of structural flags determine number of MFs in which each input variable is fuzzified and accordingly the peaks are also organized [2, 4, 5, 8].

Figure 3.4 has already shown the flowchart form of presentation of SOA-based approach (SOABA) algorithm for the FLC with supervisory control action. Here, an initial pool of candidate solution vectors has been created and the system is simulated for each candidate controller by using the candidate controller simulation (CCS) algorithm, as shown in Fig. 3.2, and individual evaluation of the fitness function has been performed. In this case also, the fitness function is chosen as the integral absolute error (*IAE*) where $IAE = \sum_{n=0}^{PST} e(n)\,\Delta t_c$, and PST = plant simulation time, Δt_c = step size or sampling time. Based on the fitness value, each candidate solution vector will generate its updated version according to the improvisation rule of the stochastic optimization algorithm chosen [5].

Hybrid Design Methodology

In these design methodologies, the complementary features of Lyapunov-based approach and the SOA-based approach are combined to obtain a superior systematic design procedure for direct AFLCs, for a class of nonlinear SISO systems. A generic schematic architecture of hybrid adaptive fuzzy control action, augmented by supervisory control, is shown in Fig. 5.1.

Within the scope of this book, the optimization process of the controller is derived in three different ways to explore the possibility of obtaining superior results [4]. These three hybrid strategies are described now.

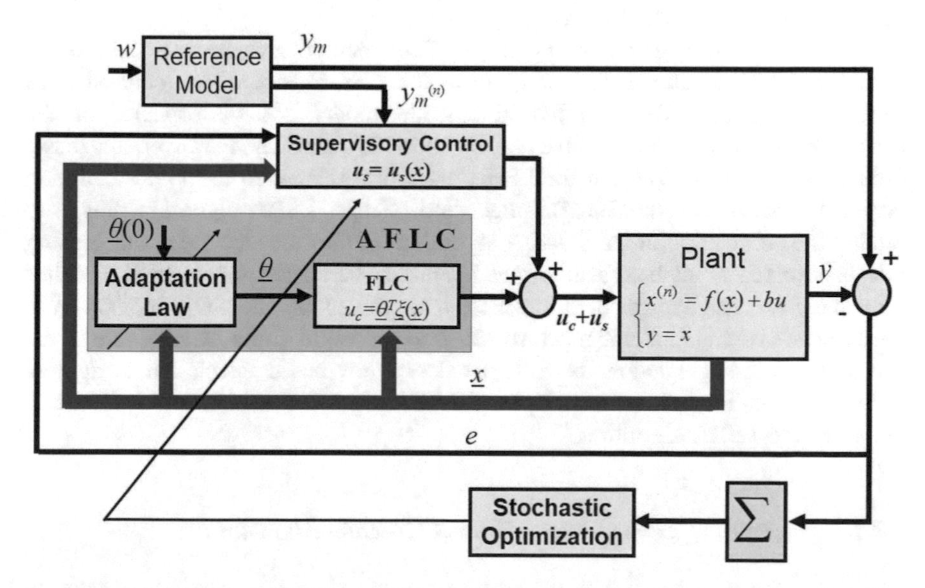

Fig. 5.1 A schematic architecture of hybrid adaptive fuzzy control action, augmented by supervisory control

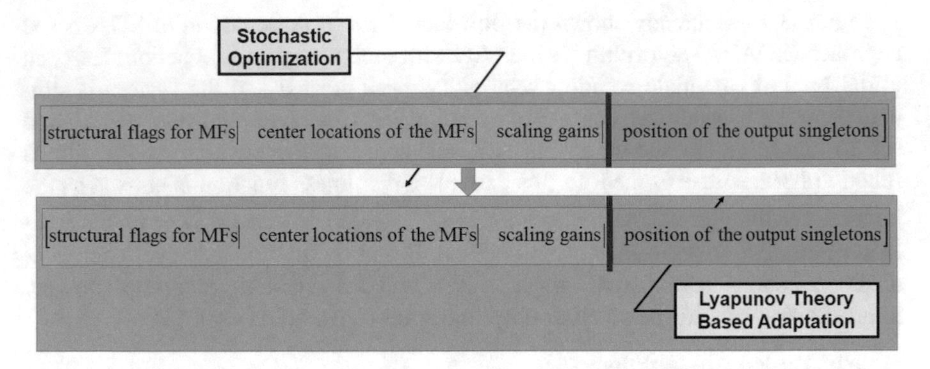

Fig. 5.2 Lyapunov theory-based hybrid cascade adaptation scheme

5.2.1 Lyapunov Theory-Based Hybrid Cascade Model

In this model, a stochastic optimization algorithm and Lyapunov-based optimization are used in cascade fashion and therefore named as hybrid cascade model (Hybd-Cas Model). Therefore, in this model, the structure and the free parameters of the controller are optimized by the stochastic optimization-based approach, as well as the positions of the output singletons are primarily tuned in a global search space by the SOA method [2, 4].

So, after a primary global tuning of $\underline{\theta}$ by SOA method, Lyapunov-based method optimizes the $\underline{\theta}$ locally to further improve the performance of the controller, as shown in Fig. 5.2. Thus, in hybrid cascade model, the performance of the Lyapunov-based method is enhanced by the SOA method. However, the basic difference lies in the fact that the LSBA algorithm is applied to the best solution vector determined by the SOABA algorithm. Hence, LSBA algorithm runs only once at the completion of the SOABA algorithm. In the cascade model, the learning rate is adjusted in such a fashion that Lyapunov-based adaptation (5.8) searches only the near neighborhood of the best $\underline{\theta}$ produced by the SOA method. The performance of this model is much sensitive to the value of the learning rate of the Lyapunov-based adaptation, because the Lyapunov-based adaptation is applied only to the solution $\underline{\theta}$ determined by the SOABA algorithm, which acts as the initial value for the LSBA algorithm.

5.2.1.1 Lyapunov Theory-Based Hybrid Cascade Algorithm

The outline of the controller design algorithm using Lyapunov theory-based hybrid cascade model is given as follows [2, 4]:

(i) Select the number of candidate solution vectors (N_{CSV}).
(ii) Select the parameters of the stochastic optimization algorithm (SOA).

(iii) The stopping criterion can be set in two ways, either by choosing the maximum iteration ($iter_{max}$) or by using a minimum error criteria, i.e., continue the SOA adaptation until a maximum pre-specified allowable error is attained by the controller.

(iv) Initialize N_{CSV} number of randomly generated candidate solution vectors (CSV).

(v) Determine the structure of the candidate controller based on structural flags as incorporated in the candidate solution vector $\underline{Z}$ (see (5.9)).

(vi) Calculate the fitness value, e.g., *IAE*, of the controller for each CSV that acts as a candidate controller in the problem under consideration, by using CCS algorithm as shown in the flowchart in Fig. 3.2.

(vii) Improvise a new CSV based on the updating rules of the SOA.

(viii) Repeat steps (v)–(vii) until the termination criterion is fulfilled.

(ix) Obtain an optimal/near-optimal CSV after the termination of the SOABA method.

(x) Determine the optimal structure of the candidate controller based on structural flags as incorporated in $\underline{Z}$ (see (5.9)).

(xi) Initialize other LSBA parameters, like learning rate, Lyapunov matrix.

(xii) Calculate the outputs of fuzzy basis functions.

(xiii) Update the positions of the output singletons ($\underline{\theta}$), based on Lyapunov theory, using (5.8).

(xiv) Calculate the feedback control u as in (5.4) and applied to the plant (5.1), where u_c is calculated from (5.5) and u_s from (5.7).

(xv) Repeat steps (xii)–(xiv) until the expiry of the plant simulation time.

The flowchart form of presentation of the hybrid cascade model is shown in Fig. 5.3 [4].

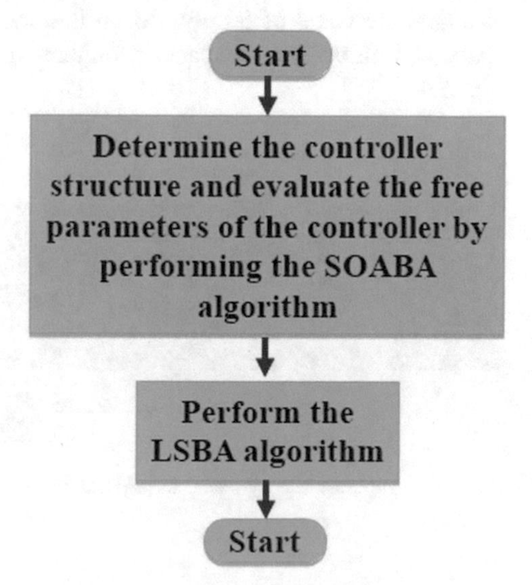

Fig. 5.3 Lyapunov theory-based hybrid cascade algorithm, in flowchart form

5.2.2 Lyapunov Theory-Based Hybrid Concurrent Model

In this stable adaptive fuzzy controller design process, the Lyapunov-based approach and the SOA-based approach run concurrently or in parallel to optimize the (i) structure of the controller, (ii) scaling gains, and (iii) the positions of the output singletons [2, 4, 5, 8] and thus, named as hybrid concurrent model (Hybd-Con Model).

In this method, a particle $\underline{Z}$ is divided into two subgroups, unlike in SOABA design methodology and is given as [2, 4]:

$$\underline{Z} = \left[\underline{\psi}|\underline{\theta}\right] \tag{5.10}$$

where

$$\left.\begin{aligned} \underline{\psi} &= [\text{structural flags for MFs}|\ldots \\ &\quad |\text{center locations of the MFs}|\text{scaling gains}] \\ \underline{\theta} &= [\text{position of the output singletons}] \end{aligned}\right\} \tag{5.11}$$

The partition of $\underline{Z}$ vector separates the parameters in such a fashion that $\underline{\psi}$ has nonlinear influence and $\underline{\theta}$ has linear influence on u_c, respectively [3, 4]. In LSBA, the values of the free parameters in the vector $\underline{\psi}$ are defined a priori and the value of the corresponding $\underline{\theta}$ minimizing the tracking error can be obtained by the adaptation law, as given in (5.8). So, in LSBA, the $\underline{\psi}$ is set by hand-tuned trial-and-error method. In this hybrid concurrent method, SOA is applied to optimize $\underline{\psi}$ and $\underline{\theta}$ in tandem by a method of CSV improvisation in optimization iteration. In addition to that, the adaptation law, as in (5.8), is applied to every updated candidate controller to adjust the value of $\underline{\theta}$ only. So, in this method, $\underline{\theta}$ explores both the local search space and global search space, simultaneously, in a twofold manner, as shown in Fig. 5.4.

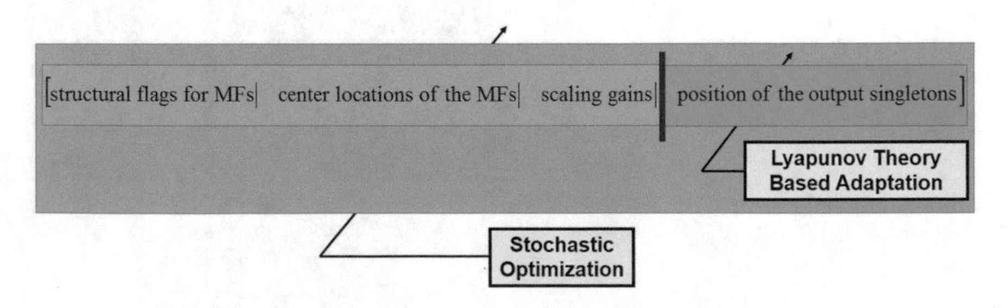

Fig. 5.4 Lyapunov theory-based hybrid concurrent adaptation scheme

5.2.2.1 Lyapunov Theory-Based Hybrid Concurrent Algorithm

The outline of the controller design algorithm using Lyapunov theory-based hybrid concurrent model is given as follows [2, 4, 5, 8]:

(i) Select the number of candidate solution vectors (N_{CSV}).
(ii) Select the parameters of the stochastic optimization algorithm (SOA).
(iii) Initialize other LSBA parameters, like learning rate, Lyapunov matrix.
(iv) The stopping criterion can be set in two ways, either by choosing the maximum iteration ($iter_{\max}$) or by using a minimum error criteria, i.e., continue the SOA adaptation until a maximum pre-specified allowable error is attained by the controller.
(v) Initialize N_{CSV} number of randomly generated candidate solution vectors (CSV).
(vi) Determine the structure of the candidate controller based on structural flags as incorporated in the candidate solution vector $\underline{Z}$ (see (5.9)).
(vii) Calculate the outputs of fuzzy basis functions.
(viii) Update the positions of the output singletons ($\underline{\theta}$) based on Lyapunov theory using (5.8).
(ix) Calculate the feedback control u as in (5.4) and applied to the plant (5.1), where u_c is calculated from (5.5) and u_s from (5.7).
(x) Repeat steps (vii)–(ix) until the expiry of the plant simulation time.
(xi) Calculate the fitness value, e.g., *IAE*, of the controller, for each CSV that acts as a candidate controller.
(xii) Replace $\underline{\theta}$ of $\underline{Z}$ by θ_{LSBA} obtained after step (x).
(xiii) Improvise a new CSV based on the updating rules of the SOA.
(xiv) Repeat steps (vi)–(xiii) until the termination criterion is fulfilled.

The flowchart form of presentation of Lyapunov theory-based hybrid concurrent model is shown in Fig. 5.5.

5.2.3 *Lyapunov Theory-Based Hybrid Preferential Model*

In this model, the SOABA method and the hybrid concurrent method both are employed to keep the tracking error as minimum as possible [2, 4] and so named as hybrid preferential model (Hybd-Pref Model). Here, each CSV in the population, i.e., each candidate controller, evaluates the CCS algorithm and the LSBA algorithm separately during the same iteration. The fitness values of the candidate controller, evaluated by employing both the CCS algorithm and the LSBA algorithm independently, are then compared, and then the fitness value of the better

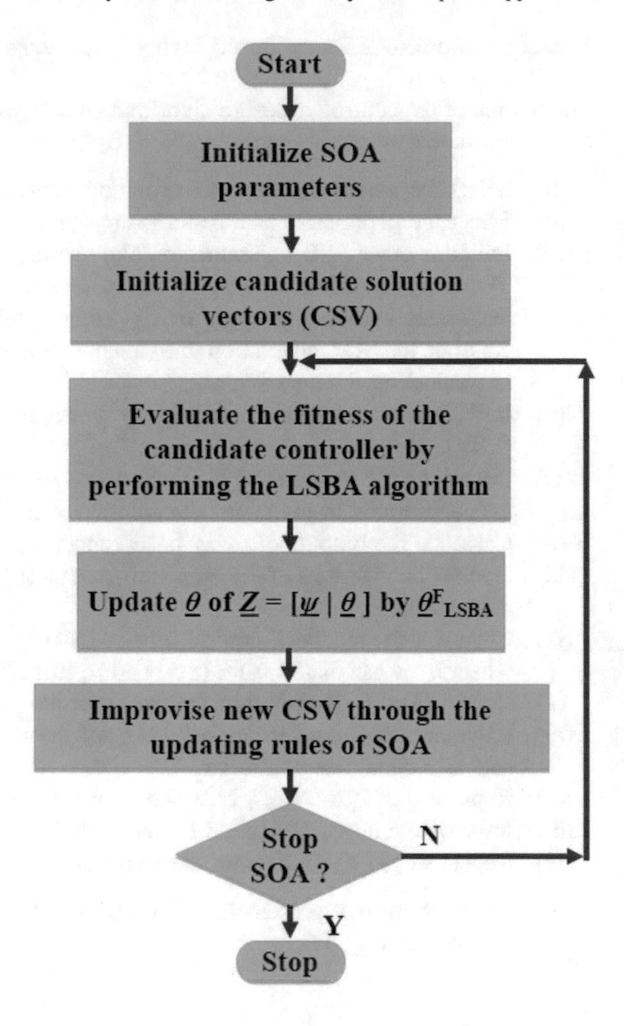

Fig. 5.5 Lyapunov theory-based hybrid concurrent algorithm, in flowchart form

algorithm is taken as the guiding factor for the update operation of the CSV in that iteration of the chosen SOA. If the LSBA algorithm is proven better for that CSV in that iteration, then the process will follow the hybrid concurrent model, otherwise, the process will follow the SOABA method in that iteration. Thus, in this process, each CSV in a population does not follow the same rule, even within the same iteration, to search the solution space. In a particular iteration of evaluation, some CSVs follow the SOABA method and some others may follow the hybrid concurrent method. This process of evaluation is performed in each iteration for each CSV, so that the same CSV may follow different update rules in different iterations, as described in Fig. 5.6. This method is proposed with an objective of avoiding getting stuck in local minima [5, 8].

Candidate Solution Vectors →

Iterations ↓

LSBA	CCS	LSBA	CCS	CCS	LSBA	LSBA	CCS	CCS	LSBA
CCS	LSBA	CCS	LSBA	LSBA	CCS	LSBA	LSBA	CCS	LSBA
CCS	LSBA	LSBA	CCS	CCS	CCS	CCS	LSBA	LSBA	CCS
LSBA	CCS	CCS	CCS	LSBA	LSBA	LSBA	CCS	CCS	LSBA
LSBA	CCS	LSBA	LSBA	CCS	LSBA	CCS	LSBA	LSBA	CCS
CCS	LSBA	LSBA	CCS	CCS	LSBA	CCS	CCS	LSBA	CCS
LSBA	LSBA	CCS	LSBA	LSBA	CCS	CCS	LSBA	CCS	LSBA
LSBA	CCS	LSBA	LSBA	CCS	LSBA	LSBA	CCS	LSBA	CCS
CCS	CCS	CCS	LSBA	LSBA	CCS	LSBA	LSBA	CCS	LSBA
LSBA	CCS	LSBA	CCS	LSBA	CCS	CCS	CCS	LSBA	CCS

Fig. 5.6 Movements of CSVs in Lyapunov theory-based hybrid preferential algorithm

5.2.3.1 Lyapunov Theory-Based Hybrid Preferential Algorithm

The outline of the controller design algorithm using Lyapunov theory-based hybrid preferential model is given as follows [2, 4, 5, 8]:

(i) Select the number of candidate solution vectors (N_{CSV}).

(ii) Select the parameters of the stochastic optimization algorithm (SOA).

(iii) Initialize other LSBA parameters, like learning rate, Lyapunov matrix.

(iv) The stopping criterion can be set in two ways, either by choosing the maximum iteration ($iter_{\max}$) or by using a minimum error criteria, i.e., continue the SOA adaptation until a maximum pre-specified allowable error is attained by the controller.

(v) Initialize N_{CSV} number of randomly generated candidate solution vectors (CSV).

(vi) Determine the structure of the candidate controller, based on structural flags as incorporated in the candidate solution vector $\underline{Z}$ (see (5.9)).

(vii) Independently evaluate:

a. CSS algorithm and calculate the $\mathrm{IAE_{CCS}}$ as shown in flowchart in Fig. 3.2.

b. LSBA algorithm and calculate the $\mathrm{IAE_{LSBA}}$ as shown in flowchart in Fig. 4.2.

(viii) If ($\mathrm{IAE_{LSBA}} < \mathrm{IAE_{CCS}}$), then update the values of the output singletons $\underline{\theta}$ by the final adapted value from the LSBA algorithm and set $\mathrm{IAE} = \mathrm{IAE_{LSBA}}$, otherwise, keep the $\underline{\theta}$ unchanged and set $\mathrm{IAE} = \mathrm{IAE_{CCS}}$.

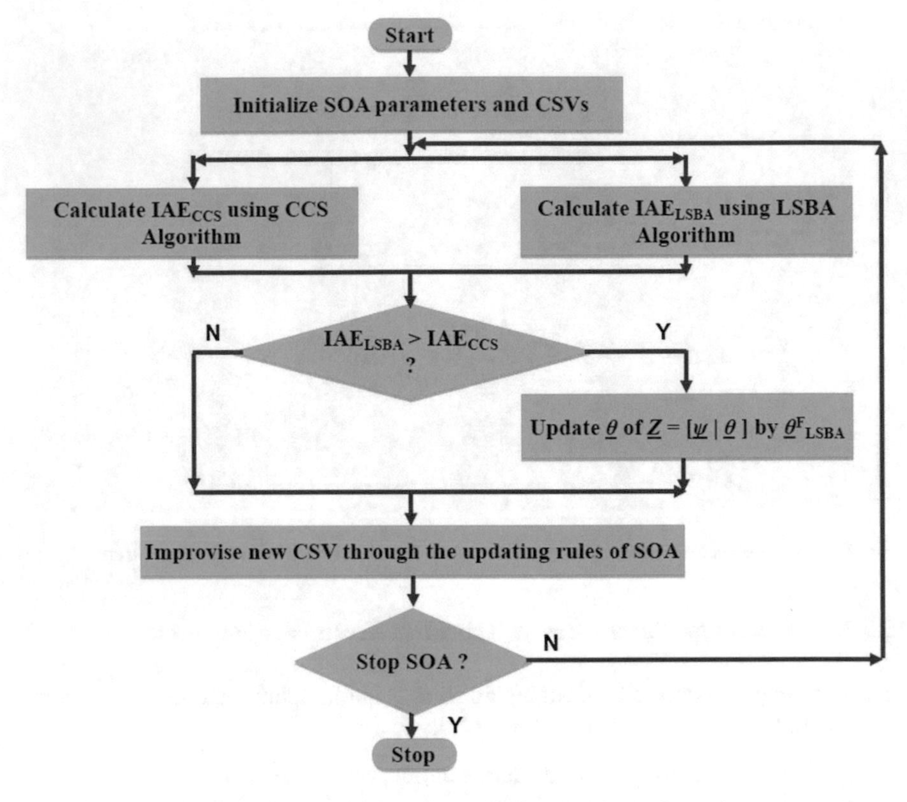

Fig. 5.7 Lyapunov theory-based hybrid preferential algorithm, in flowchart form

(ix) Improvise a new CSV based on the updating rules of the SOA.
(x) Repeat steps (vi)–(ix) until the termination criterion is fulfilled.

The flowchart form of presentation of Lyapunov theory-based hybrid preferential model is shown in Fig. 5.7.

5.3 Case Studies

To demonstrate the effectiveness of the hybrid adaptive control schemes, in this section also the same two benchmark case studies are considered, which were previously utilized in otherchapters of this book. To study the performances in terms of *IAE*, nonlinear systems with fixed structure configurations and a variable structure configuration, with minimum 2 and maximum 7 numbers of MFs for each input, are simulated, respectively. The process models are simulated each using a fixed-step fourth-order Runge–Kutta method with sampling time $\Delta t_c = 0.01$ s. PSO and HSA are the two potential contemporary SOAs employed in this case study to hybridize with Lyapunov-based adaptation strategy. In all simulation case studies, except the hybrid cascade model, the plant is evaluated for 10 s, after the training is completed (200 iterations of CSV improvisations) by different optimization strategies, to compare the results. At the outset, in case of hybrid cascade model, 200 iterations of SOA are used to train the CSVs and then the Lyapunov theory-based adaptation is performed on the resulting optimal CSV for 200 s. Finally, the trained controller is evaluated for 10 s [2].

5.3.1 *Case Study I: DC Motor Containing Nonlinear Friction Characteristics*

As mentioned before, the controlled plant under consideration is a second-order DC motor containing nonlinear friction characteristics described by the following model [4, 6]:

$$\left.\begin{aligned} \dot{x}_1 &= x_2 \\ \dot{x}_2 &= -\tfrac{f(x_2)}{J} + \tfrac{C_T}{J}u \\ y &= x_1 \end{aligned}\right\} \tag{5.12}$$

where $y = x_1$ is the angular position of the rotor (in rad), x_2 is the angular speed (in rad/s) and u is the current fed to the motor (in A). The plant parameters are $C_T = 10$ Nm/A, $J = 0.1$ kgm^2, and the nonlinear friction torque is defined as:

$$f(x_2) = 5\tan^{-1}(5x_2)\,\text{Nm} \tag{5.13}$$

The control objective is to make the angular position y follow a reference signal given by

$$\ddot{y}_m = 400w(t) - 400y_m - 40\dot{y}_m \tag{5.14}$$

The detailed discussions about the plant and the reference model are discussed before in Sect. 3.6.1 of Chap. 3.

The control signal obtained as the output from the fuzzy controller, in response to input signals, is given as:

$$u_c = u_c(\underline{x}, w|\underline{Z}) \tag{5.15}$$

Here also the controller is trained by applying a signal $w(t)$, which changes its value randomly in the interval (−1.5, 1.5) in every 0.5 s as shown in Fig. 3.6a, to the reference model, which gives an output $y_m(t)$ as shown in Fig. 3.6b. As it has been already mentioned that this reference signal $y_m(t)$ contains a huge number of frequencies and leads to good exploration of controller input space [6].

Different design processes as discussed before are implemented for this case study, and the observations are presented in detail now.

5.3.1.1 Performance Analysis of PSO-Based Hybrid Controllers

In PSO-based hybrid control schemes, all the parameters of $\underline{Z}$ are optimized by the PSO algorithm. To perform this simulation, population size of ten particles is taken. For each simulation, 200 iterations of PSO are set. In fixed structure controller design, a fixed number of evenly distributed input MFs are used for initial population of PSO. PSO algorithm further tunes the free parameters and the positions of the output singletons to minimize the tracking error. In variable structure controller design, the structural flags are utilized to automatically vary the structure of the candidate controller, in each iteration. The input MFs and the corresponding positions of the output singletons are set according to the active structural flags, as detailed in Sect. 3.3. In this case study too, the *IAE* between the reference signal y_m and the motor output y is taken as the fitness function for optimization. To respect the stochastic behavior of the optimization process, twenty simulations are performed and the average *IAE* is calculated for each control strategy.

In PSO-based hybrid cascade (PSO-Hybd-Cas) model, the simulation process initializes with the PSOBA algorithm. The results obtained from PSOBA are then passed through the Lyapunov-based adaptation strategy to further adjust the positions of the output singletons $\underline{\theta}$. The adaptation gain ν during the LSBA is set to 5 to explore the near neighborhood of the best $\underline{\theta}$ as determined by the PSOBA.

In PSO-based hybrid concurrent (PSO-Hybd-Con) model, PSOBA-based optimization of parameters for $Z = [\underline{\psi}|\underline{\theta}]$ is carried out and LSBA algorithm is employed to only adjust $\underline{\theta}$. These two processes run concurrently. The adaptation gain (ν) is varied over the range 14,000–16,000. The structural flags are set and reset in the same manner as carried out in PSOBA, and utilized to automatically determine the structure of the candidate controller.

In PSO-based hybrid preferential (PSO-Hybd-Pref) model, both hybrid concurrent model and pure PSOBA model are implemented for the same particle separately at the same time and the candidate controllers, i.e., the particles of PSO, are updated accordingly. The same particle can execute different searching algorithms to search the solution space, depending on the value of its fitness function (*IAE*), during the entire optimization process.

The results of simulation studies are shown in Table 5.1. In all these simulations, the plant is evaluated for 10 s, after the training is completed by different adaptation strategies, to compare the results. The simulations are carried out for three fixed structures of $3 \times 3 \times 3$, $5 \times 5 \times 5$, and $7 \times 7 \times 7$ input MFs and a variable structure with a potentially biggest possible structure of $7 \times 7 \times 7$ MFs. Figures 5.8, 5.11, 5.14, and 5.17 show temporal variations of system responses for the fixed structure controllers with $3 \times 3 \times 3$, $5 \times 5 \times 5$, $7 \times 7 \times 7$, and variable structure input MFs, respectively, for PSO-Hybd-Cas model control strategy. During the

Table 5.1 Comparison of simulation results for case study I (nonlinear DC motor system): PSO-based hybrid design strategies

Control strategy	No. of MFs	No. of rules	Avg. IAE	Best IAE	Std. dev.
PSO-Hybd-Cas	3 × 3 × 3	27	0.6552	0.6550	0.0003
PSO-Hybd-Con			0.5749	0.5569	0.0136
PSO-Hybd-Pref			0.6312	0.6034	0.0352
PSO-Hybd-Cas	5 × 5 × 5	125	0.3267	0.3219	0.0069
PSO-Hybd-Con			0.3119	0.2977	0.0172
PSO-Hybd-Pref			0.2931	0.2754	0.0250
PSO-Hybd-Cas	7 × 7 × 7	343	0.3426	0.3351	0.0106
PSO-Hybd-Con			0.3624	0.3368	0.0256
PSO-Hybd-Pref			0.2942	0.2872	0.0099
PSO-Hybd-Cas-V	4 × 6 × 5	120	–	0.4625	–
PSO-Hybd-Con-V	4 × 5 × 4	80	–	0.4823	–
PSO-Hybd-Pref-V	4 × 5 × 5	100	–	0.3692	–

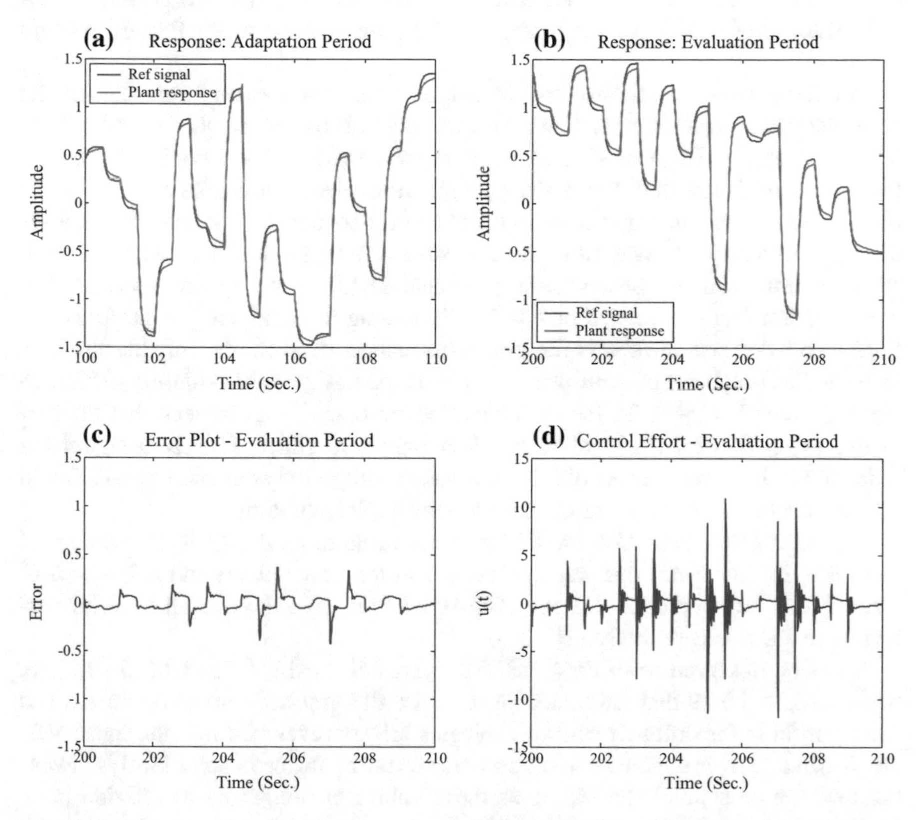

Fig. 5.8 Responses of the adaptive fuzzy controller with PSO-Hybd-Cas model, for 3 × 3 × 3 configuration of input MFs, of case study I (nonlinear DC motor system): **a** adaptation period response after 100 s, **b** evaluation period response for 200–210 s, **c** evaluation period error variation, **d** evaluation period control effort variation

implementation of the hybrid cascade model, the results of PSOBA model are used as the initial values for the LSBA model. Thus, the input MFs and the positions of the output singletons for this case are identical as in Figs. 3.7, 3.8, 3.9, and 3.10 in Chap. 3. Figures 5.8a, 5.11a, 5.14a, and 5.17a show the adaptation period responses for 100–110 s. In (b)–(d) of Figs. 5.8, 5.11, 5.14, and 5.17 the evaluation period responses, error variations and control effort variations for 200–210 s are shown, respectively. Figures 5.9, 5.12, 5.15, and 5.18 show temporal variations of system responses for the fixed structure controllers with $3 \times 3 \times 3$, $5 \times 5 \times 5$, $7 \times 7 \times 7$ and variable structure input MFs, respectively, for PSO-Hybd-Con model control strategy. In (a) and (b) of Figs. 5.9, 5.12, 5.15, and 5.18, the input MFs and positions of the output singletons, respectively, after the completion of the adaptation process are shown. Figures 5.9c, 5.12c, 5.15c, and 5.18c show the adaptation period responses after 100 PSO iterations. In (d)–(f) of Figs. 5.9, 5.12, 5.15, and 5.18, the evaluation period responses, error variations, and control effort variations are shown, respectively. Figures 5.10, 5.13, 5.16, and 5.19 show temporal variations of system responses for the fixed structure controllers with $3 \times 3 \times 3$, $5 \times 5 \times 5$, $7 \times 7 \times 7$, and variable structure input MFs, respectively, for PSO-Hybd-Pref model control strategy in the same fashion as for PSO-Hybd-Con model.

For fixed structure controllers, it can be seen that the performances of the controllers, in terms of *IAE*, improve with increase in the number of rules from $3 \times 3 \times 3$ to $5 \times 5 \times 5$ configuration. However, for $7 \times 7 \times 7$ configuration, the results are not better than $5 \times 5 \times 5$ configuration. Hence, it can be concluded that merely increasing number of rules in fuzzy controllers does not necessarily improve their performances as revealed in the case studies in earlier chapters. This is also in conformation with the observations presented in [3]. For a given controller MF configuration (e.g., $5 \times 5 \times 5$ or $7 \times 7 \times 7$), among the competing controllers, the PSO-Hybd-Pref scheme shows the best performance. Even for the variable structure scheme, PSO-Hybd-Pref controller emerges as the best possible solution. Although the *IAE* value is higher for the variable structure case, it can be seen that the best result, in terms of *IAE*, is obtained with only 100 rules. This is a significant reduction in the total number of rules and such a simplified structure of an FLC will be very useful from the point of view of real implementation.

Although PSO-Hybd-Con model could converge to an even smaller number of rules (80, in this case), this was achieved with a comparatively much larger *IAE*. Thus, it can be concluded that the PSO-Hybd-Pref gave the overall superior performance for this case study [2].

It can be observed from Figs. 5.8, 5.9, 5.10, 5.11, 5.12, 5.13, 5.14, 5.15, 5.16, 5.17, 5.18, and 5.19 that the center locations of all input MFs are so optimized and the algorithms for different control strategies are so developed that the input MFs are distributed over the universe of discourse ensuring the crosspoint level is always 0.5 and the crosspoint ratio is 1, as these values are expected to provide good performances for FLC design [18–20]. It can also be observed that the output singletons are more or less uniformly distributed over its universe of discourse in case of all PSO Hybd-Pref models, compared to the other design strategies that may

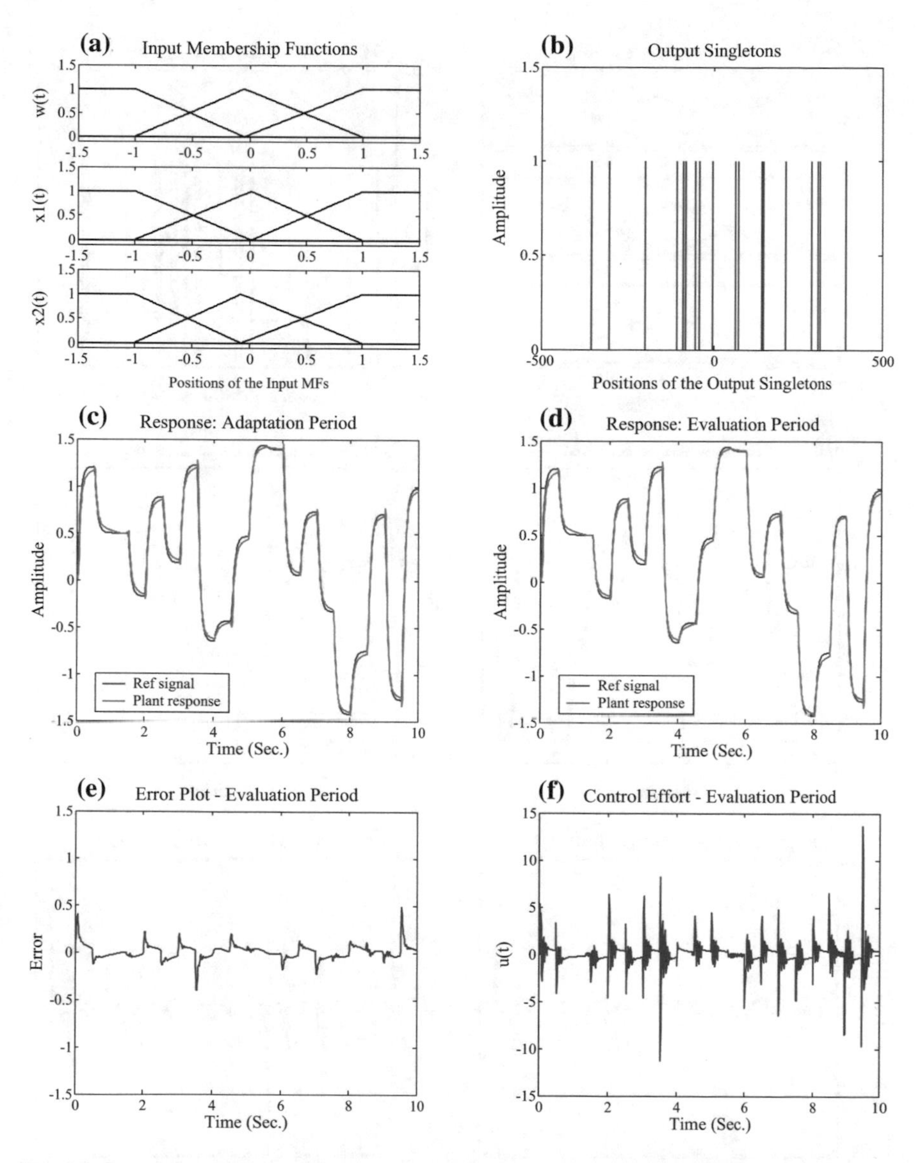

Fig. 5.9 Responses of the adaptive fuzzy controller with PSO-Hybd-Con model, for $3 \times 3 \times 3$ configuration of input MFs, of case study I (nonlinear DC motor system): **a** positions of input MFs, **b** positions of output singletons, **c** adaptation period response after 100 PSO generations, **d** evaluation period response for 0–10 s after 200 PSO generations, **e** evaluation period error variation, **f** evaluation period control effort variation

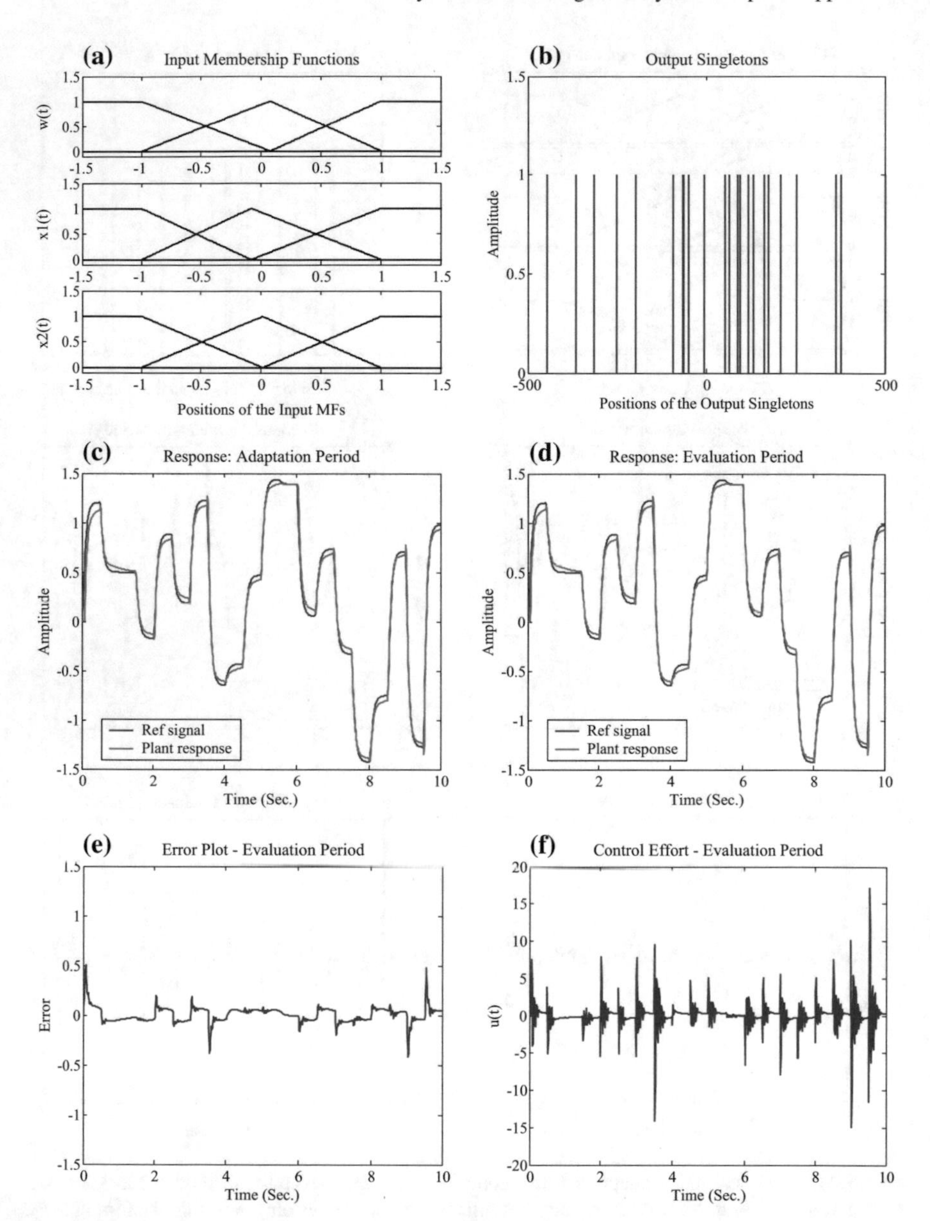

Fig. 5.10 Responses of the adaptive fuzzy controller with PSO-Hybd-Pref model, for $3 \times 3 \times 3$ configuration of input MFs, of case study I (nonlinear DC motor system): **a** positions of input MFs, **b** positions of output singletons, **c** adaptation period response after 100 PSO generations, **d** evaluation period response for 0–10 s after 200 PSO generations, **e** evaluation period error variation, **f** evaluation period control effort variation

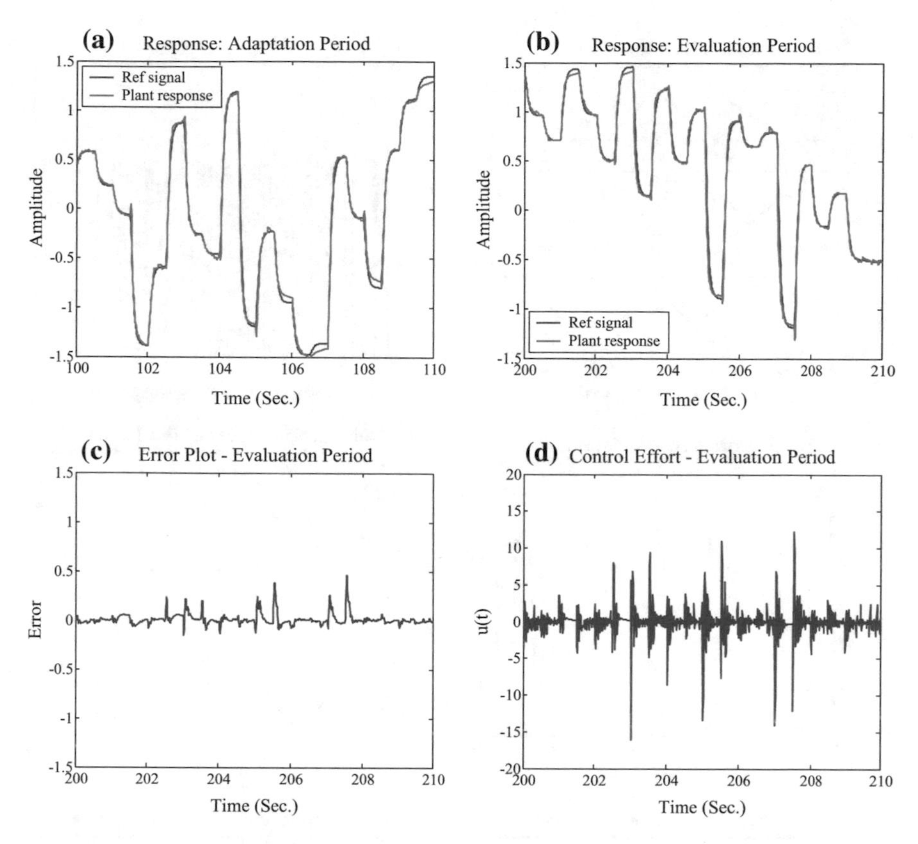

Fig. 5.11 Responses of the adaptive fuzzy controller with PSO-Hybd-Cas model, for $5 \times 5 \times 5$ configuration of input MFs, of case study I (nonlinear DC motor system): **a** adaptation period response after 100 s, **b** evaluation period response for 200–210 s, **c** evaluation period error variation, **d** evaluation period control effort variation

lead to better performances from PSO-Hybd-Pref model of controller design. Again it is noticed that there are some amount of oscillations in the responses for PSO-Hybd-Cas model and PSO-Hybd-Con model design strategies for $7 \times 7 \times 7$ input MFs configuration, but the oscillations are largely suppressed in PSO-Hybd-Pref model of design strategy [2].

5.3.1.2 Performance Analysis of HSA-Based Hybrid Controllers

To perform this case study, a harmony memory (*HM*) of ten harmonies is considered. For each simulation, 200 generations of *HM* improvisation are carried out. The other design parameters are chosen same as those for the case study I in Chap. 3. In fixed structure controller design, a fixed number of evenly distributed input MFs are used for initial *HM* formation and in variable structure controller

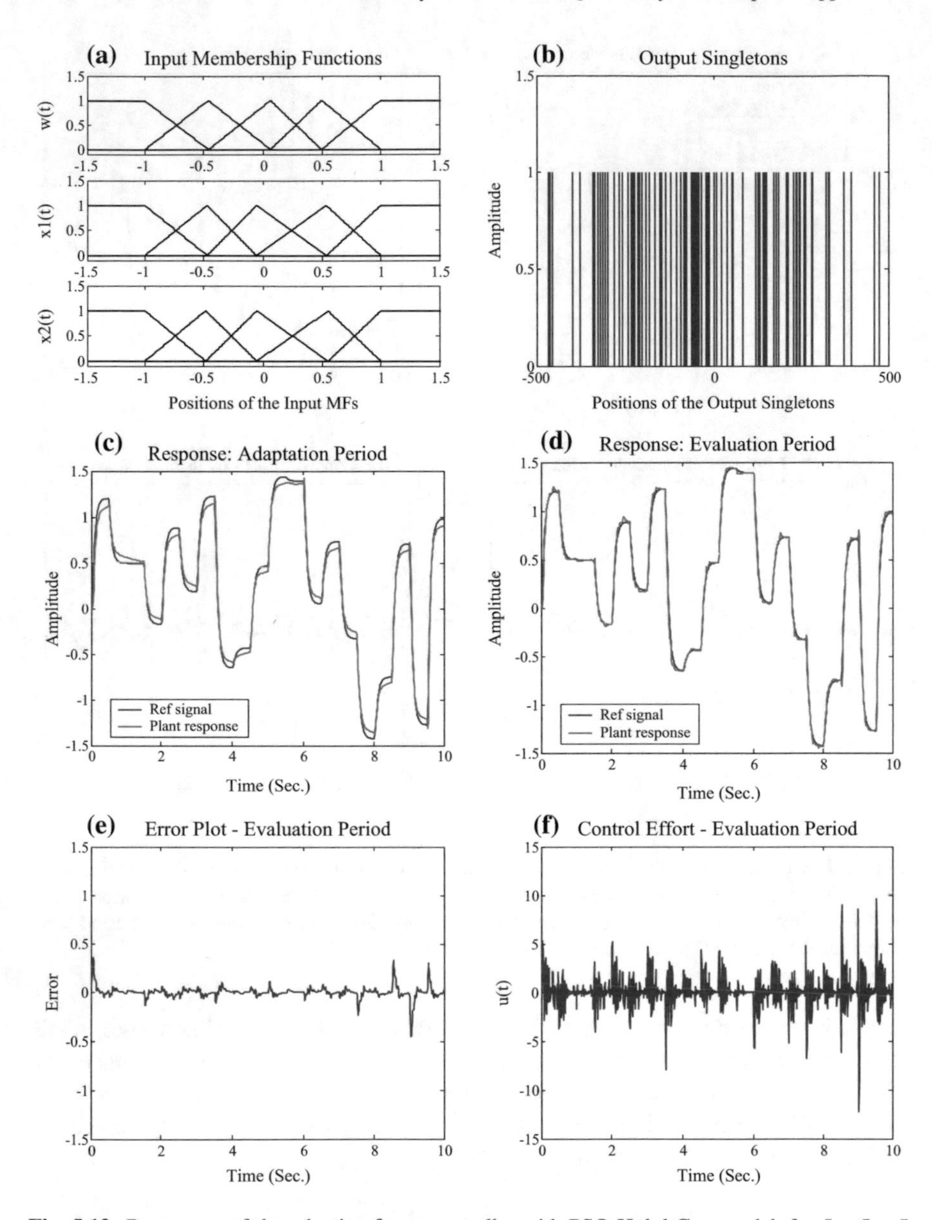

Fig. 5.12 Responses of the adaptive fuzzy controller with PSO-Hybd-Con model, for $5 \times 5 \times 5$ configuration of input MFs, of case study I (nonlinear DC motor system): **a** positions of input MFs, **b** positions of output singletons, **c** adaptation period response after 100 PSO generations, **d** evaluation period response for 0–10 s after 200 PSO generations, **e** evaluation period error variation, **f** evaluation period control effort variation

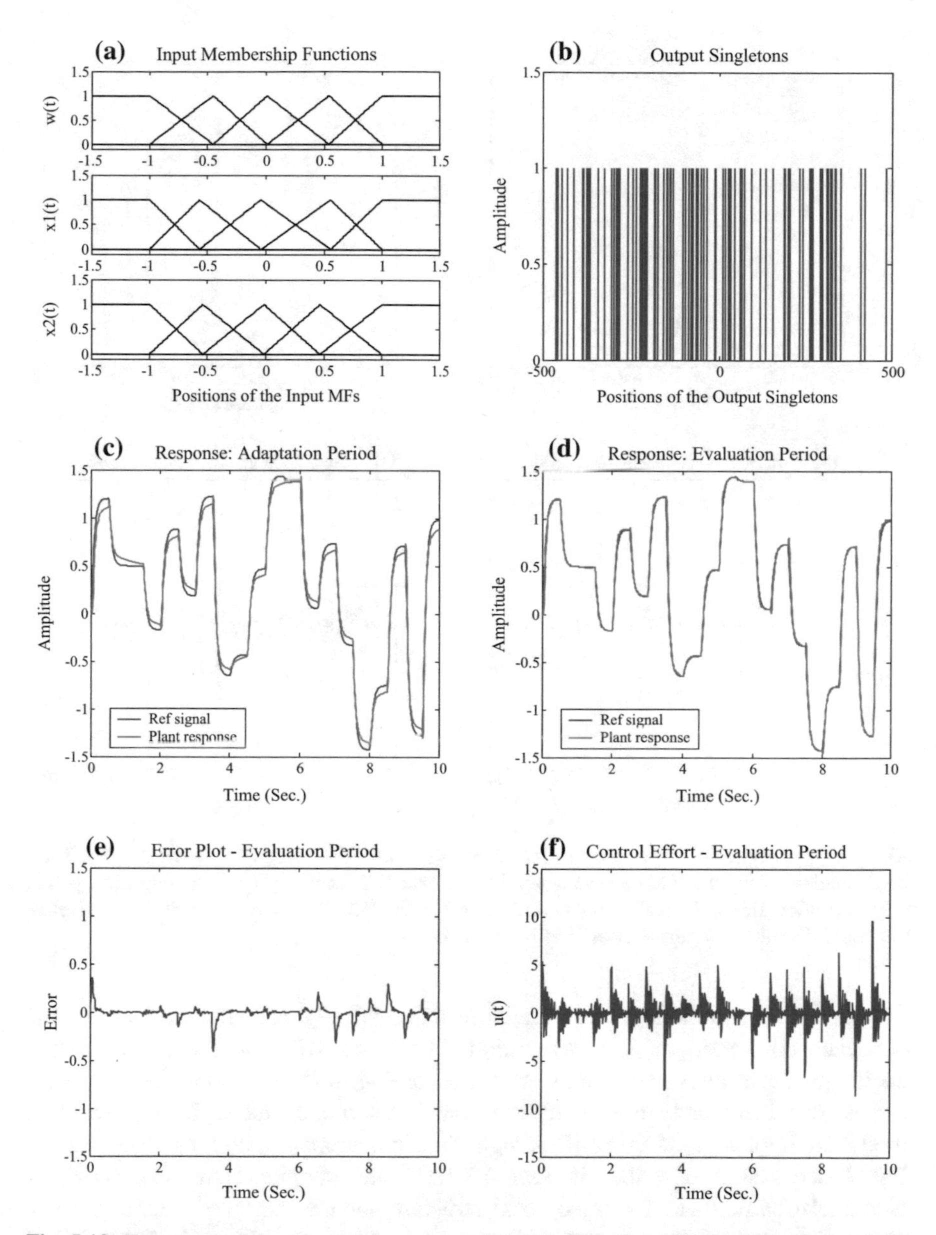

Fig. 5.13 Responses of the adaptive fuzzy controller with PSO-Hybd-Pref model, for $5 \times 5 \times 5$ configuration of input MFs, of case study I (nonlinear DC motor system): **a** positions of input MFs, **b** positions of output singletons, **c** adaptation period response after 100 PSO generations, **d** evaluation period response for 0–10 s after 200 PSO generations, **e** evaluation period error variation, **f** evaluation period control effort variation

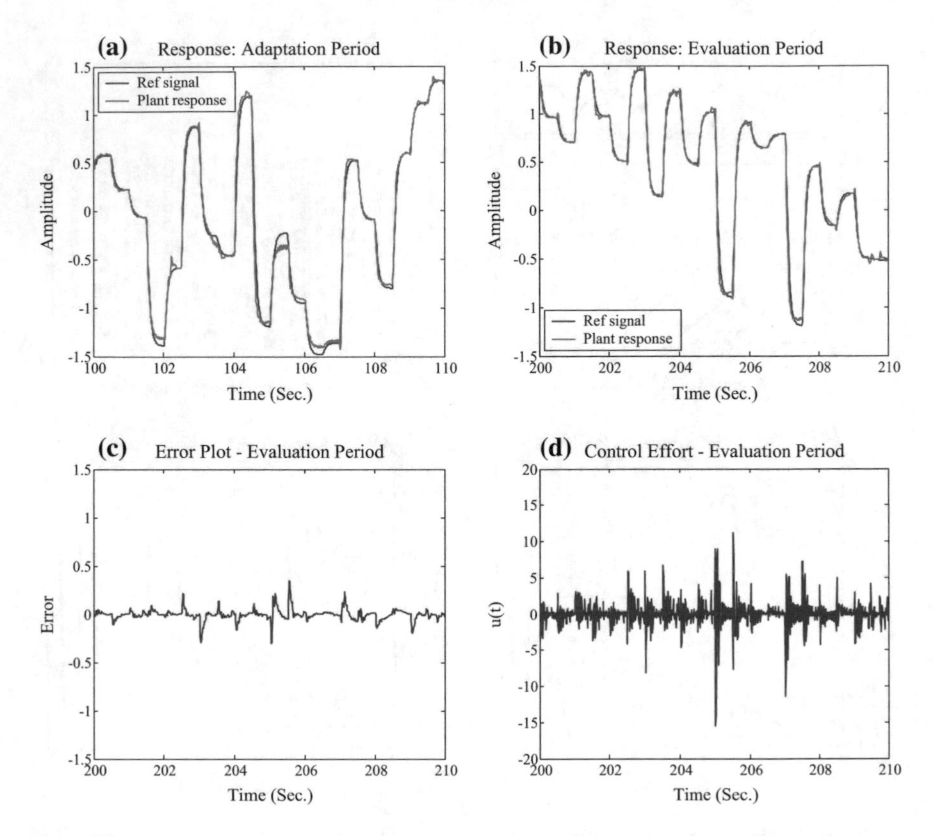

Fig. 5.14 Responses of the adaptive fuzzy controller with PSO-Hybd-Cas model, for 7 × 7 × 7 configuration of input MFs, of case study I (nonlinear DC motor system): **a** adaptation period response after 100 s, **b** evaluation period response for 200–210 s, **c** evaluation period error variation, **d** evaluation period control effort variation

design, the structural flags are utilized to automatically vary the structure of the candidate controller, in each generation. The input MFs and the corresponding positions of the output singletons are set according to the structural flags [2, 8].

The hybrid cascade model, hybrid concurrent model and hybrid preferential model are implemented using HSA algorithm in a similar fashion as these were in PSO-based design, described in Sect. 5.3.1.1 of this chapter. Here also, the simulations are carried out for three fixed structures and a variable structure with a potentially biggest possible structure of 7 × 7 × 7 MFs, as shown in Table 5.2. It can be seen from Table 5.2 that the performance of HSA-Hybd-Con model, in terms of *IAE*, is not better than the HSABA counterpart for 5 × 5 × 5 and 7 × 7 × 7 input MF configurations. This observation justifies the proposal of using HSA-Hybd-Pref model for designing AFLCs. The performances of HSA-Hybd-Pref model are superior in case of all fixed structure configurations. Among the variable structure configurations, the *IAE* value is smallest for the

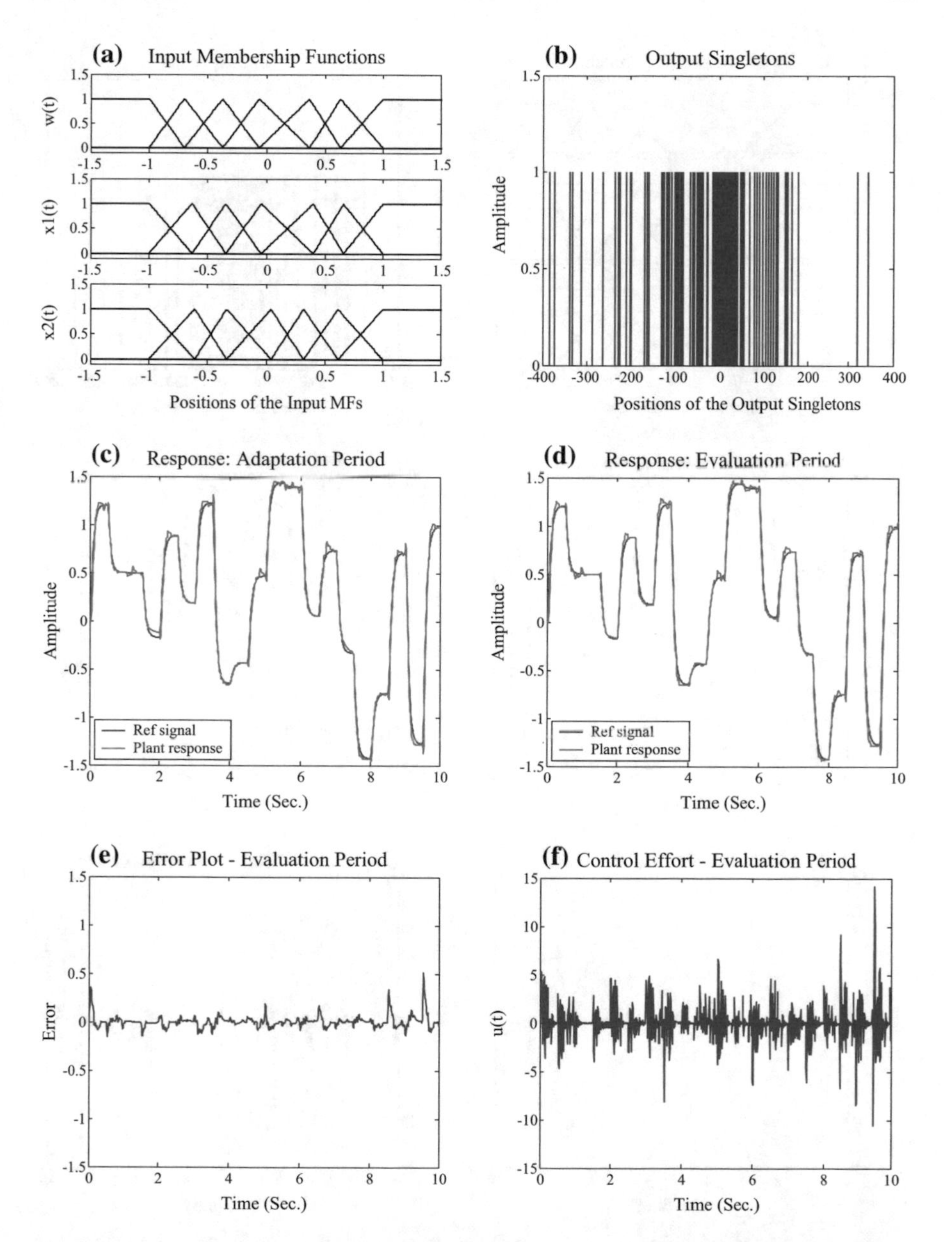

Fig. 5.15 Responses of the adaptive fuzzy controller with PSO-Hybd-Con model, for $7 \times 7 \times 7$ configuration of input MFs, of case study I (nonlinear DC motor system): **a** positions of input MFs, **b** positions of output singletons, **c** adaptation period response after 100 PSO generations, **d** evaluation period response for 0–10 s after 200 PSO generations, **e** evaluation period error variation, **f** evaluation period control effort variation

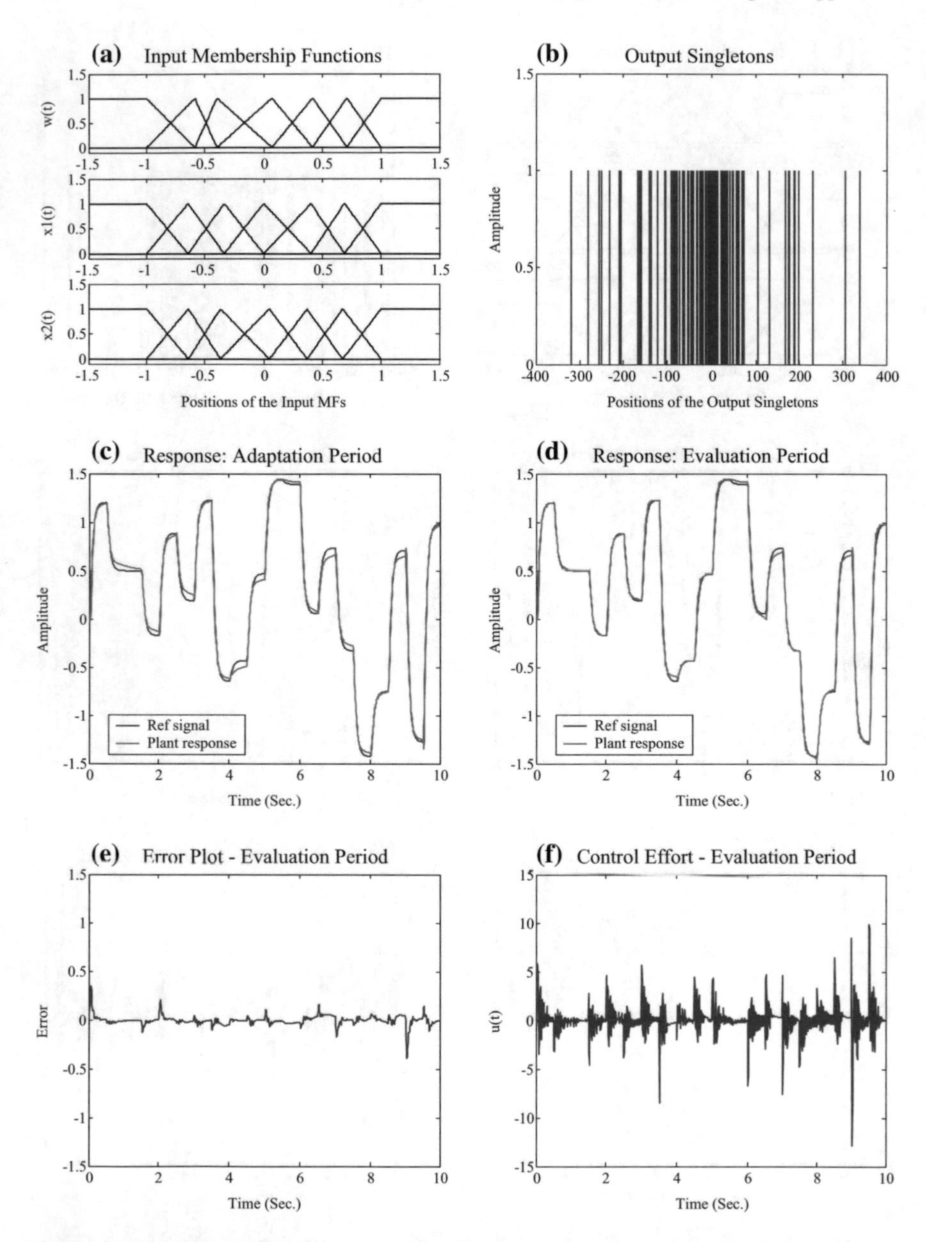

Fig. 5.16 Responses of the adaptive fuzzy controller with PSO-Hybd-Pref model, for $7 \times 7 \times 7$ configuration of input MFs, of case study I (nonlinear DC motor system): **a** positions of input MFs, **b** positions of output singletons, **c** adaptation period response after 100 PSO generations, **d** evaluation period response for 0–10 s after 200 PSO generations, **e** evaluation period error variation, **f** evaluation period control effort variation

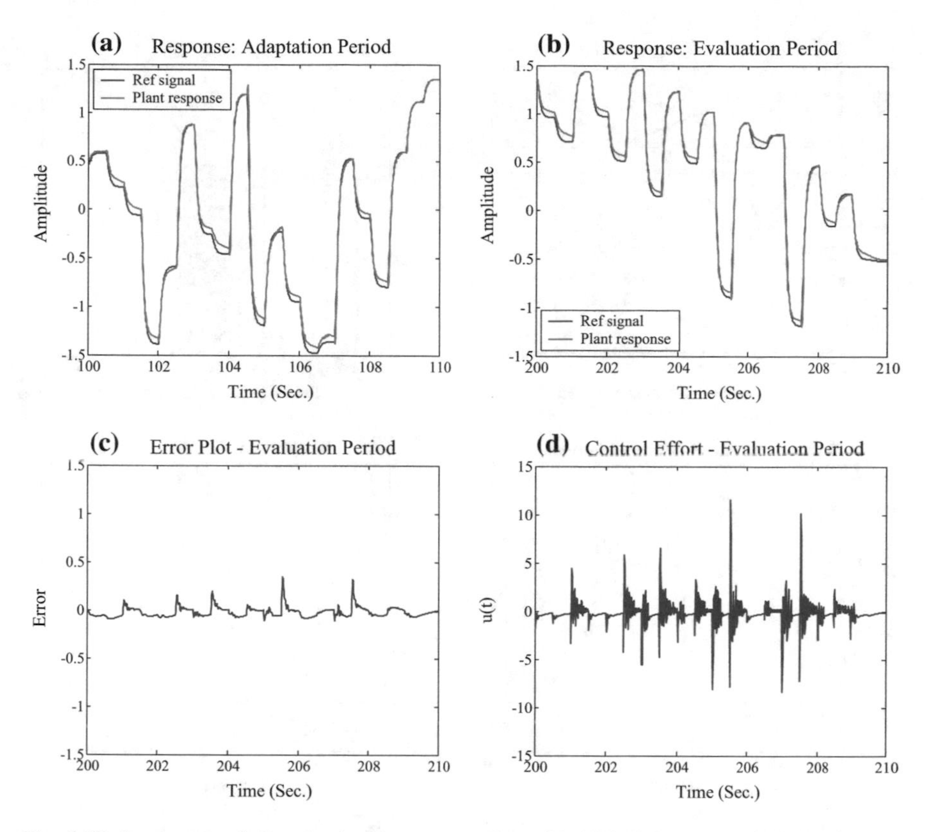

Fig. 5.17 Responses of the adaptive fuzzy controller with PSO-Hybd-Cas model, for variable structure configuration of input MFs, of case study I (nonlinear DC motor system): **a** adaptation period response after 100 s, **b** evaluation period response for 200–210 s, **c** evaluation period error variation, **d** evaluation period control effort variation

HSA-Hybd-Pref model with 30 rules. In comparison, HSA-Hybd-Con model produces a little higher value of *IAE* compared to HSA-Hybd-Pref model, but with a smaller number of rules (only 27 out of a maximum possible 343 number of rules). Although the overall *IAE* performances of all variable structure controllers are inferior compared to the fixed structure controllers, a significant reduction in the number of rules is a strong highlighting feature. Hence, for a given controller MF configuration, among the competing controllers, the HSA-Hybd-Pref model shows the overall best performance. Even for the variable structure scheme, HSA-Hybd-Pref method emerges as a suitable solution. A significant reduction in the total number of rules and such a simplified structure of an FLC will be very useful from the point of view of implementation in real life. Figures 5.20, 5.21, and 5.22 show the temporal variations of system responses for the sample case of fixed structure controllers with $5 \times 5 \times 5$ input MFs configurations for HSA-based hybrid cascade, hybrid concurrent, and hybrid preferential schemes, respectively,

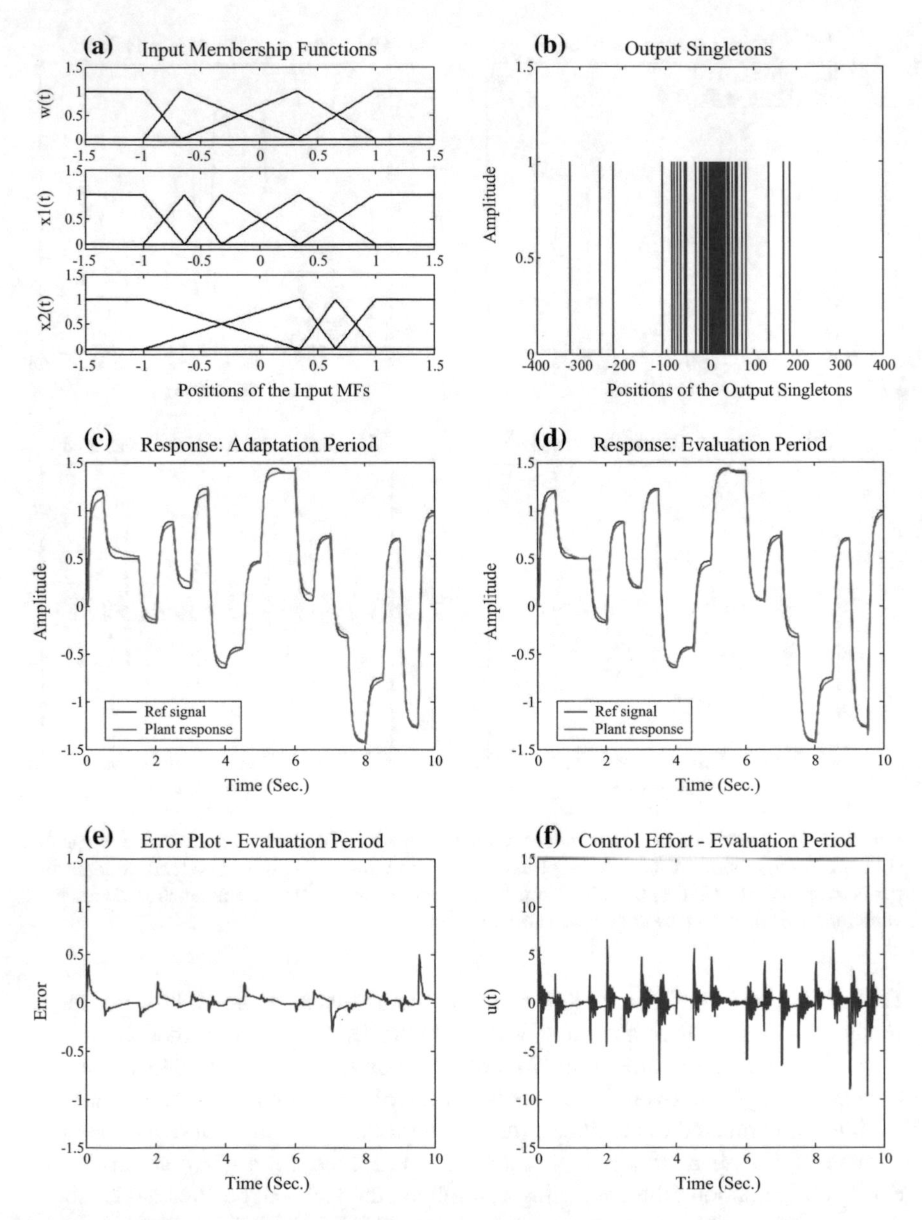

Fig. 5.18 Responses of the adaptive fuzzy controller with PSO-Hybd-Con model, for variable structure configuration of input MFs, of case study I (nonlinear DC motor system): **a** positions of input MFs, **b** positions of output singletons, **c** adaptation period response after 100 PSO generations, **d** evaluation period response for 0–10 s after 200 PSO generations, **e** evaluation period error variation, **f** evaluation period control effort variation

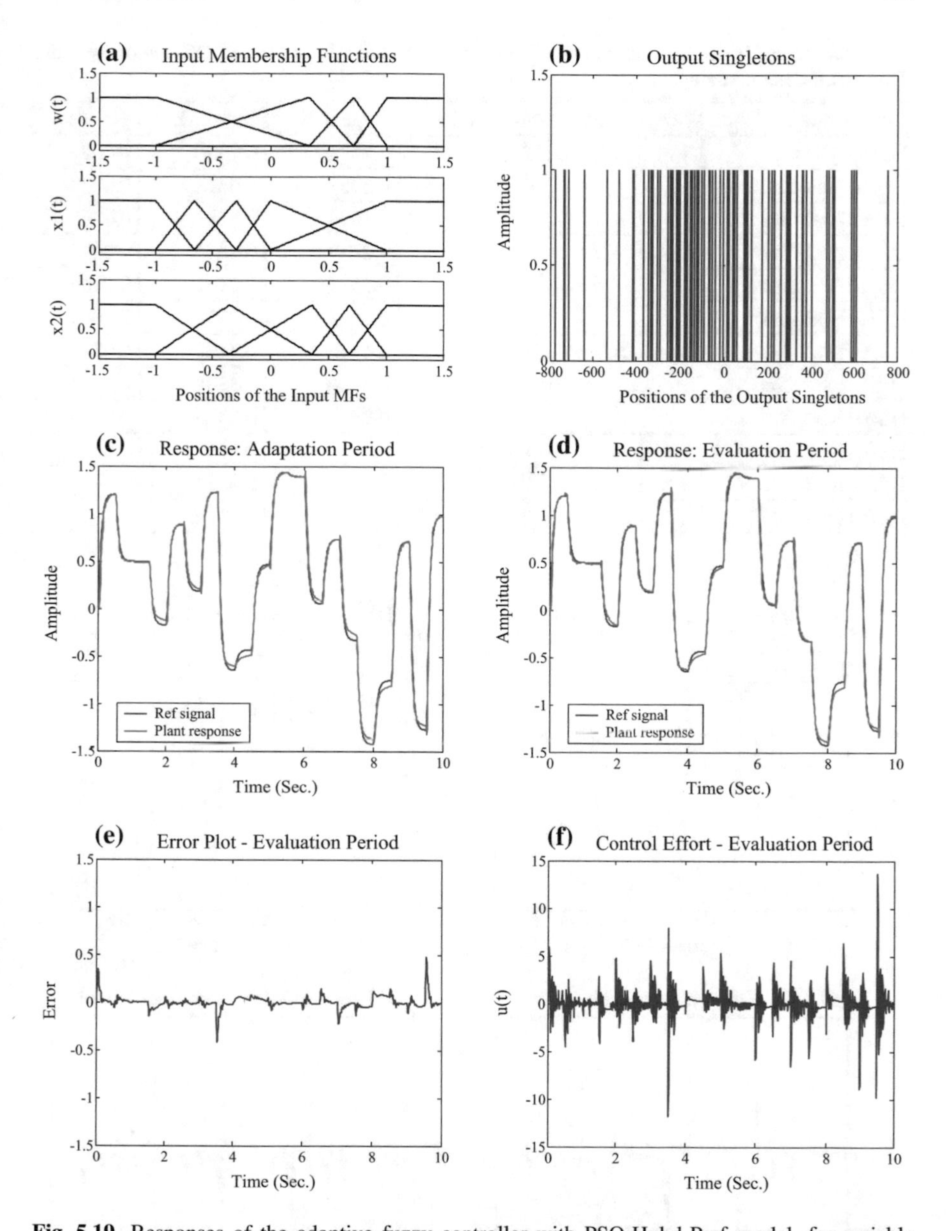

Fig. 5.19 Responses of the adaptive fuzzy controller with PSO-Hybd-Pref model, for variable structure configuration of input MFs, of case study I (nonlinear DC motor system): **a** positions of input MFs, **b** positions of output singletons, **c** adaptation period response after 100 PSO generations, **d** evaluation period response for 0–10 s after 200 PSO generations, **e** evaluation period error variation, **f** evaluation period control effort variation

Table 5.2 Comparison of simulation results for case study I (nonlinear DC motor system): HSA-based design strategies

Control strategy	No. of MFs	No. of rules	Avg. IAE	Best IAE	Std. dev.
HSA-Hybd-Cas	3 × 3 × 3	27	0.7573	0.7466	0.0094
HSA-Hybd-Con			0.6757	0.6671	0.0073
HSA-Hybd-Pref			0.6639	0.6583	0.0068
HSA-Hybd-Cas	5 × 5 × 5	125	0.2728	0.2646	0.0095
HSA-Hybd-Con			0.2513	0.2435	0.0104
HSA-Hybd-Pref			0.2301	0.2257	0.0044
HSA-Hybd-Cas	7 × 7 × 7	343	0.4446	0.4168	0.0273
HSA-Hybd-Con			0.3406	0.3361	0.0034
HSA-Hybd-Pref			0.2542	0.2485	0.0048
HSA-Hybd-Cas-V	4 × 3 × 5	60		0.6863	
HSA-Hybd-Con-V	3 × 3 × 3	27	–	0.6584	–
HSA-Hybd-Pref-V	2 × 3 × 5	30	–	0.6456	–

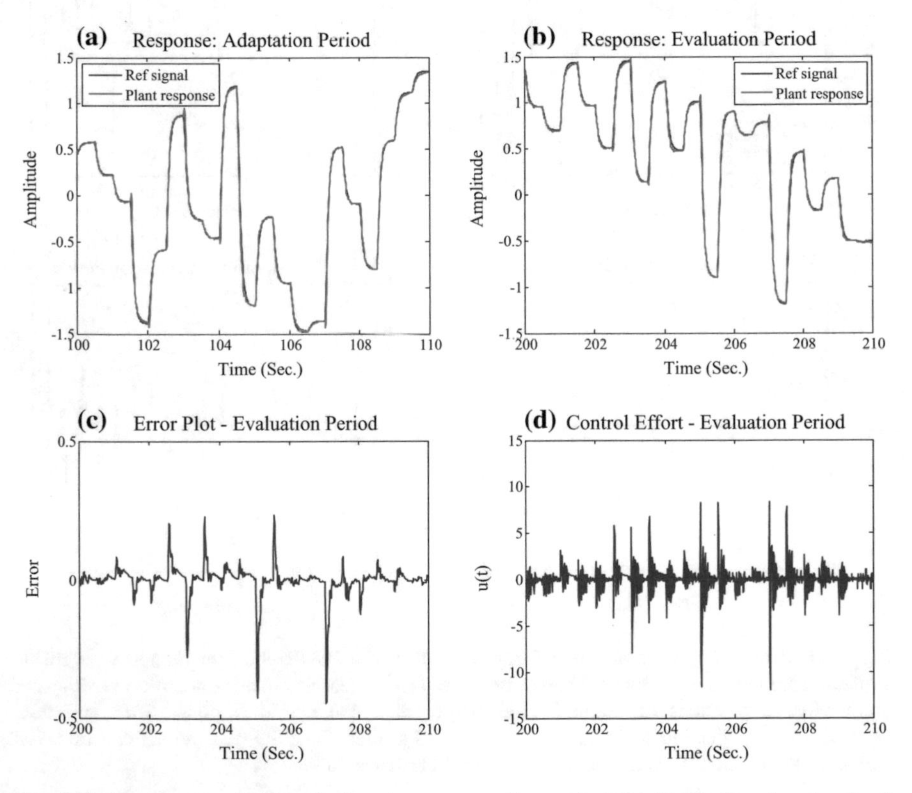

Fig. 5.20 Responses of the adaptive fuzzy controller with HSA-Hybd-Cas, for 5 × 5 × 5 configuration of input MFs, of case study I (nonlinear DC motor system): **a** adaptation period response after 100 s, **b** evaluation period response for 200–210 s, **c** evaluation period error variation, **d** evaluation period control effort variation

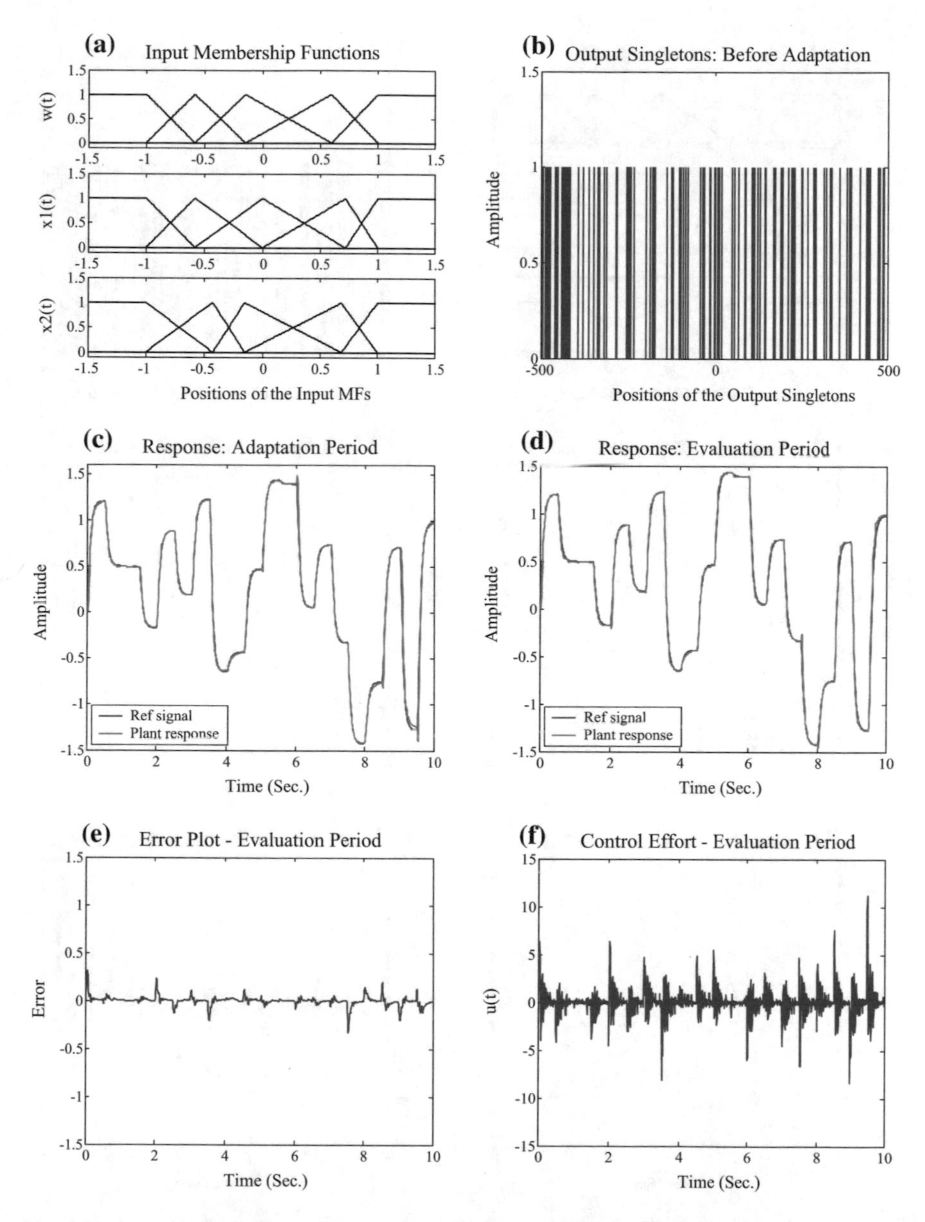

Fig. 5.21 Responses of the adaptive fuzzy controller with HSA-Hybd-Con model, for $5 \times 5 \times 5$ configuration of input MFs, of case study I (nonlinear DC motor system): **a** positions of input MFs, **b** positions of output singletons, **c** adaptation period response after 100 HSA generations, **d** evaluation period response for 0–10 s after 200 HSA generations, **e** evaluation period error variation, **f** evaluation period control effort variation

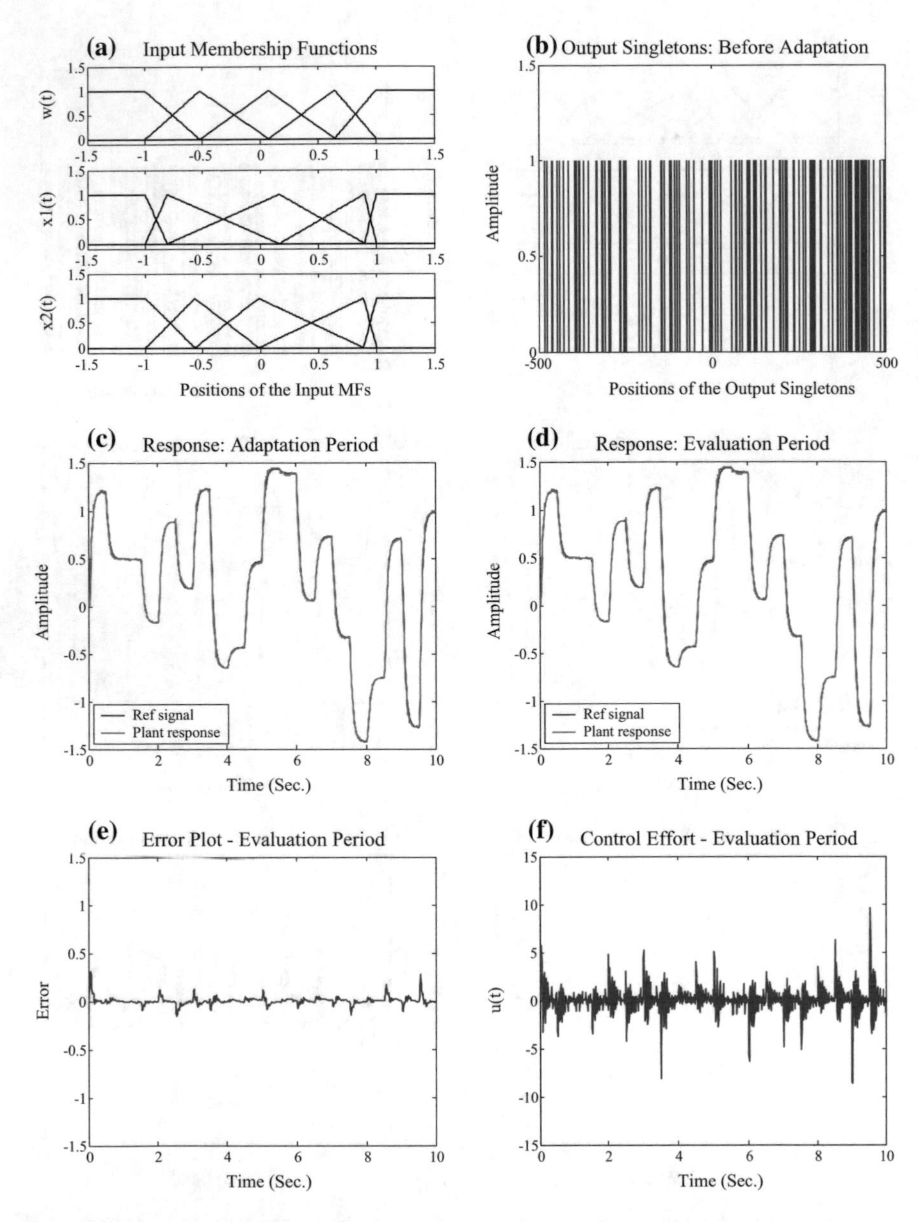

Fig. 5.22 Responses of the adaptive fuzzy controller with HSA-Hybd-Pref model, for $5 \times 5 \times 5$ configuration of input MFs, of case study I (nonlinear DC motor system): **a** positions of input MFs, **b** positions of output singletons, **c** adaptation period response after 100 HSA generations, **d** evaluation period response for 0–10 s after 200 HSA generations, **e** evaluation period error variation, **f** evaluation period control effort variation

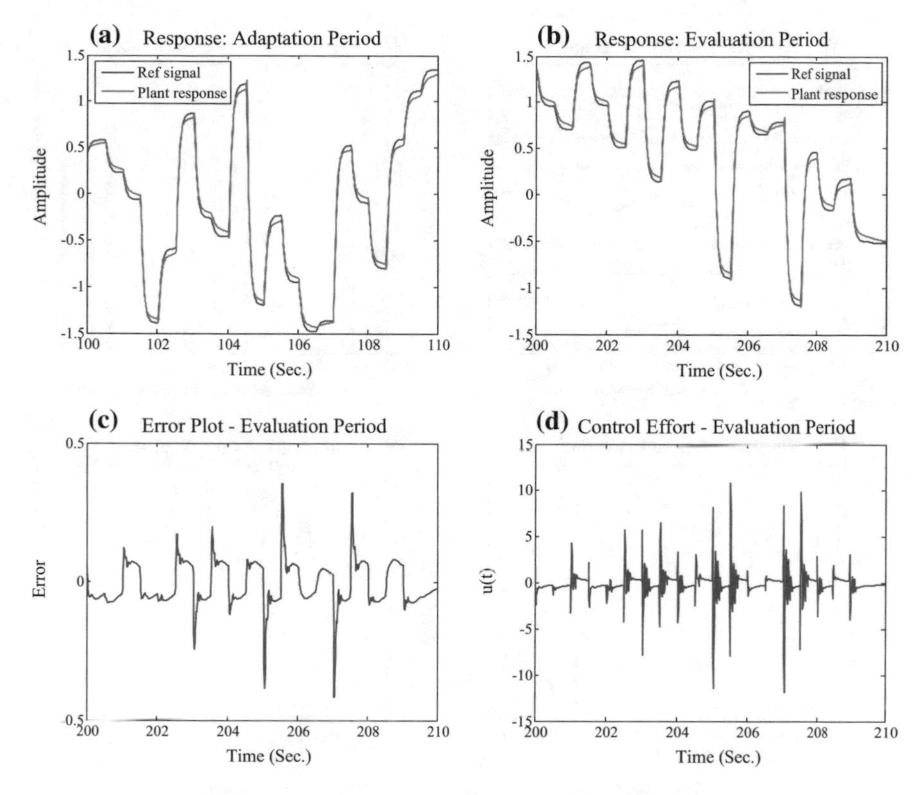

Fig. 5.23 Responses of the adaptive fuzzy controller with HSA-Hybd-Cas, for variable structure configuration of input MFs, of case study I (nonlinear DC motor system): **a** adaptation period response after 100 s, **b** evaluation period response for 200–210 s, **c** evaluation period error variation, **d** evaluation period control effort variation

and Figs. 5.23, 5.24, and 5.25 show the same for variable structure AFLCs. Each of these figures shows the different features of the case study as in the same manner as in PSO-based hybrid schemes.

5.3.2 *Case Study II: Duffing's Oscillatory System*

In this case study, once more we consider Duffing's oscillatory system for demonstrating the usefulness of the proposed hybrid adaptive control strategies. In this section, first the Duffing's forced oscillation system is considered and then the Duffing's forced oscillation system with disturbance is considered for the detail study.

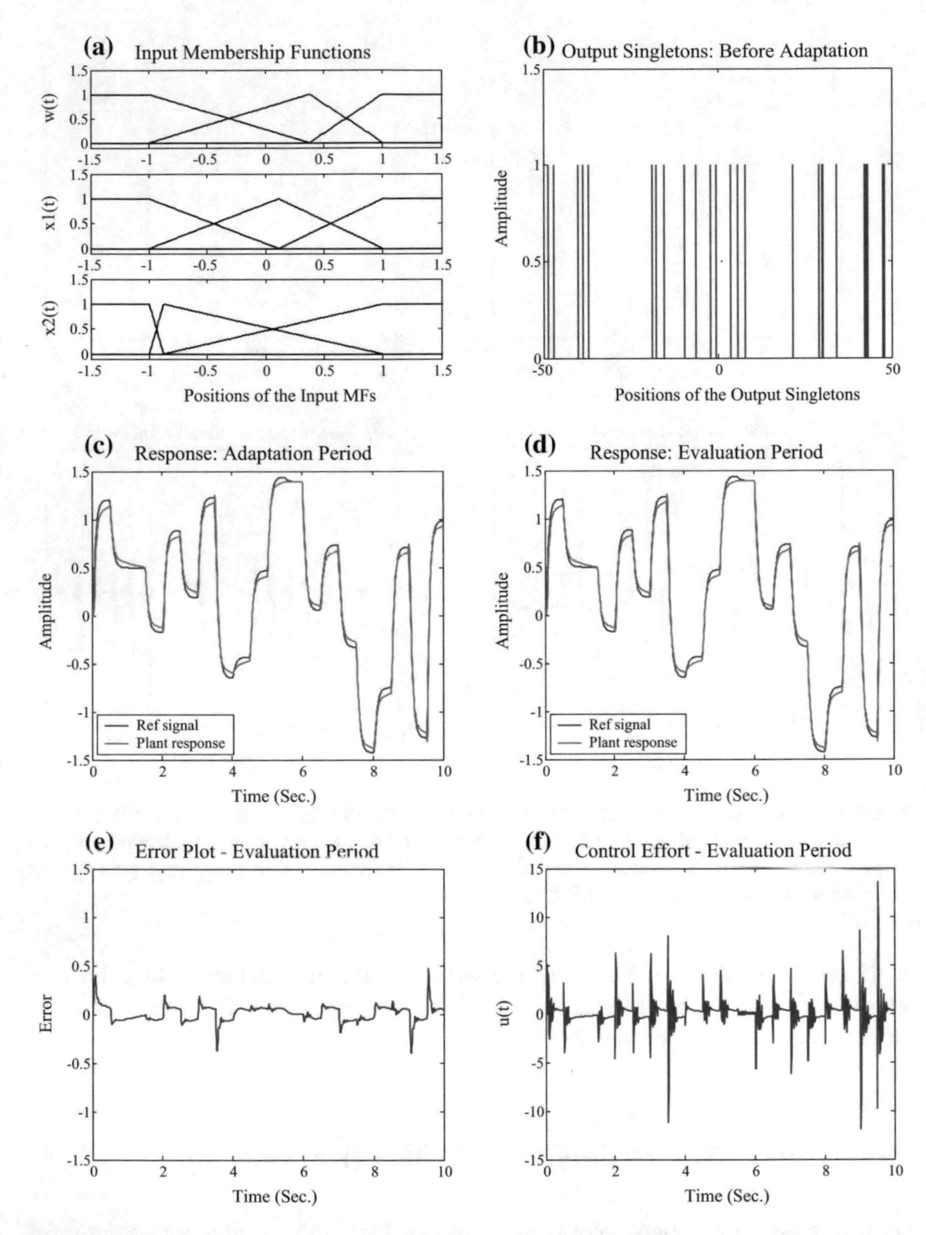

Fig. 5.24 Responses of the adaptive fuzzy controller with HSA-Hybd-Con model, for variable structure configuration of input MFs, of case study I (nonlinear DC motor system): **a** positions of input MFs, **b** positions of output singletons, **c** adaptation period response after 100 HSA generations, **d** evaluation period response for 0–10 s after 200 HSA generations, **e** evaluation period error variation, **f** evaluation period control effort variation

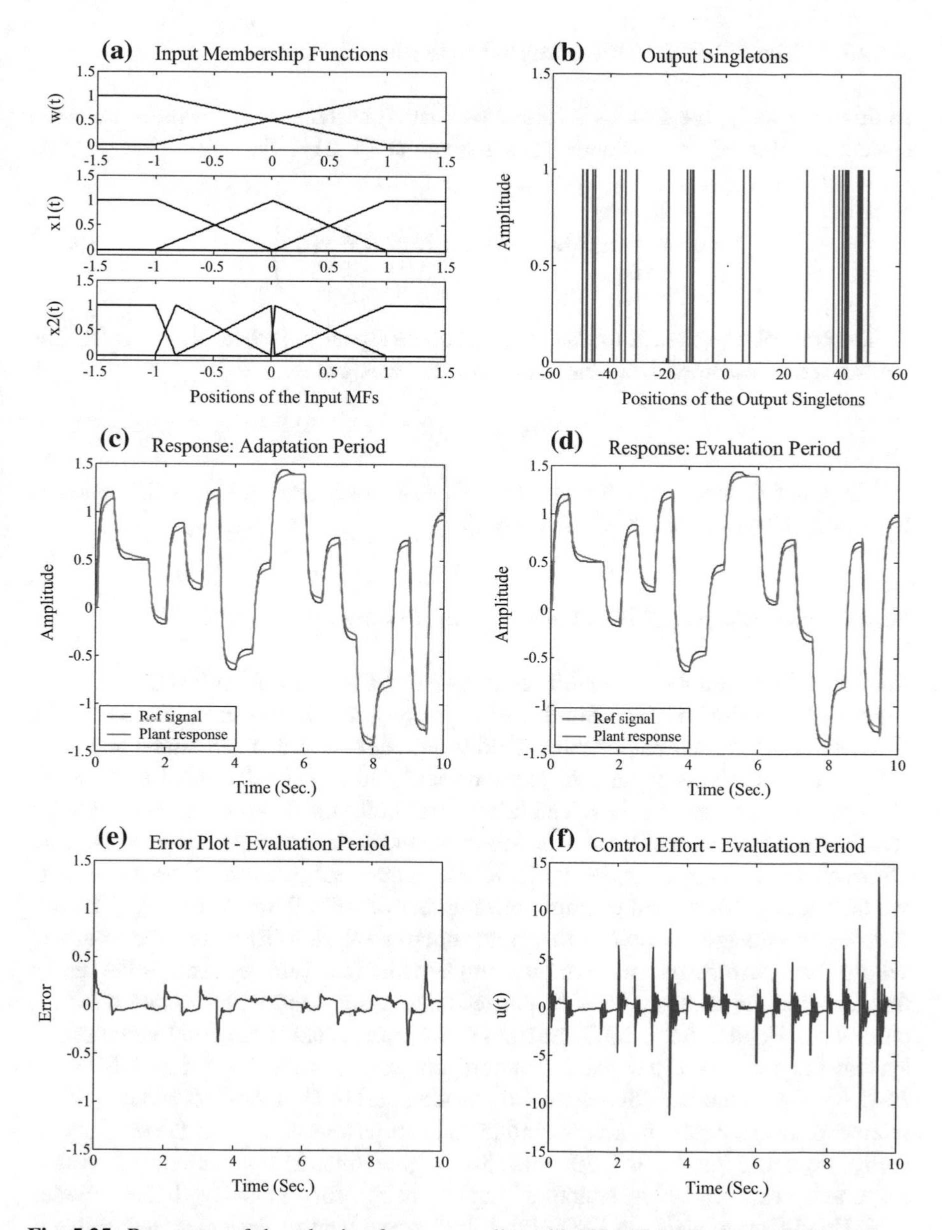

Fig. 5.25 Responses of the adaptive fuzzy controller with HSA-Hybd-Pref model, for variable structure configuration of input MFs, of case study I (nonlinear DC motor system): **a** positions of input MFs, **b** positions of output singletons, **c** adaptation period response after 100 HSA generations, **d** evaluation period response for 0–10 s after 200 HSA generations, **e** evaluation period error variation, **f** evaluation period control effort variation

5.3.2.1 Case Study II(A): Duffing's Oscillatory System

In this case study, the Duffing's forced oscillation system, which is itself a chaotic system if unforced, is considered and is given as [4, 21–23]:

$$\left.\begin{aligned} \dot{x}_1 &= x_2 \\ \dot{x}_2 &= -0.1x_2 - x_1^3 + 12\cos(t) + u(t) \\ y &= x_1 \end{aligned}\right\} \tag{5.16}$$

The control objective is to track the reference signal y_m, where y_m = sin (t). The control signal obtained from the fuzzy controller is as:

$$u_c = u_c(\underline{x}|\underline{Z}) \tag{5.17}$$

The other design considerations for this case study are similar as discussed in case study II(A) in Sect. 3.6.2 of Chap. 3.

Performance Analysis of PSO-Based Hybrid Controllers

The adaptation gain v is set to 0.5 for PSO-Hybd-Cas, 2 for PSO-Hybd-Con and 1 for PSO-Hybd-Pref models. Similar to the approach adopted in case study I, the simulations are carried out for three fixed structures of 3 × 3, 5 × 5, and 7 × 7 input MFs and a variable structure. In this case study also, the PSO-Hybd-Pref design strategy emerges as the best candidate, for both fixed structure and variable structure configurations. In fact, for this case study, the best performance result for PSO-Hybd-Pref controller for variable structure configuration is obtained with simultaneous achievement of minimum number of rules (here it is 15) [2]. Hence, this design strategy evolves as the most superior solution from both the points of view of best performance and simplest implementation. Table 5.3 shows the results obtained for the case study II(A) carried out, using fixed and variable structure controllers. Figures 5.26, 5.27, and 5.28 show representative temporal variations of system responses for the fixed structure controllers with 5 × 5 input MFs for PSO-Hybd-Cas model, PSO-Hybd-Con model, and PSO-Hybd-Pref model control strategies, respectively, in a very similar manner as was utilized in the earlier case study. Again, Figs. 5.29, 5.30, and 5.31 show temporal variations of system responses for variable structure input MFs for PSO-Hybd-Cas model, PSO-Hybd-Con model, and PSO-Hybd-Pref model control strategies, respectively, in a similar manner as was shown in the fixed structure case.

In this case study also, the values of crosspoint level and crosspoint ratio are chosen as 0.5 and 1, respectively, to achieve better performances from the FLCs designed [18, 20]. It is observed from the pictorial representations of this case study that the oscillations during the adaptation period are fairly suppressed during the evaluation period.

Table 5.3 Comparison of simulation results for case study II (A) (Duffing's system): PSO-based design strategies

Control strategy	No. of MFs	No. of rules	Avg. IAE	Best IAE	Std. dev.
PSO-Hybd-Cas	3 × 3	9	0.2501	0.2383	0.0168
PSO-Hybd-Con			0.3121	0.2922	0.0136
PSO-Hybd-Pref			0.2274	0.1962	0.0441
PSO-Hybd-Cas	5 × 5	25	0.2376	0.2354	0.0031
PSO-Hybd-Con			0.2805	0.2271	0.0751
PSO-Hybd-Pref			0.2004	0.1960	0.0062
PSO-Hybd-Cas	7 × 7	49	0.2531	0.2407	0.0175
PSO-Hybd-Con			0.2505	0.2165	0.0258
PSO-Hybd-Pref			0.1982	0.1756	0.0320
PSO-Hybd-Cas-V	4 × 7	28	–	0.2339	–
PSO-Hybd-Con-V	3 × 5	15	–	0.3164	–
PSO-Hybd-Pref-V	5 × 3	15	–	0.2197	–

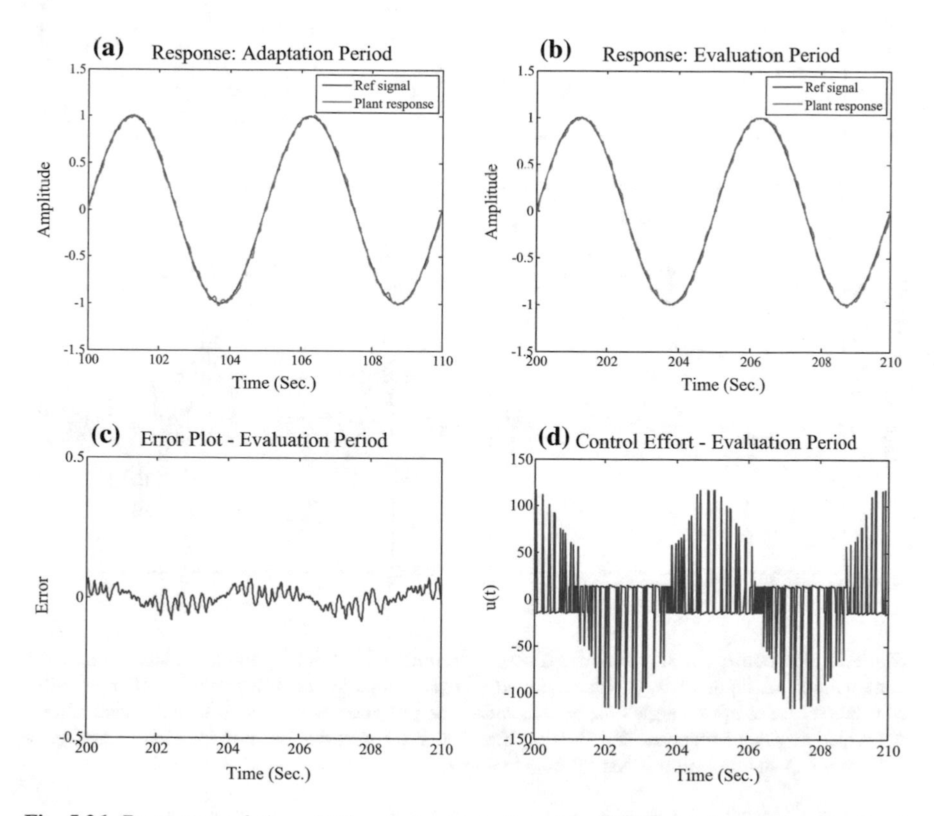

Fig. 5.26 Responses of the adaptive fuzzy controller with PSO-Hybd-Cas model, for 5 × 5 configuration of input MFs, of case study II (A) (Duffing's system): **a** adaptation period response after 100 s, **b** evaluation period response for 200–210 s, **c** evaluation period error variation, **d** evaluation period control effort variation

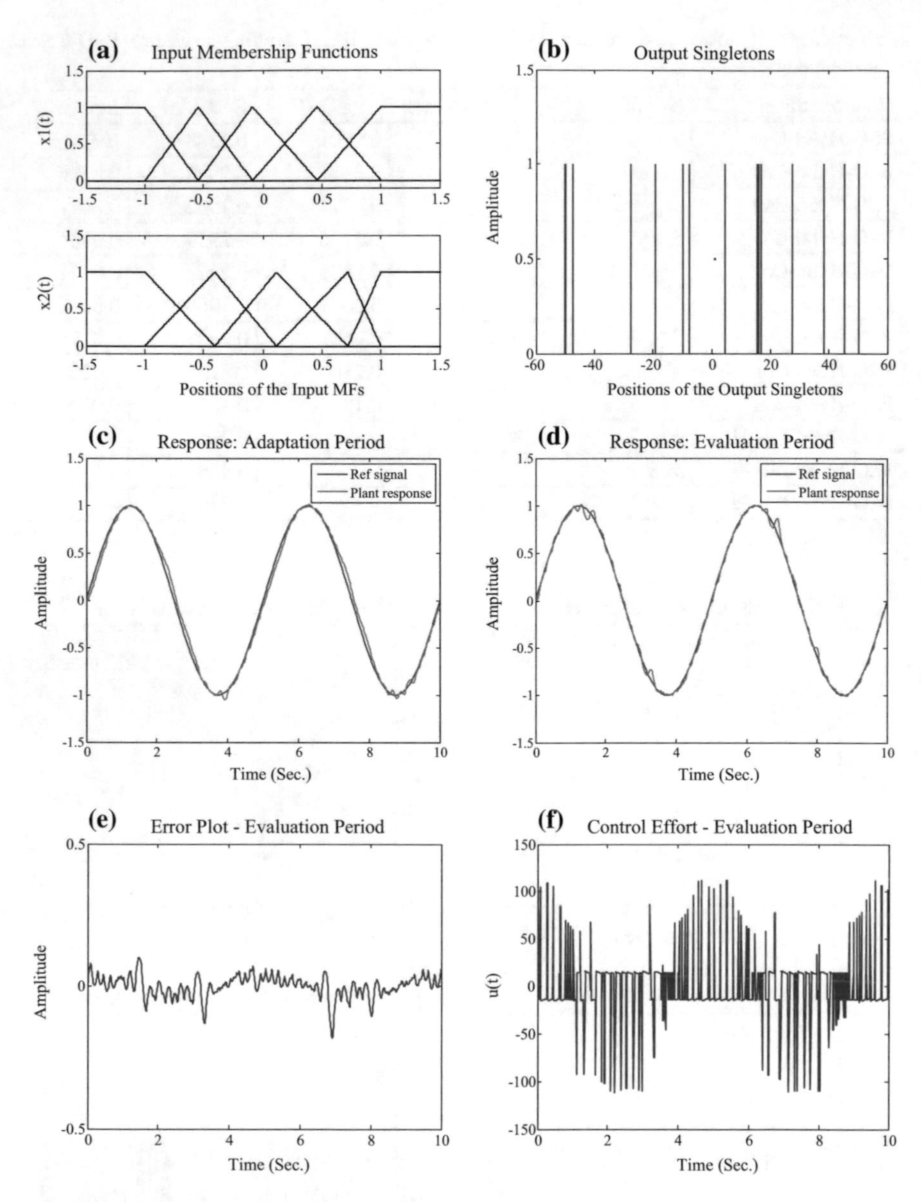

Fig. 5.27 Responses of the adaptive fuzzy controller with PSO-Hybd-Con model, for 5×5 configuration of input MFs, of case study II (A) (Duffing's system): **a** positions of input MFs, **b** positions of output singletons, **c** adaptation period response after 100 PSO generations, **d** evaluation period response for 0–10 s after 200 PSO generations, **e** evaluation period error variation, **f** evaluation period control effort variation

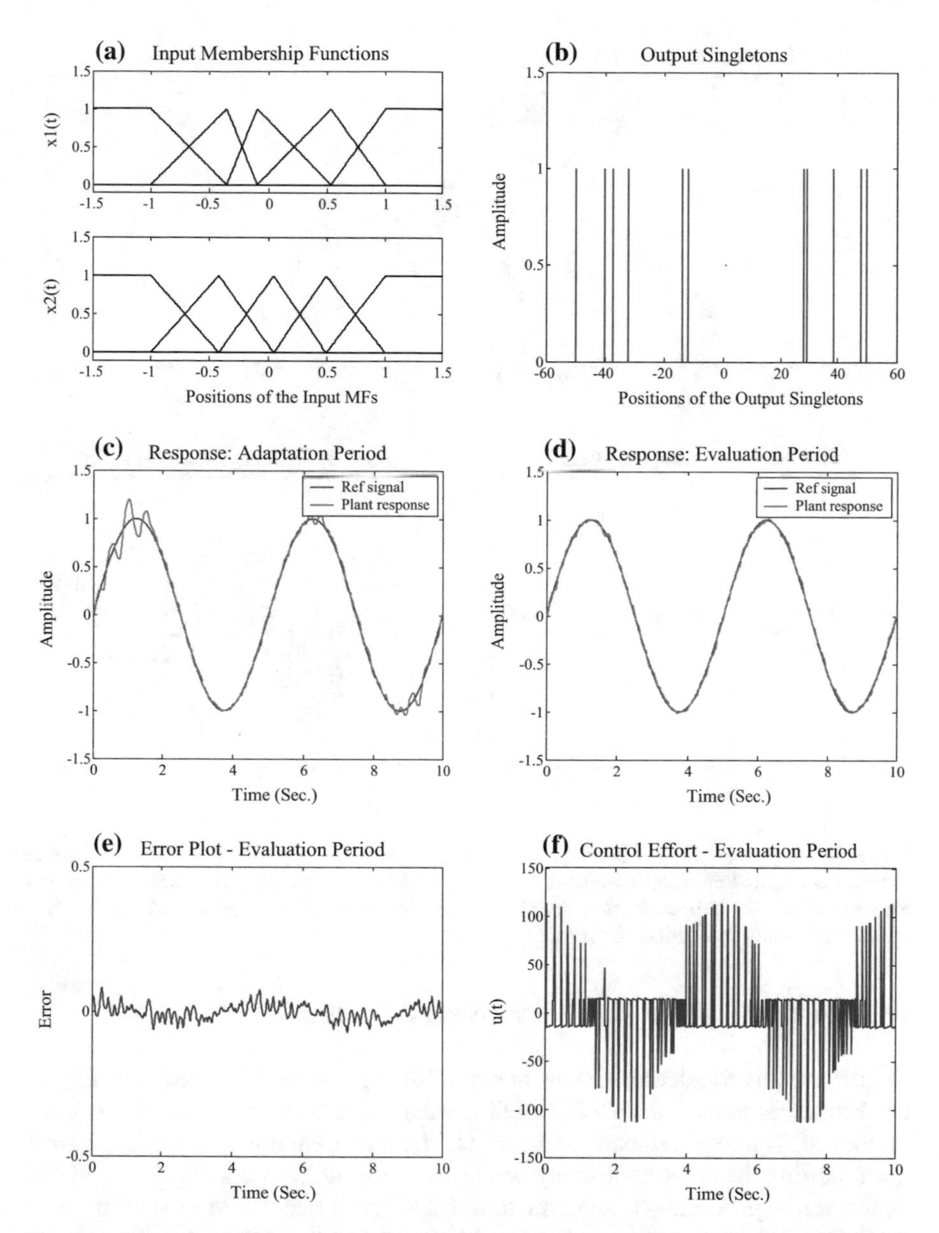

Fig. 5.28 Responses of the adaptive fuzzy controller with PSO-Hybd-Pref model, for 5 × 5 configuration of input MFs, of case study II (A) (Duffing's system): **a** positions of input MFs, **b** positions of output singletons, **c** adaptation period response after 100 PSO generations, **d** evaluation period response for 0–10 s after 200 PSO generations, **e** evaluation period error variation, **f** evaluation period control effort variation

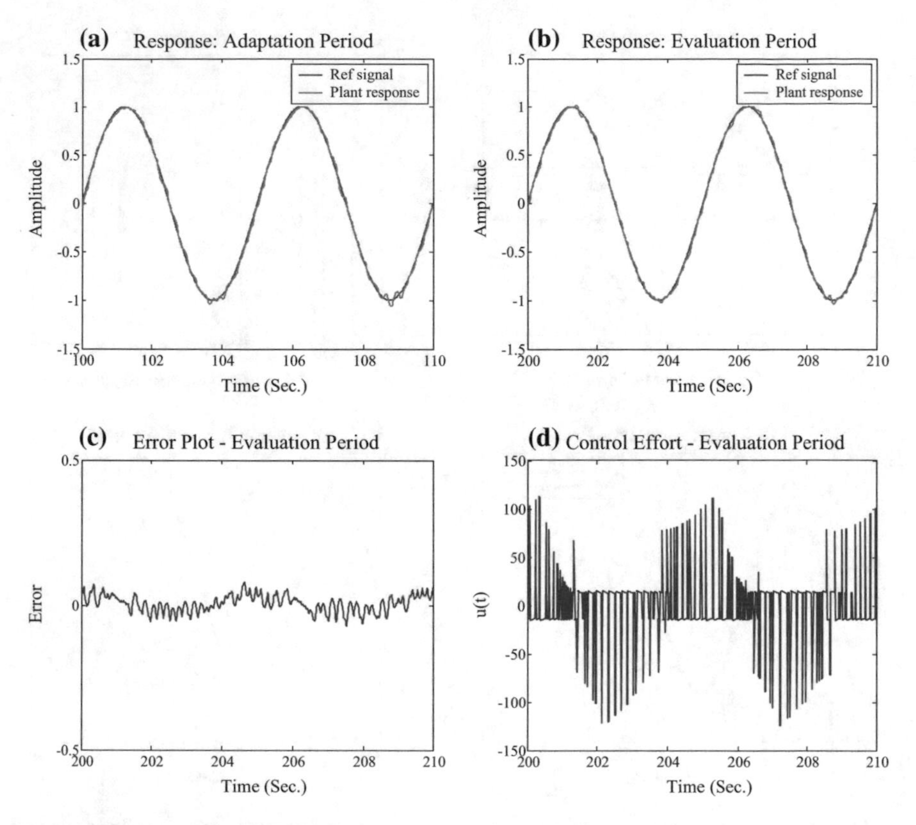

Fig. 5.29 Responses of the adaptive fuzzy controller with PSO-Hybd-Cas model, for variable structure configuration of input MFs, of case study II (A) (Duffing's system): **a** adaptation period response after 100 s, **b** evaluation period response for 200–210 s, **c** evaluation period error variation, **d** evaluation period control effort variation

Performance Analysis of HSA-Based Hybrid Controllers

To perform this simulation also, as in case of the case study I, harmony memory of ten harmonies is considered. For each simulation, 200 iterations of HSABA are performed. The fixed structure and variable structure design procedures are carried out following the same philosophy which was adopted in case study I. In variable structure design technique, the structural flags bear a major role to stochastically determine the number of input MFs and its corresponding output singletons.

In HSA-based hybrid design process, the adaptation gain v is set as 0.01, 1.5, and 1, for cascade model, concurrent model and preferential model, respectively [2]. In this case study also, the simulations are carried out for three fixed structures of 3×3, 5×5, and 7×7 input MFs and a variable structure with a potentially biggest possible structure of 7×7 MFs. The results of simulation studies are shown in Table 5.4. For fixed structure controllers, it can be seen that the performances of

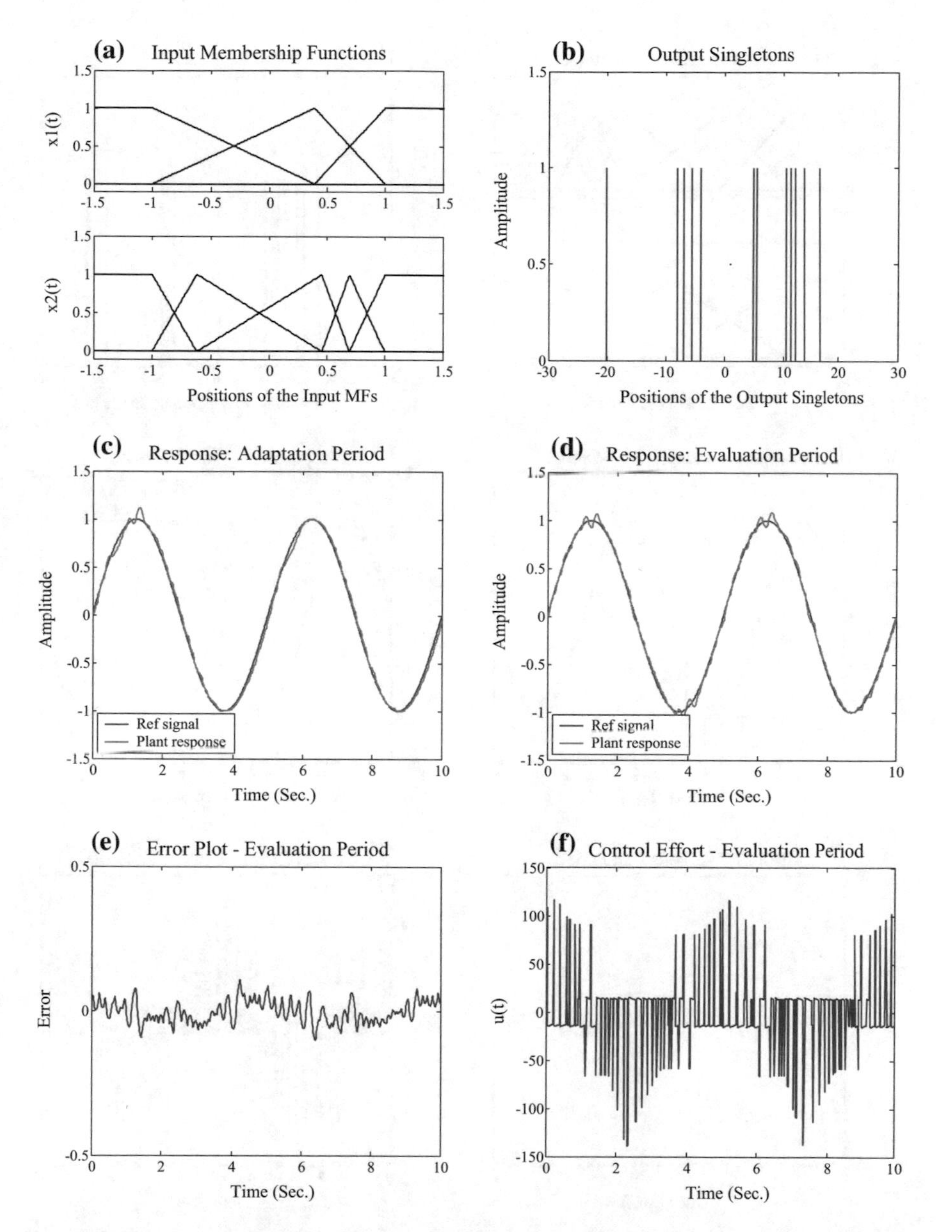

Fig. 5.30 Responses of the adaptive fuzzy controller with PSO-Hybd-Con model, for variable structure configuration of input MFs, of case study II (A) (Duffing's system): **a** positions of input MFs, **b** positions of output singletons, **c** adaptation period response after 100 PSO generations, **d** evaluation period response for 0–10 s after 200 PSO generations, **e** evaluation period error variation, **f** evaluation period control effort variation

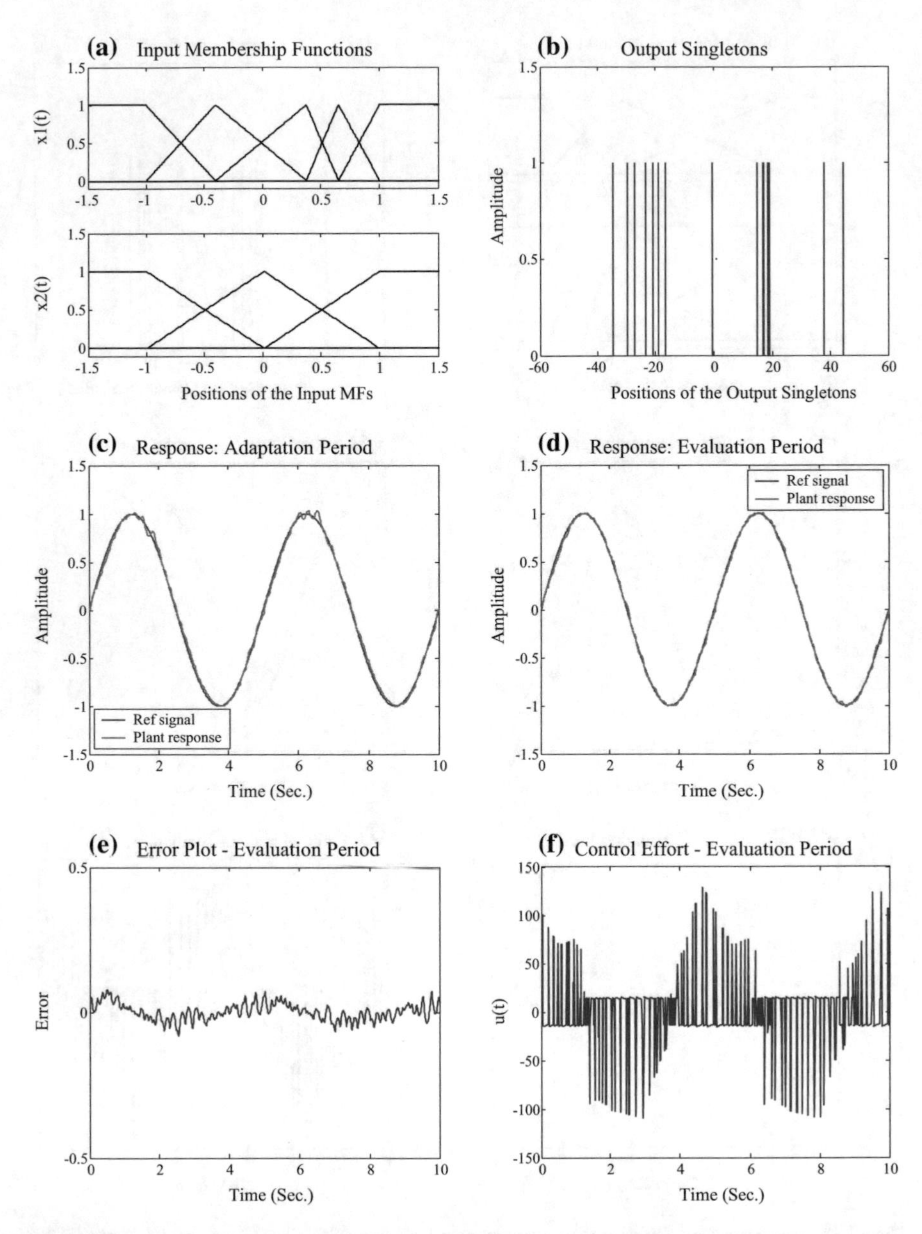

Fig. 5.31 Responses of the adaptive fuzzy controller with PSO-Hybd-Pref model, for variable structure configuration of input MFs, of case study II (A) (Duffing's system): **a** positions of input MFs, **b** positions of output singletons, **c** adaptation period response after 100 PSO generations, **d** evaluation period response for 0–10 s after 200 PSO generations, **e** evaluation period error variation, **f** evaluation period control effort variation

Table 5.4 Comparison of simulation results for case study II (A) (Duffing's system): HSA-based design strategies

Control strategy	No. of MFs	No. of rules	Avg. IAE	Best IAE	Std. dev.
HSA-Hybd-Cas	3 × 3	9	0.3582	0.3418	0.0129
HSA-Hybd-Con			0.2889	0.2610	0.0206
HSA-Hybd-Pref			0.2882	0.2700	0.0120
HSA-Hybd-Cas	5 × 5	25	0.2495	0.2356	0.0127
HSA-Hybd-Con			0.2347	0.2127	0.0157
HSA-Hybd-Pref			0.2058	0.1894	0.0130
HSA-Hybd-Cas	7 × 7	49	0.2537	0.2340	0.0187
HSA-Hybd-Con			0.2645	0.2562	0.0117
HSA-Hybd-Pref			0.2127	0.2080	0.0067
HSA-Hybd-Cas-V	4 × 6	24	–	0.2428	–
HSA-Hybd-Con-V	4 × 5	20	–	0.4823	–
HSA-Hybd-Pref-V	3 × 5	15	–	0.2169	–
	3 × 4	12	–	0.3102	–

the controllers (demonstrated by the performance index of *IAE*) improve with increase in the number of rules from 3 × 3 to 5 × 5 configuration. However, for 7 × 7 configuration, the results obtained are not better than those obtained for 5 × 5 configuration. Therefore, it can be concluded again that, merely increasing the number of rules in fuzzy controllers does not necessarily improve their performance [2, 3], as the operation of a fuzzy controller depends on many complex relationships between the variables associated with the operation. For a given controller MF configuration (e.g., 5 × 5 or 7 × 7), among the competing controllers, the HSA-Hybd-Pref scheme showed the best performance. Even for the variable structure scheme, HSA-Hybd-Pref controller emerged as the best possible solution in terms of the number of rules. Although the *IAE* value is higher for the variable structure case, it can be seen that the best result, in terms of *IAE*, is obtained with only 12 rules. This is a significant reduction in the total number of rules and such a simplified structure of an FLC will be very useful for real implementation of such control strategies. So, from the detail analysis of the results obtained for all competing controller design strategies, it can be concluded that the HSA-Hybd-Pref algorithm emerge as the overall superior design methodology for this case study also. Figures 5.32 and 5.33 show the sample presentations of evaluation phase responses of the controller designed according to hybrid preferential model strategy, with fixed structure 5 × 5 MFs and variable structure MFs, respectively.

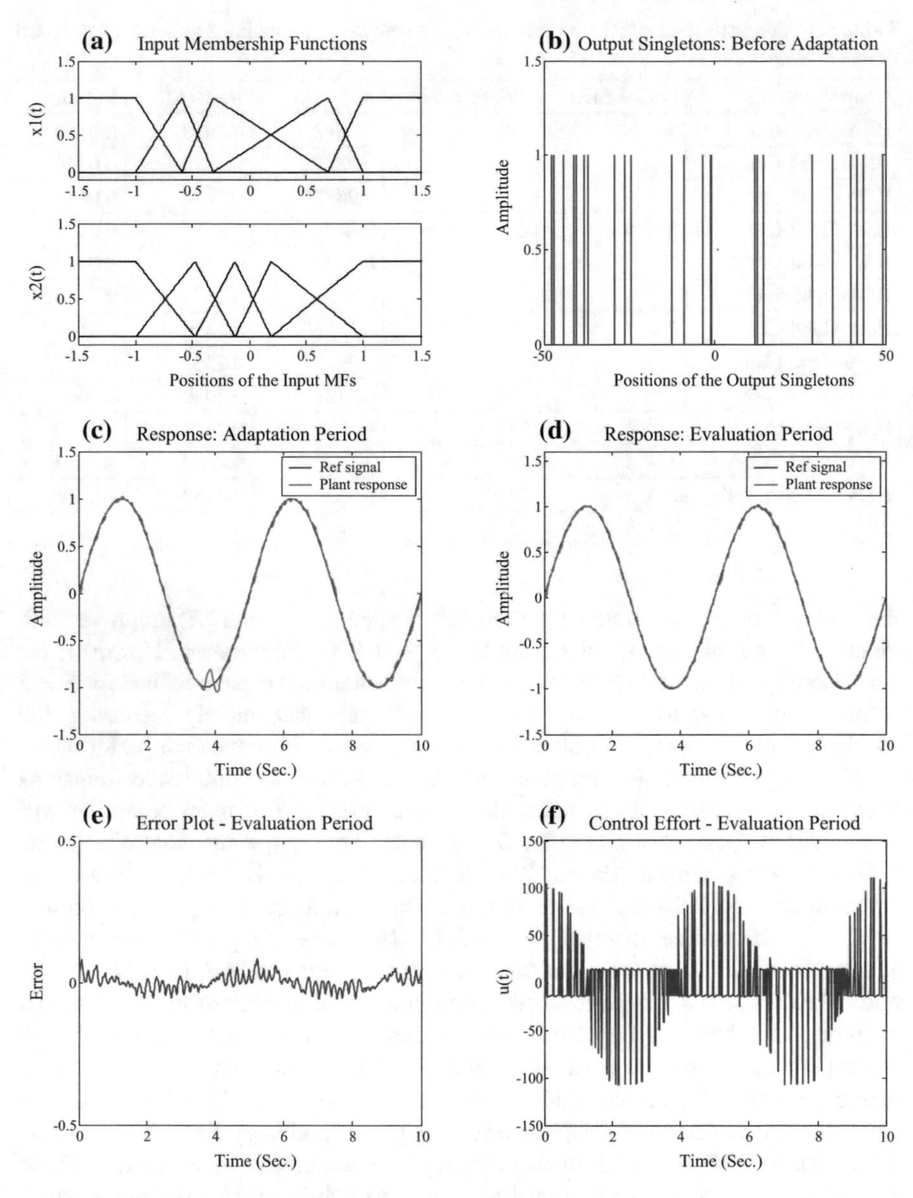

Fig. 5.32 Responses of the adaptive fuzzy controller with HSA-Hybd-Pref model, for 5×5 configuration of input MFs, of case study II (A) (Duffing's system): **a** positions of input MFs, **b** positions of output singletons, **c** adaptation period response after 100 HSA generations, **d** evaluation period response for 0–10 s after 200 HSA generations, **e** evaluation period error variation, **f** evaluation period control effort variation

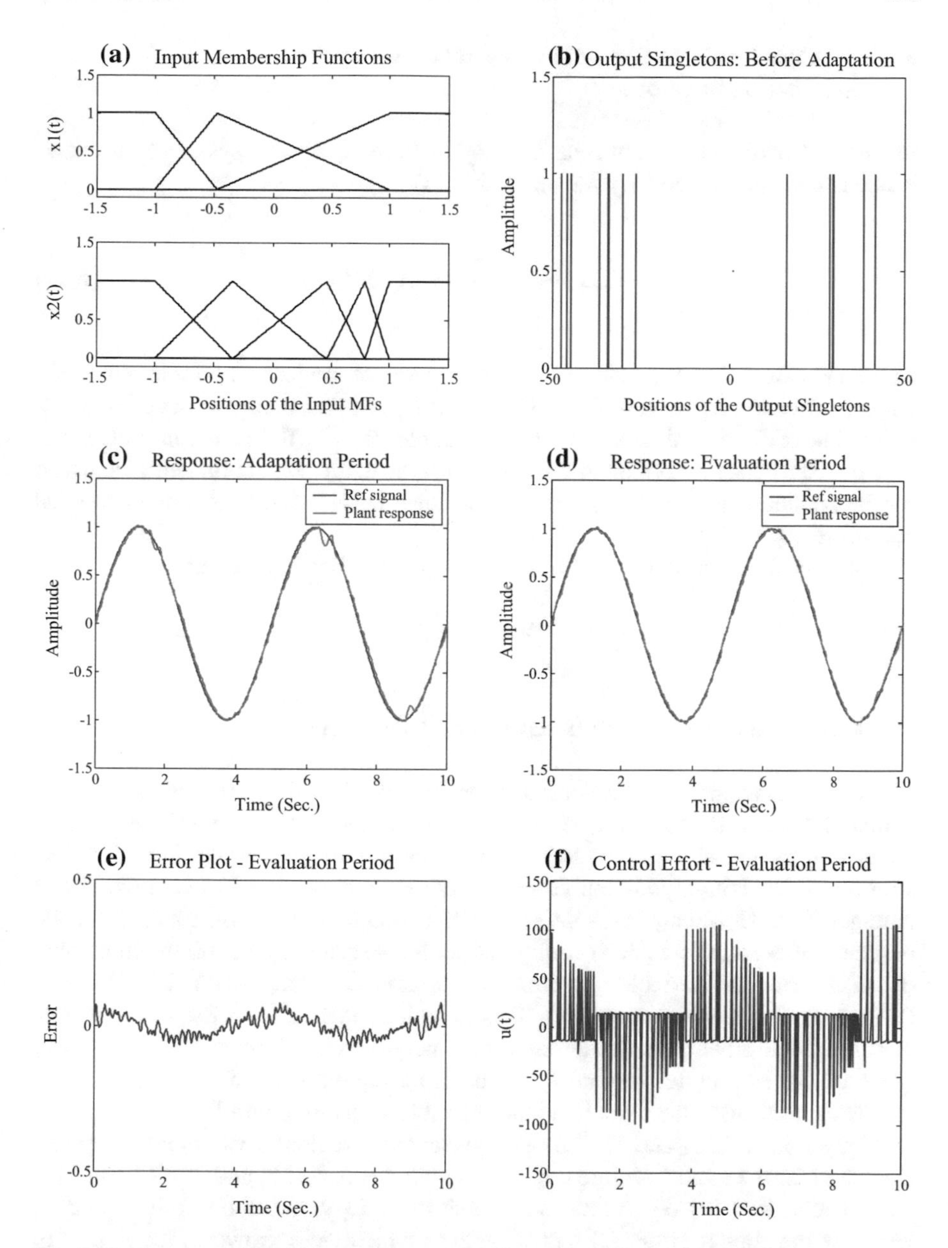

Fig. 5.33 Responses of the adaptive fuzzy controller with HSA-Hybd-Pref model, for variable structure configuration of input MFs, of case study II (A) (Duffing's system): **a** positions of input MFs, **b** positions of output singletons, **c** adaptation period response after 100 HSA generations, **d** evaluation period response for 0–10 s after 200 HSA generations, **e** evaluation period error variation, **f** evaluation period control effort variation

5.3.2.2 Case Study II(B): Duffing's Oscillatory System with Disturbance

In this part of the case study, the Duffing's forced oscillation system with disturbance is considered and is given as [4, 21–23]:

$$\left.\begin{aligned} \dot{x}_1 &= x_2 \\ \dot{x}_2 &= -0.1x_2 - x_1^3 + 12\cos(t) + u(t) + d \\ y &= x_1 \end{aligned}\right\} \quad (5.18)$$

where the external disturbance d is a square wave of random amplitude within the range $[-1, 1]$ and a period of 0.5 s. Figure 3.17 shows the nature of disturbance for a duration of 0–10 s, demonstrated as a sample case. In the case study also, this type of disturbance is applied to the system for the entire adaptation period and also for the evaluation period [2, 4]. The control objective is to track the reference signal y_m, where $y_m = \sin(t)$.

The control signal obtained from the fuzzy controller is given as:

$$u_c = u_c(\underline{x}|\underline{Z}) \quad (5.19)$$

Performance Analysis of PSO-Based Hybrid Controllers

The design strategies of PSO-based different hybrid methods are simulated in a similar fashion as in case study II(A) and the same parameters and adaptation gains are chosen for this simulation process. In this case study, among the fixed structure controllers, the PSO-Hybd-Con version achieves least average *IAE* value for 7×7 configuration. However, PSO-HASBA-Pref configuration emerges as the second best controller solution in each configuration. It also achieves the minimum number of rules for the variable structure configuration (along with PSOBA and PSO-Hybd-Cas model controllers). This is understandable from the fact that very rarely a single controller design strategy emerges as the best solution for many types of systems under several types of input conditions. Table 5.5 shows the results for this case study for fixed and variable structure controllers.

Figures 5.34, 5.35, and 5.36 show representative temporal variations of system responses for the fixed structure controllers with 3×3, 5×5, and 7×7 input MFs respectively, for PSO-Hybd-Pref model control strategy. In (a)–(f) of these figures, we show the similar graphical representations which were shown in the figures of the case study II(A). Similarly, Figs. 5.37, 5.38, and 5.39 show temporal variations of system responses for the variable structure input MFs for PSO-Hybd-Cas model, PSO-Hybd-Con model and PSO-Hybd-Pref model control strategies, respectively, in a similar fashion as was presented in case study II(A). In this case study, the evaluation period responses show that a randomly varying disturbance has been engulfed by the plant with the proper design parameters.

Table 5.5 Comparison of simulation results for case study II (B) (Duffing's system with disturbance): PSO-based design strategies

Control strategy	No. of MFs	No. of rules	Avg. IAE	Best IAE	Std. dev.
PSO-Hybd-Cas	3 × 3	9	0.2699	0.2566	0.0190
PSO-Hybd-Con			0.2568	0.2566	0.0140
PSO-Hybd-Pref			0.2462	0.2073	0.0550
PSO-Hybd-Cas	5 × 5	25	0.2700	0.2639	0.0090
PSO-Hybd-Con			0.2380	0.2310	0.0100
PSO-Hybd-Pref			0.2194	0.2109	0.0120
PSO-Hybd-Cas	7 × 7	49	0.2640	0.2639	0.0001
PSO-Hybd-Con			0.2125	0.1972	0.0220
PSO-Hybd-Pref			0.2215	0.2095	0.0170
PSO-Hybd-Cas-V	4 × 3	12	–	0.4002	–
PSO-Hybd-Con-V	5 × 4	20	–	0.2837	–
PSO-Hybd-Pref-V	3 × 4	12	–	0.2708	–

Performance Analysis of HSA-Based Hybrid Controllers

In this case study also, we have chosen the same parameters for this simulation process, as were chosen in case study II(A). The design strategies of different hybrid methods are simulated in a similar fashion as in case study II(A). In this case study, among the fixed structure controllers, the HSA-Hybd-Pref model achieved lower average *IAE* values than the HSA-Hybd-Con model, overall. Also the HSA-Hybd-Pref model showed best performance in terms of best *IAE* value for variable configuration, although with highest number of rules, 21 to be exact. This is understandable from the fact that very rarely a single controller design strategy emerge as the best solution from every point of view of the performances for many types of systems, under several types of input conditions. Table 5.6 shows the results for the fixed structure and the variable structure controllers, for the system under consideration. Figures 5.40 and 5.41 show sample evaluation phase responses for fixed structure HSA-Hybd-Pref controller and variable structure HSA-Hybd-Pref controller, respectively, for case study II (B).

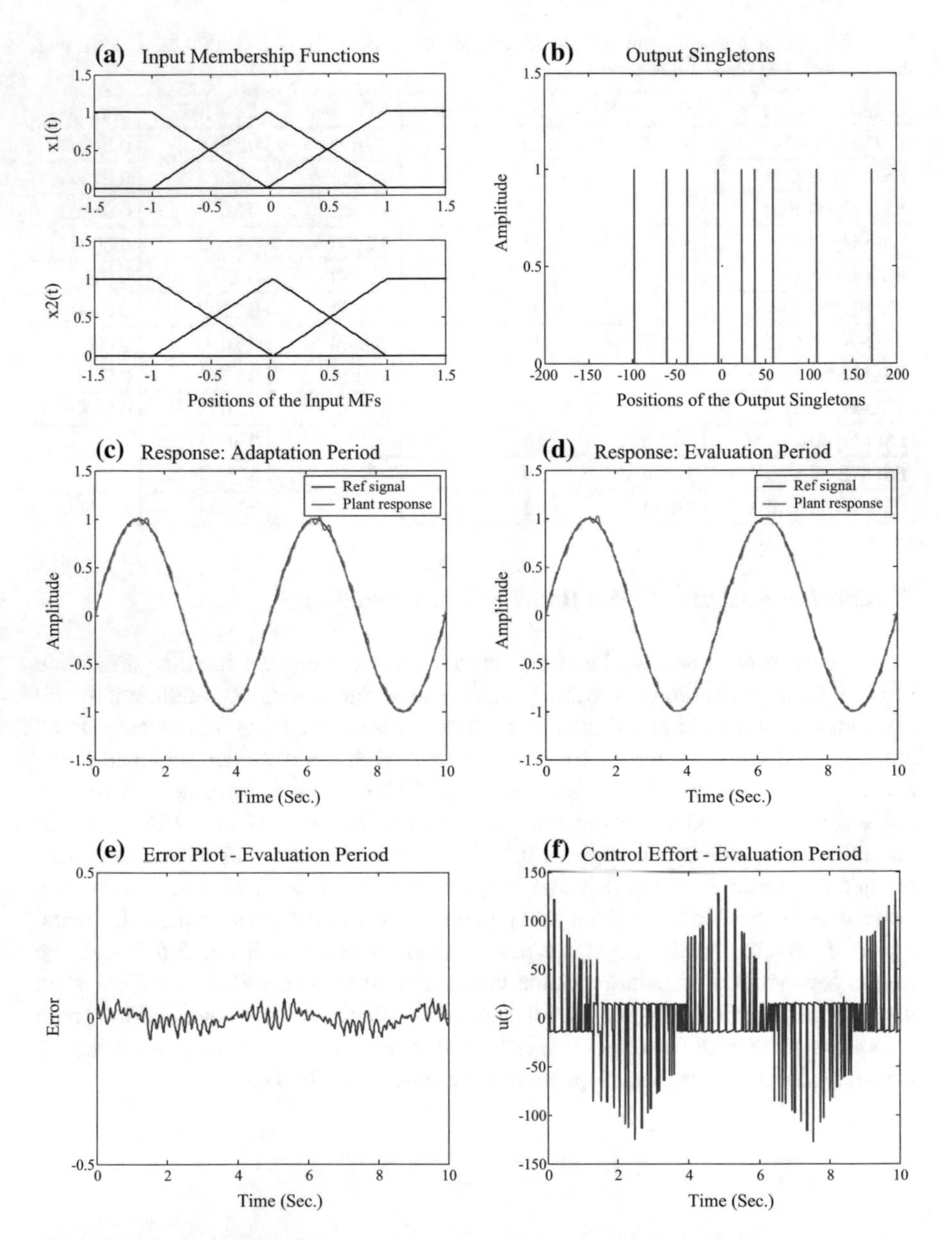

Fig. 5.34 Responses of the adaptive fuzzy controller with PSO-Hybd-Pref model, for 3×3 configuration of input MFs, of case study II (B) (Duffing's system with disturbance): **a** positions of input MFs, **b** positions of output singletons, **c** adaptation period response after 100 PSO generations, **d** evaluation period response for 0–10 s after 200 PSO generations, **e** evaluation period error variation, **f** evaluation period control effort variation

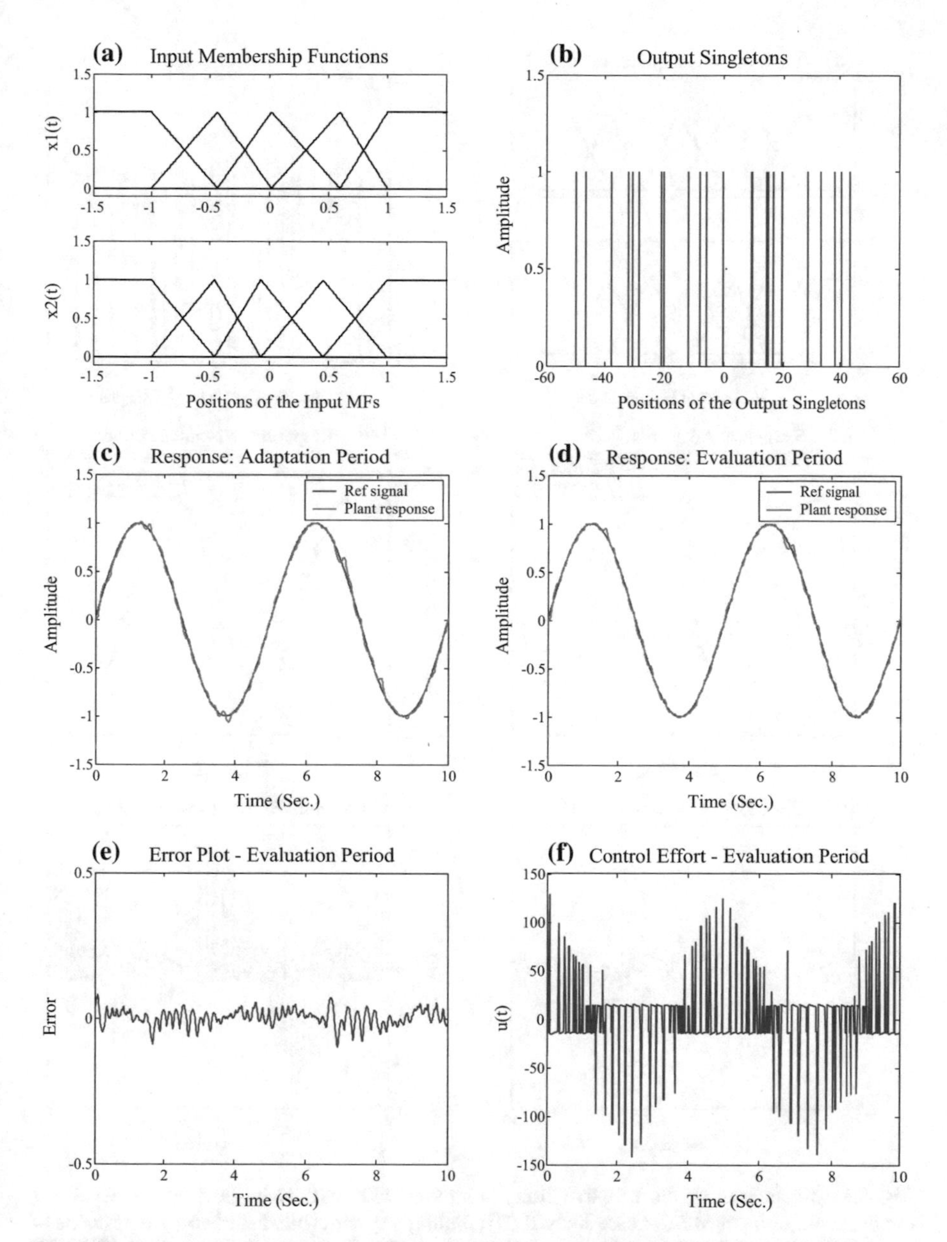

Fig. 5.35 Responses of the adaptive fuzzy controller with PSO-Hybd-Pref model, for 5 × 5 configuration of input MFs, of case study II (B) (Duffing's system with disturbance): **a** positions of input MFs, **b** positions of output singletons, **c** adaptation period response after 100 PSO generations, **d** evaluation period response for 0–10 s after 200 PSO generations, **e** evaluation period error variation, **f** evaluation period control effort variation

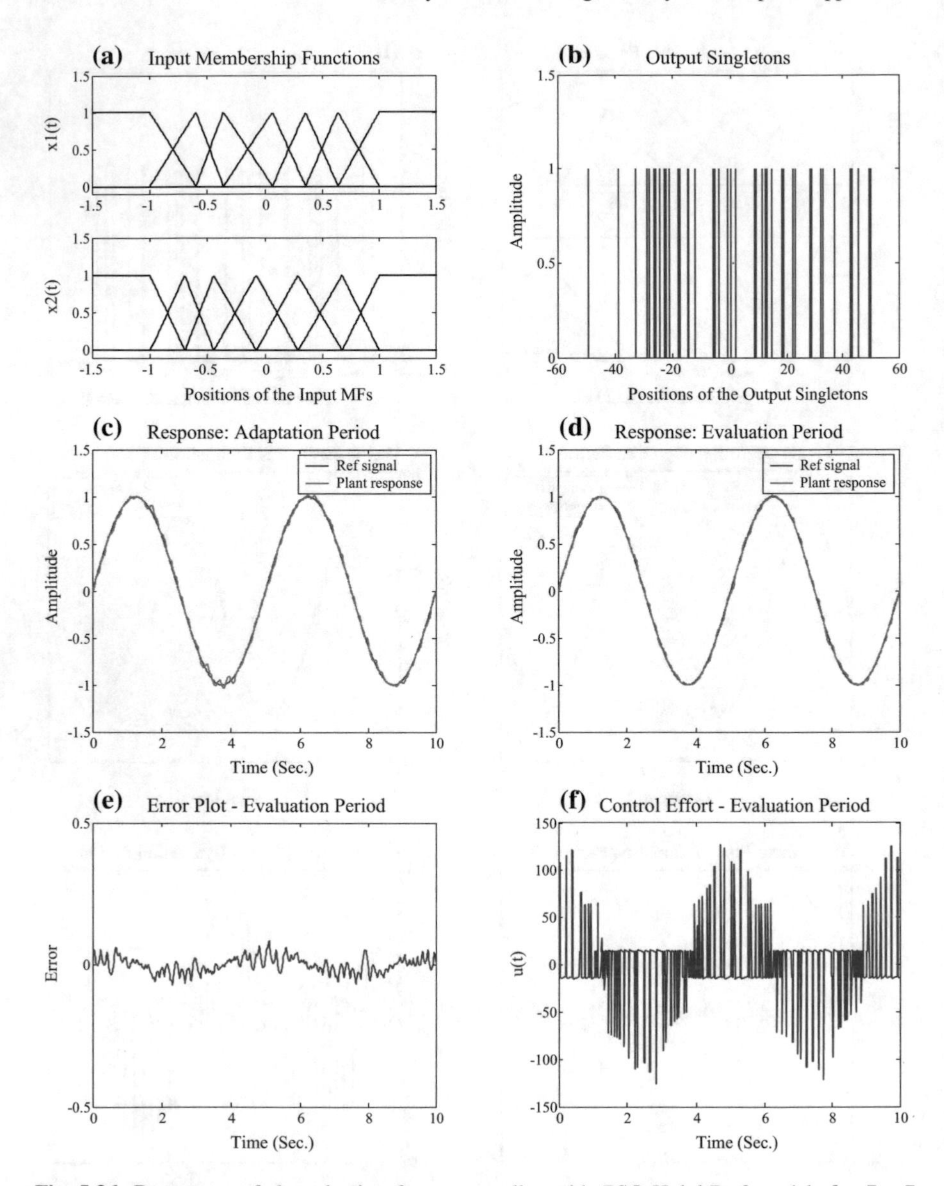

Fig. 5.36 Responses of the adaptive fuzzy controller with PSO-Hybd-Pref model, for 7×7 configuration of input MFs, of case study II (B) (Duffing's system with disturbance): **a** positions of input MFs, **b** positions of output singletons, **c** adaptation period response after 100 PSO generations, **d** evaluation period response for 0–10 s after 200 PSO generations, **e** evaluation period error variation, **f** evaluation period control effort variation

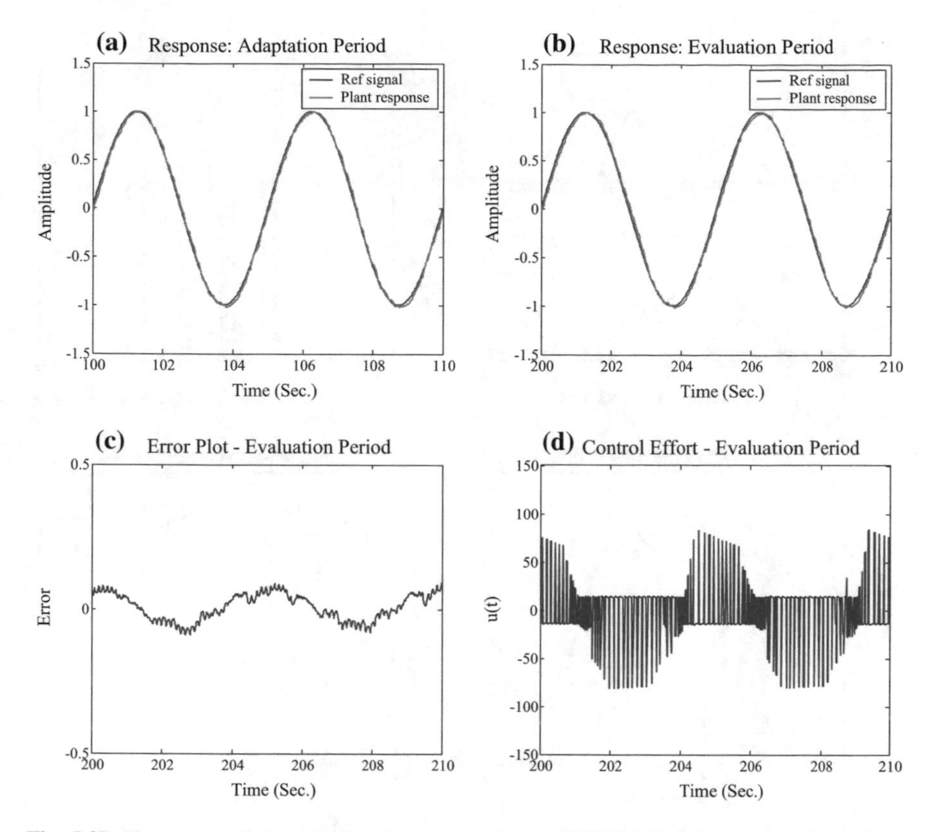

Fig. 5.37 Responses of the adaptive fuzzy controller with PSO-Hybd-Cas model, for variable structure configuration of input MFs, of case study II (B) (Duffing's system with disturbance) **a** adaptation period response after 100 s, **b** evaluation period response for 200–210 s, **c** evaluation period error variation, **d** evaluation period control effort variation

5.4 Summary

The present chapter elaborated a new hybrid approach for designing the AFLCs employing the conventional Lyapunov theory, to locally optimize the positions of the output singletons, and a stochastic optimization technique. In this hybridization process, while LSBA optimizes the positions of the output singletons, SOABA is utilized to design the structure of the AFLC and also evaluate the free parameters of it. Three variants of the hybrid models are detailed and extensively simulated for two benchmark case studies. The hybrid preferential model largely evolved as a

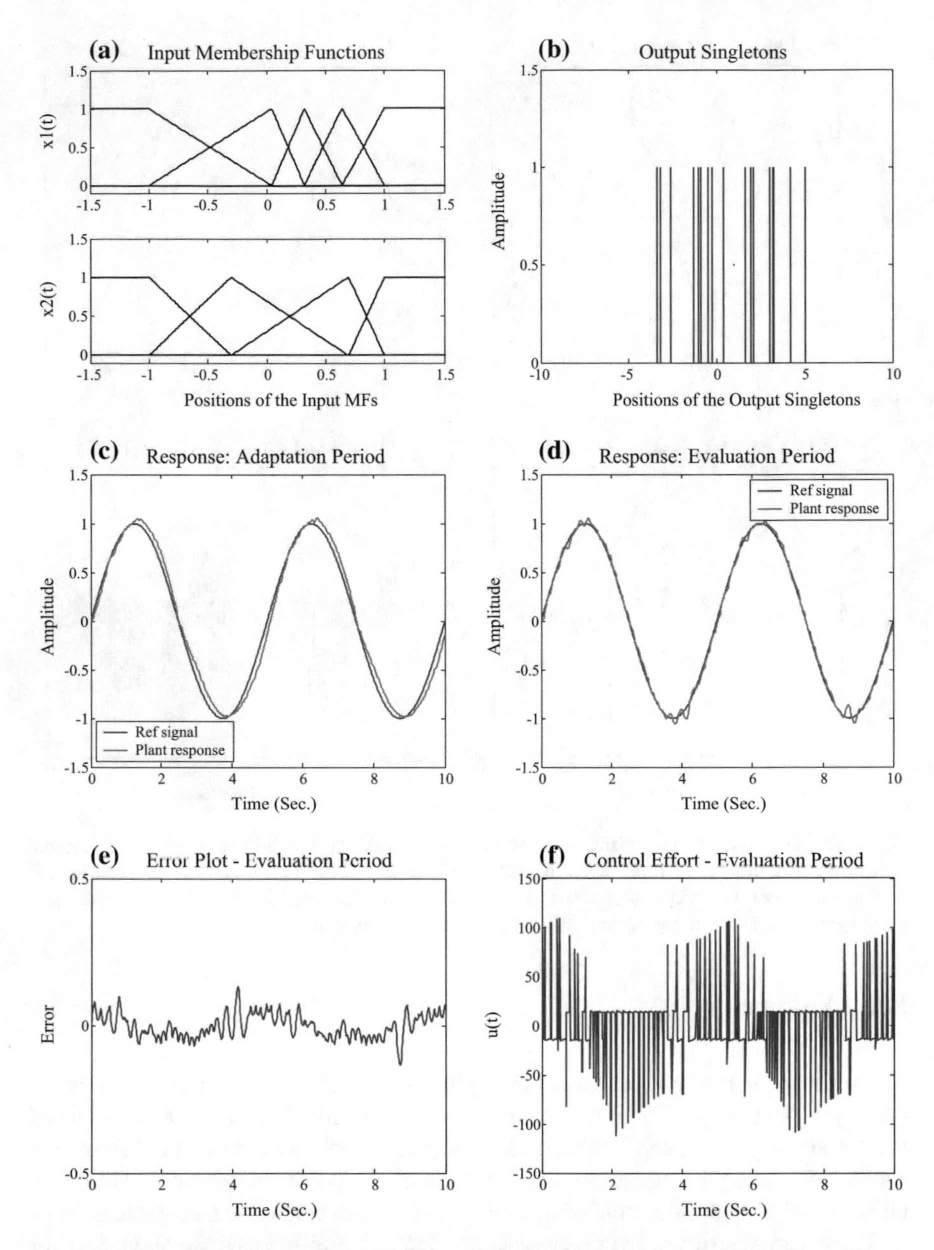

Fig. 5.38 Responses of the adaptive fuzzy controller with PSO-Hybd-Con model, for variable structure configuration of input MFs, of case study II (B) (Duffing's system with disturbance): **a** positions of input MFs, **b** positions of output singletons, **c** adaptation period response after 100 PSO generations, **d** evaluation period response for 0–10 s after 200 PSO generations, **e** evaluation period error variation, **f** evaluation period control effort variation

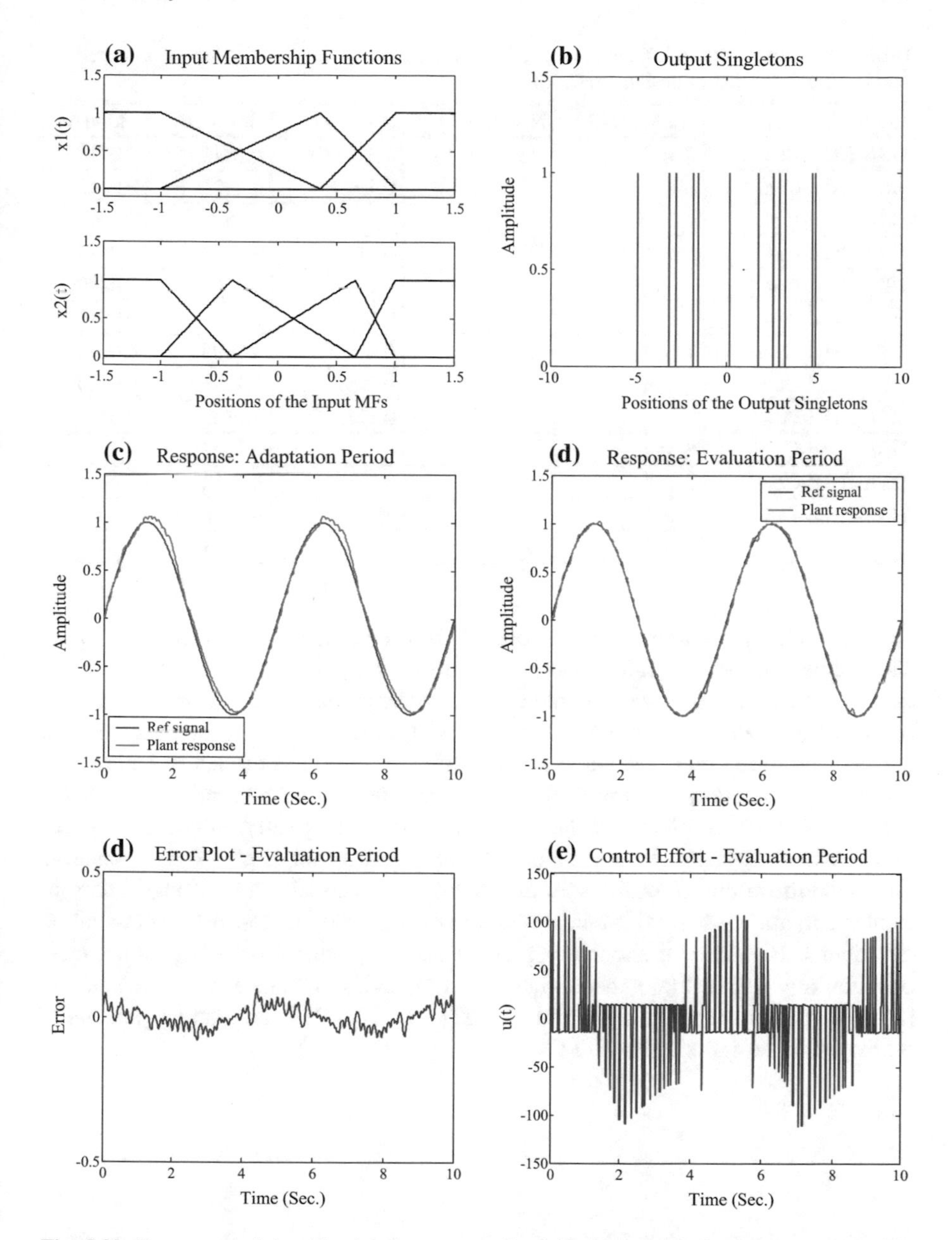

Fig. 5.39 Responses of the adaptive fuzzy controller with PSO-Hybd-Pref model, for variable structure configuration of input MFs, of case study II (B) (Duffing's system with disturbance): **a** positions of input MFs, **b** positions of output singletons, **c** adaptation period response after 100 PSO generations, **d** evaluation period response for 0–10 s after 200 PSO generations, **e** evaluation period error variation, **f** evaluation period control effort variation

Table 5.6 Comparison of simulation results for case study II (B) (Duffing's system with disturbance): HSA-based design strategies

Control strategy	No. of MFs	No. of rules	Avg. IAE	Best IAE	Std. dev.
HSA-Hybd-Cas	3 × 3	9	0.3108	0.2856	0.0218
HSA-Hybd-Con			0.2920	0.2871	0.0042
HSA-Hybd-Pref			0.2577	0.2533	0.0046
HSA-Hybd-Cas	5 × 5	25	0.3171	0.2957	0.0183
HSA-Hybd-Con			0.2176	0.2068	0.0089
HSA-Hybd-Pref			0.2166	0.2151	0.0012
HSA-Hybd-Cas	7 × 7	49	0.2794	0.2604	0.0143
HSA-Hybd-Con			0.2257	0.2163	0.0120
HSA-Hybd-Pref			0.2235	0.2129	0.0086
HSA-Hybd-Cas-V	3 × 5	15	–	0.3027	–
HSA-Hybd-Con-V	3 × 4	12	–	0.3902	–
HSA-Hybd-Pref-V	3 × 5	15	–	0.2718	–
	3 × 7	21	–	0.2353	–

superior technique compared to the other hybrid models described in this chapter. The stability of the closed-loop system and the convergence of the plant output to a desired reference signal are guaranteed with greater precision in this case. It has also been demonstrated that another hybrid model, called hybrid concurrent model, emerged as the overall second best alternative among the competing controllers. The main advantages of these methods are that they require no a priori knowledge about the controlled plant and the performance error is greatly reduced, in almost every case. The rationale behind the choice of PSO and HSA as representative global optimization algorithms is that these are relatively simple algorithms to implement, among several candidates within the category of stochastic optimization algorithms. However, it should be pointed out that, other stochastic, global optimization algorithms like genetic algorithm [23, 24], covariance matrix adaptation [25], gravitational search algorithm [26, 27], etc. [28, 29] can also be potentially employed to design similar AFLCs.

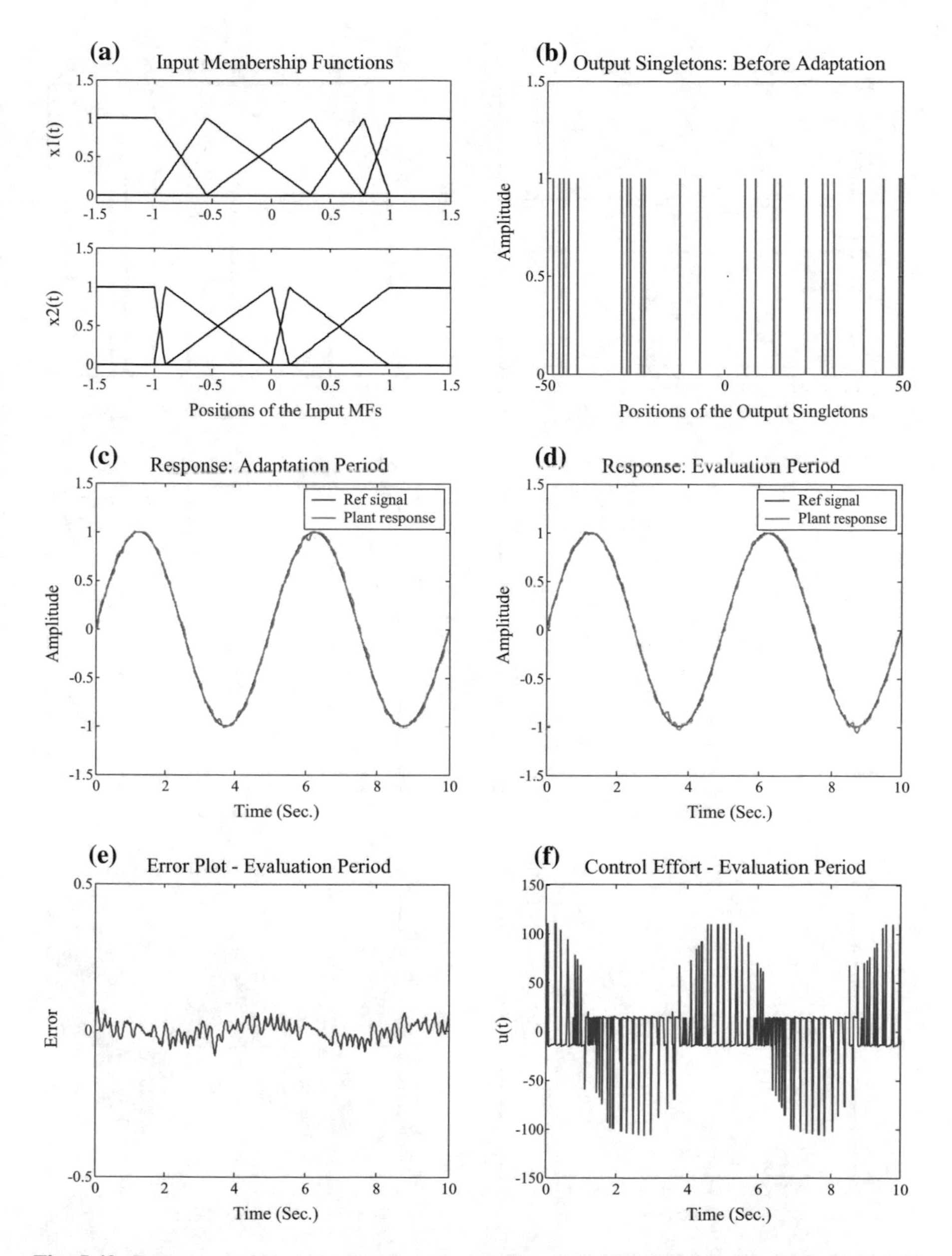

Fig. 5.40 Responses of the adaptive fuzzy controller with HSA-Hybd-Pref model, for 5 × 5 configuration of input MFs, of case study II (B) (Duffing's system with disturbance): **a** positions of input MFs, **b** positions of output singletons, **c** adaptation period response after 100 HSA generations, **d** evaluation period response for 0–10 s after 200 HSA generations, **e** evaluation period error variation, **f** evaluation period control effort variation

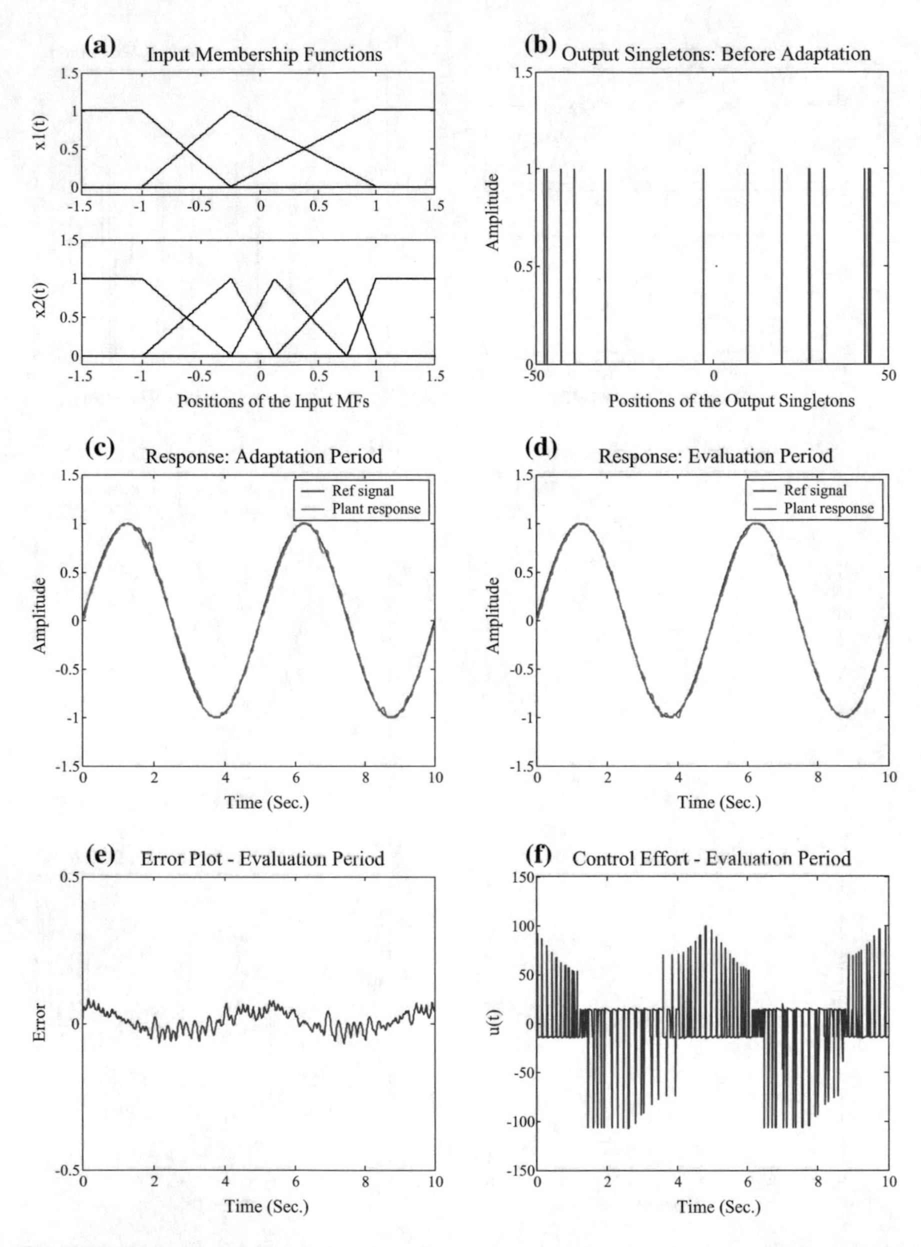

Fig. 5.41 Responses of the adaptive fuzzy controller with HSA-Hybd-Pref model, for variable structure configuration of input MFs, of case study II (B) (Duffing's system with disturbance): **a** positions of input MFs, **b** positions of output singletons, **c** adaptation period response after 100 HSA generations, **d** evaluation period response for 0–10 s after 200 HSA generations, **e** evaluation period error variation, **f** evaluation period control effort variation

References

1. L.-X. Wang, *Adaptive Fuzzy Systems and Control: Design and Stability Analysis* (Prentice-Hall, Englewood Cliffs, NJ, 1994)
2. K. Das Sharma, *Hybrid Methodologies for Stable Adaptive Fuzzy Control*, Doctoral Dissertation, Jadavpur University, India, 2012
3. F. Cupertino, E. Mininno, D. Naso, B. Turchiano, L. Salvatore, On-line genetic design of anti-windup unstructured controllers for electric drives with variable load. IEEE Trans. Evol. Comput. **8**(4), 347–364 (2004)
4. K. Das Sharma, A. Chatterjee, A. Rakshit, A hybrid approach for design of stable adaptive fuzzy controllers employing Lyapunov theory and particle swarm optimization. IEEE Trans. Fuzzy Syst. **17**(2), 329–342 (2009)
5. K. Das Sharma, A. Chatterjee, A. Rakshit, Harmony search algorithm and Lyapunov theory based hybrid adaptive fuzzy controller for temperature control of air heater system with transport-delay. Appl. Soft Comput. **25**, 40–50, Dec (2014)
6. K. Fischle, D. Schroder, An improved stable adaptive fuzzy control method. IEEE Trans. Fuzzy Syst. **7**(1), 27–40 (1999)
7. K. Das Sharma, A. Chatterjee, F. Matsuno, A Lyapunov theory and stochastic optimization based stable adaptive fuzzy control methodology, in *Proceedings of SICE Annual Conference 2008*, Japan, 20–22 Aug, pp. 1839–1844
8. K. Das Sharma, A. Chatterjee, A. Rakshit, Design of a hybrid stable adaptive fuzzy controller employing Lyapunov theory and harmony search algorithm. IEEE Trans. Contr. Syst. Technol. **18**(6), 1440–1447, Nov (2010)
9. R.C. Eberhart, J. Kennedy, A new optimizer using particle swarm theory, in *Proceedings of the 6th International Symposium on Micro Machine and Human Science*, Nagoya, Japan, 1995, pp. 39–43
10. J. Kennedy, R.C. Eberhart, Particle swarm optimization. Proc. IEEE Int. Conf. Neural Netw. **4**, 1942–1948 (1995)
11. G. Venter, J.S. Sobieski, Particle swarm optimization, in *43rd AIAA Conference*, Colorado, 2002
12. Y. Shi, R.C. Eberhart, A modified particle swarm optimizer, in *Proceedings of the IEEE International Conference on Evolutionary Computation*, Alaska, May, 1998
13. Z.W. Geem, J.H. Kim, G.V. Loganathan, A new heuristic optimization algorithm: harmony search. Simulation **76**(2), 60–68 (2001)
14. K.S. Lee, Z.W. Geem, A new structural optimization method based on the harmony search algorithm. Comput. Struct. **82**, 781–798 (2004)
15. M. Mahdavi, M. Fesanghary, E. Damangir, An improved harmony search algorithm for solving optimization problems. Appl. Math. Comput. **188**, 1567–1679 (2007)
16. S.H. Zak, *Systems and Control* (Oxford University Press, New York, 2003)
17. C.-T. Chen, *Control System Design, Orlando* (Saunders College Publishing, Florida, 1993)
18. D. Driankov, H. Hellendoorn, M.M. Reinfrank, *An Introduction to Fuzzy Control* (Springer-Verlag, Heidelberg, Germany, 1993)
19. K.M. Passino, S. Yurkovich, Fuzzy control, in *Handbook on Control*, ed. by W. Levine (CRC, Boca Raton, FL, 1996), pp. 1001–1017
20. D. Chakroborty, K. Das Sharma, A study on the design methodologies of fuzzy logic controller, in *Proceedings of 17th State Science & Technology Congress, WBUAFS*, Kolkata, India, March 2010, p. 69
21. L.X. Wang, Stable adaptive fuzzy control of nonlinear system. IEEE Trans. Fuzzy Syst. **1**(2), 146–155, May (1993)
22. K. Das Sharma, Stable fuzzy controller design employing group improvisation based harmony search algorithm. Int. J. Contr. Autom. Syst. **11**(5), 1046–1052 (2013), Springer
23. A. Roy, K. Das Sharma, GA and Lyapunov theory based hybrid adaptive fuzzy controller for nonlinear systems. Int. J. Electron. **120**(2), 312–325 (2015), Taylor & Francis

24. D.E. Goldberg, *Genetic Algorithms in Search, Optimization and Machine Learning* (Kluwer Academic Publishers, Boston, 1989)
25. A. Auger, N. Hansen, A restart CMA evolution strategy with increasing population size. Cong. Evol. Comp. **2**, 1769–1776 (2005)
26. E. Rashedi, H. Nezamabadi-pour, S. Saryazdi, GSA: a gravitational search algorithm. Inf. Sci. **179**(13), 2232–2248 (2009)
27. A. Roy, K. Das Sharma, Gravitational search algorithm and Lyapunov theory based stable adaptive fuzzy logic controller, in *Proceedings of 1st International Conference on Computational Intelligence: Modelling, Techniques and Applications (CIMTA-2013)*, Procedia Technol. **10**, 581–586 (2013), Elsevier
28. K.M. Passino, Biomimicry of bacterial foraging for distributed optimization and control, in *IEEE Control Systems Magazine*, 2002, pp. 52–67
29. R. Toscano, P. Lyonnet, Stabilization of systems by static output feedback via heuristic Kalman algorithm. J. Appl. Comput. Math. **5**(2), 154–165 (2006)

Part III
H^{∞} Strategy Based Design Methodologies

Chapter 6
Fuzzy Controller Design IV: H^{∞} Strategy-Based Robust Approach

6.1 Introduction

In Chap. 4, we have described the design of stable adaptive fuzzy controllers for the nonlinear plants employing Lyapunov theory-based adaptation rules. An additional supervisory control law was also utilized to ensure the boundedness of the states of the plants in course of tracking the reference signal. In addition to the stable design of the fuzzy controller, achieving the robust behavior is also a very important feature during the implementation of the plant in practice [1–17]. In [1], the authors presented one of the earliest, most notable methodologies of designing H^{∞} tracking control-based robust adaptive fuzzy controllers. Here, matrix Riccati-like theorem had been utilized to ensure the attenuation of tracking error in the presence of external disturbances applied to the plant. Lyapunov theory had also been utilized to derive the adaptation law and the robust control law additional to the fuzzy control law, obtained from a T-S-type fuzzy controller configuration. Thus, the stability and robustness, along with H^{∞} control-based tracking performance, were achieved for a class of nonlinear plants with external disturbances. In [2, 3], authors further showed the applicability of the H^{∞} tracking control-based robust adaptive fuzzy controllers with some other mathematical treatments, but the final formulation of the adaptive law and robust control action was identical with [1].

In this chapter, we shall discuss the developments of the stable robust H^{∞} tracking control-based adaptive fuzzy controller which can achieve the simultaneous tracking control and disturbance rejection objectives. Some case studies with benchmark systems are also presented to evaluate the performances of the designed controllers.

K. Das Sharma et al., *Intelligent Control*, Cognitive Intelligence and Robotics,
https://doi.org/10.1007/978-981-13-1298-4_6

6.2 Robust Adaptive Fuzzy Controllers

Let us consider an *n*th-order nonlinear plant given as [1–3]:

$$\left.\begin{aligned} x^{(n)} &= f(\underline{x}) + bu(t) + d(t) \\ y &= x \end{aligned}\right\} \tag{6.1}$$

where $f(\cdot)$ is an unknown continuous function, $u \in R$ and $y \in R$ are the input and output of the plant, b is an unknown positive constant, and d is unknown, but bounded external disturbances. It is assumed that the state vector is given as $\underline{x} = (x_1, x_2, \ldots, x_n)^{\mathrm{T}} = (x, \dot{x}, \ldots x^{(n-1)})^{\mathrm{T}} \in R^n$. In order that (6.1) be controllable in certain controllability region $U_{\underline{x}} \subset R^n$, we require that $b \neq 0$. Thus, without loss of generality, we can assume that $b > 0$.

The control objective is to track the reference signal $y_m(t)$, and thus, the tracking error is $e = y - y_m$. The objective is to design a stable AFLC for the system described in (6.1). Here, we need to find a feedback control law $u = u_c(\underline{x}|\underline{\theta})$, using fuzzy logic system and an adaptive law for adjusting the parameter vector $\underline{\theta}$ such the following conditions are satisfied:

(i) The closed-loop system must be globally stable in the sense that all variables, $\underline{x}(t), \underline{\theta}(t)$ and $u(\underline{x}|\underline{\theta})$ must be uniformly bounded, i.e.,$|\underline{x}(t)| \leq M_x < \infty$, $|\underline{\theta}(t)| \leq M_\theta < \infty$, and $|u(\underline{x}|\underline{\theta})| \leq M_u < \infty$, where M_x, M_θ, and M_u are set by the designer [1, 4, 5].

(ii) The tracking error $e(t)$ should be as small as possible under the constraints in (i), and the $\underline{e} - \underline{\theta}$ space should be stable in the large for the system [1, 18–20].

Let the error vector be $\underline{e} = (e, \dot{e}, \ldots, e^{(n-1)})^{\mathrm{T}}$ and $\underline{k} = (k_1, k_2, \ldots, k_n)^{\mathrm{T}} \in R^n$ be such that all the roots of the Hurwitz polynomial $s^n + k_n s^{n-1} + \cdots + k_2 s + k_1$ are in the left half of s-plane.

Now, the ideal control law for the system in (6.1) with the introduction of H^∞ control action is given as [1, 10]:

$$u^* = \frac{1}{b}\left[-f(\underline{x}) + y_m^{(n)} + \underline{k}^{\mathrm{T}}\underline{e} - u_r\right] \tag{6.2}$$

where $y_m^{(n)}$ is the *n*th derivative of the reference signal and the control signal u_r is applied to attenuate the external disturbance and the error due to fuzzy approximation of the AFLC [1, 10].

Thus, from (6.1) and (6.2), the error differential equation becomes

$$\begin{aligned}
bu^* &= -f(\underline{x}) + y_m^{(n)} + \underline{k}^{\mathrm{T}}\underline{e} - u_r \\
y_m^{(n)} - f(\underline{x}) - bu^* &= -\underline{k}^{\mathrm{T}}\underline{e} + u_r \\
y_m^{(n)} - f(\underline{x}) - bu^* - d &= -\underline{k}^{\mathrm{T}}\underline{e} + u_r - d \\
y_m^{(n)} - x^{(n)} &= -\underline{k}^{\mathrm{T}}\underline{e} + u_r - d \\
e^{(n)} &= -\underline{k}^{\mathrm{T}}\underline{e} + u_r - d \\
e^{(n)} &= -k_1 e - k_2 \dot{e} - \cdots - k_n e^{(n-1)} + u_r - d
\end{aligned} \tag{6.3}$$

This definition implies that u^* guarantees perfect tracking, i.e., $y(t) \equiv y_m(t)$ if $\underset{t\to\infty}{Lt}\ e(t) = 0$. As f and b are not known precisely, the ideal u^* of (6.2) cannot be implemented in practice. The design objective of asymptotically stable tracking may be achieved if ω, the summation of the fuzzy approximation error (ε) and the external disturbance (d), is zero; otherwise, the H^{∞} tracking performance will come into play [1–10]:

$$\begin{aligned}
&\int_0^{\mathrm{T}} \underline{e}^{\mathrm{T}} Q \underline{e}\,\mathrm{d}t \le \underline{e}^{\mathrm{T}}(0) P \underline{e}(0) + \tfrac{1}{v}\underline{\tilde{\theta}}^{\mathrm{T}}(0)\underline{\tilde{\theta}}(0) + \rho^2 \int_0^{\mathrm{T}} \omega^{\mathrm{T}}\omega\,\mathrm{d}t \\
&\quad \forall T \in [0, \infty)\omega \in L_2[0, T]
\end{aligned} \tag{6.4}$$

for given weighting matrices $Q = Q^{\mathrm{T}} \ge 0$, $P = P^{\mathrm{T}} \ge 0$, an adaptation gain $v > 0$, and prescribed attenuation level ρ.

Let us assume that the AFLC is constructed using a zero-order Takagi–Sugeno (T-S) fuzzy system. Then, $u_c(\underline{x}|\underline{\theta})$ for the AFLC is given in the form [1, 4, 11–18]:

$$u_c(\underline{x}|\underline{\theta}) = \sum_{l=1}^{N} \theta_l * \alpha_l(\underline{x}) \Big/ \left(\sum_{l=1}^{N} \alpha_l(\underline{x}) \right) = \underline{\theta}^{\mathrm{T}} * \underline{\xi}(\underline{x}) \tag{6.5}$$

where $\underline{\theta} = [\theta_1\ \theta_2\ \ldots\ \theta_N]^{\mathrm{T}}$ = the vector of the output singletons, $\alpha_l(\underline{x}) = \prod_{i=1}^{r} \mu_i^l(x_i)$ = the firing degree of rule "l", N = the total number of rules, $\mu_i^l(x_i)$ = the membership value of the ith input membership function (MF) in the activated lth rule, $\underline{\xi}(\underline{x})$ = vector containing normalized firing strength of all fuzzy IF-THEN rules $= (\xi_1(\underline{x}), \xi_2(\underline{x}), \ldots, \xi_N(\underline{x}))^{\mathrm{T}}$ and

$$\xi_l(\underline{x}) = \alpha_l(\underline{x}) \Big/ \left(\sum_{l=1}^{N} \alpha_l(\underline{x}) \right) \tag{6.6}$$

Design of Robust Control Law and Adaptation Law

Now, to ensure stability, it is assumed that the control law $u(t)$ is actually the summation of the fuzzy control $u_c(\underline{x}|\underline{\theta})$ and a robust control term, given as:

$$u(t) = u_c(\underline{x}|\underline{\theta}) - \frac{1}{b}u_r(\underline{x}) = \underline{\theta}^{\mathrm{T}} * \xi(\underline{x}) - \frac{1}{b}u_r(\underline{x}) \quad (6.7)$$

where $u_r(\underline{x})$ is the H^{∞} robust term to compensate for the fuzzy approximation error and the external disturbance d and it also ensures the stability of the closed-loop system.

Now the closed-loop error equation, in light of fuzzy approximation error and external disturbance, with the help of (6.3), becomes

$$\underline{\dot{e}} = A\underline{e} + Bu_r - bB[u_c(\underline{x}|\underline{\theta}) - u_1^*] - Bd \quad (6.8)$$

where

$$A = \begin{bmatrix} 0 & 1 & 0 & 0 & \cdots & 0 & 0 \\ 0 & 0 & 1 & 0 & \cdots & 0 & 0 \\ \vdots & \vdots & \vdots & \vdots & \ddots & \vdots & \vdots \\ 0 & 0 & 0 & 0 & \cdots & 0 & 1 \\ -k_1 & -k_2 & \cdots & \cdots & \cdots & -k_{n-1} & -k_n \end{bmatrix} \text{ and } B = \begin{bmatrix} 0 \\ \vdots \\ 0 \\ 1 \end{bmatrix} \quad (6.9)$$

and $u_1^* = \frac{1}{b}\left[-f(\underline{x}) + y_m^{(n)} + \underline{k}^{\mathrm{T}}\underline{e}\right]$ is the ideal control law when external disturbance (d) is zero; i.e., the H^{∞} control action has not been initiated.

Let us define the optimal parameter vector $\underline{\theta}^*$ for which optimal u_1^* will be approximated by the fuzzy logic system, as [18, 19]:

$$\underline{\theta}^* = \arg \min_{\underline{\theta} \in \Re^{\prod_{i=1}^{n} m_i}} \left[\sup_{\underline{x} \in \Re^n} \left| u_c(\underline{x}|\underline{\theta}) - u_1^* \right| \right] \quad (6.10)$$

and the minimum fuzzy approximation error as [1, 10]:

$$\varepsilon = -b\left[u_c(\underline{x}|\underline{\theta}^*) - u_1^*\right] \quad (6.11)$$

Using (6.5) and (6.11), the error dynamical Eq. (6.8) becomes:

$$\begin{aligned} \underline{\dot{e}} &= \Lambda_c\underline{e} + Bu_r - Bb\left[u_c(\underline{x}|\underline{\theta}^*) - u_1^*\right] - Bd + B\varepsilon \\ &= \Lambda_c\underline{e} + Bu_r - Bb\underline{\xi}^{\mathrm{T}}(\underline{x}) * [\underline{\theta} - \underline{\theta}^*] + B(\varepsilon - d) \\ &= \Lambda_c\underline{e} + Bu_r - Bb\underline{\xi}^{\mathrm{T}}(\underline{x}) * \underline{\tilde{\theta}} + B\omega \end{aligned} \quad (6.12)$$

where $\omega = (\varepsilon - d)$ and $\underline{\tilde{\theta}} = \underline{\theta} - \underline{\theta}^*$. As ε in (6.11) is defined as a negative quantity, therefore ω effectively denotes the summation of the fuzzy approximation error and the external disturbances.

Now, let us define the Lyapunov function as

$$V_e = \frac{1}{2}\underline{e}^{\mathrm{T}}P\underline{e} + \frac{1}{2v}\underline{\tilde{\theta}}^{\mathrm{T}}\underline{\tilde{\theta}} \tag{6.13}$$

where v is a positive constant and P is a symmetric positive definite matrix, satisfying the Riccati-like equation [1]:

$$\Lambda_c^{\mathrm{T}}P + P\Lambda_c - PB\left(\frac{2}{r} - \frac{1}{\rho^2}\right)B^{\mathrm{T}}P = -Q \tag{6.14}$$

Here, Q is a positive definite matrix, V_e in (6.13) is positive, and r is positive scalar constant. Using (6.12) and (6.14), we have

$$\dot{V}_e = \frac{1}{2}\underline{\dot{e}}^{\mathrm{T}}P\underline{e} + \frac{1}{2}\underline{e}^{\mathrm{T}}P\underline{\dot{e}} + \frac{1}{2v}\underline{\dot{\tilde{\theta}}}^{\mathrm{T}}\underline{\tilde{\theta}} + \frac{1}{2v}\underline{\tilde{\theta}}^{\mathrm{T}}\underline{\dot{\tilde{\theta}}} \tag{6.15}$$

Now, $\underline{\dot{\tilde{\theta}}} = \underline{\dot{\theta}}$, so, (6.15) can be rewritten as,

$$\begin{aligned}\dot{V}_e = \tfrac{1}{2}[&\underline{e}^{\mathrm{T}}\Lambda_c^{\mathrm{T}}P\underline{e} + u_r^{\mathrm{T}}B^{\mathrm{T}}P\underline{e} - \underline{\tilde{\theta}}^{\mathrm{T}}\underline{\xi}(\underline{x})bB^{\mathrm{T}}P\underline{e} + \omega^{\mathrm{T}}B^{\mathrm{T}}P\underline{e} \\ &+ \tfrac{1}{v}\underline{\dot{\tilde{\theta}}}^{\mathrm{T}}\underline{\tilde{\theta}} + \underline{e}^{\mathrm{T}}P\Lambda_c\underline{e} + \underline{e}^{\mathrm{T}}PBu_r - \underline{e}^{\mathrm{T}}PBb\underline{\xi}^{\mathrm{T}}(\underline{x})\underline{\dot{\tilde{\theta}}} \\ &+ \underline{e}^{\mathrm{T}}PB\omega + \tfrac{1}{v}\underline{\tilde{\theta}}^{\mathrm{T}}\underline{\dot{\tilde{\theta}}}]\end{aligned} \tag{6.16}$$

Now, if we choose

$$u_r = -\frac{1}{r}B^{\mathrm{T}}P\underline{e} \tag{6.17}$$

Then, (6.16) can be rewritten as

$$\begin{aligned}\dot{V}_e = &\tfrac{1}{2}\underline{e}^{\mathrm{T}}\left[P\Lambda_c + \Lambda_c^{\mathrm{T}}P - \tfrac{2}{r}PBB^{\mathrm{T}}P\right]\underline{e} \\ &-\underline{\tilde{\theta}}^{\mathrm{T}}[\underline{\xi}(\underline{x})bB^{\mathrm{T}}P\underline{e} - \tfrac{1}{v}\underline{\dot{\theta}}] + \tfrac{1}{2}\omega^{\mathrm{T}}B^{\mathrm{T}}P\underline{e} + \tfrac{1}{2}\underline{e}^{\mathrm{T}}PB\omega\end{aligned} \tag{6.18}$$

Thus, the parameter update law becomes [1, 10]:

$$\underline{\dot{\theta}} = v\underline{\xi}(\underline{x})bB^{\mathrm{T}}P\underline{e} \tag{6.19}$$

where v is treated as adaptation gain or learning rate.

Now, using (6.14) and (6.19), (6.18) becomes

$$\begin{aligned}\dot{V}_e &= -\frac{1}{2}\underline{e}^{\mathrm{T}}Q\underline{e} - \frac{1}{2\rho^2}\underline{e}^{\mathrm{T}}PBB^{\mathrm{T}}P\underline{e} + \frac{1}{2}\left[\omega^{\mathrm{T}}B^{\mathrm{T}}P\underline{e} + \underline{e}^{\mathrm{T}}PB\omega\right] \\ &= -\frac{1}{2}\underline{e}^{\mathrm{T}}Q\underline{e} - \frac{1}{2}\left[\frac{1}{\rho}B^{\mathrm{T}}P\underline{e} - \rho\omega\right]^{\mathrm{T}}\left[\frac{1}{\rho}B^{\mathrm{T}}P\underline{e} - \rho\omega\right] + \frac{1}{2}\rho^2\omega^{\mathrm{T}}\omega\end{aligned} \tag{6.20}$$

Therefore,

$$\dot{V}_e \le -\frac{1}{2}\underline{e}^{\mathrm{T}}Q\underline{e} + \frac{1}{2}\rho^2\omega^{\mathrm{T}}\omega \tag{6.21}$$

Using (6.13) and (6.21)

$$\frac{1}{2}\int_0^T \underline{e}^{\mathrm{T}}Q\underline{e}dt \le \frac{1}{2}\underline{e}^{\mathrm{T}}(0)P\underline{e}(0) + \frac{1}{2v}\underline{\tilde{\theta}}^{\mathrm{T}}(0)\underline{\tilde{\theta}}(0) + \frac{1}{2}\rho^2\int_0^{\mathrm{T}}\omega^{\mathrm{T}}\omega \tag{6.22}$$

Equation (6.22) is identical to that of (6.4), i.e., the condition for the H^∞ tracking control. Thus, using the control law $u_r(\underline{x})$, one can guarantee the boundedness in $\underline{x}$.

Further, the boundedness condition [1, 10, 18] in $\underline{\theta}$ results in:

$$\dot{\underline{\theta}} = \begin{cases} v\underline{\xi}(\underline{x})bB^{\mathrm{T}}P\underline{e} & \text{if } (|\underline{\theta}| < M_\theta) \text{ or} \\ & \left(|\underline{\theta}| = M_\theta \text{ and } \underline{e}^{\mathrm{T}}PBb\underline{\xi}^{\mathrm{T}}(\underline{x})\underline{\theta} \le 0\right) \\ P\{v\underline{\xi}(\underline{x})bB^{\mathrm{T}}P\underline{e}\} & \text{if } \left(|\underline{\theta}| = M_\theta \text{ and } \underline{e}^{\mathrm{T}}PBb\underline{\xi}^{T}(\underline{x})\underline{\theta} > 0\right) \end{cases} \tag{6.23}$$

where $P\{*\}$ is the projection operator and is given by)

$$P\{v\underline{\xi}(\underline{x})bB^{\mathrm{T}}P\underline{e}\} = v\underline{\xi}(\underline{x})bB^{\mathrm{T}}P\underline{e} - v\frac{\underline{e}^{\mathrm{T}}PBb\underline{\xi}^{\mathrm{T}}(\underline{x})\underline{\theta}}{|\underline{\theta}|^2}\underline{\theta} \tag{6.24}$$

Hence, the closed-loop stability is guaranteed [1, 10]. A schematic diagram of H^∞-based approach for adaptive control action is shown in Fig. 6.1.

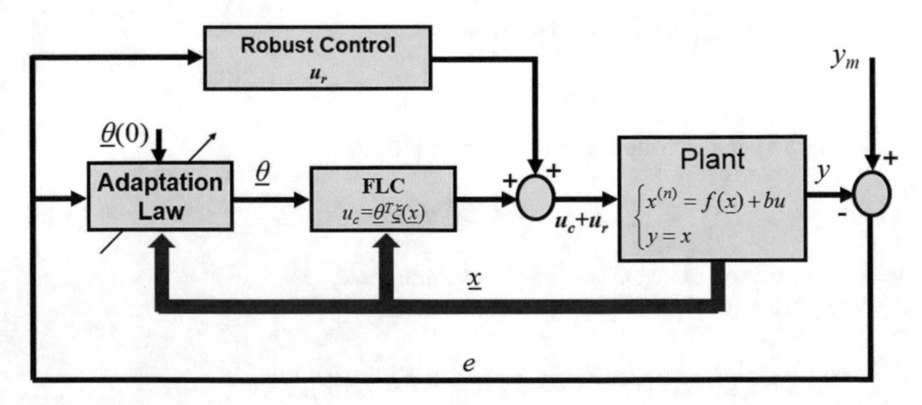

Fig. 6.1 A schematic representation of H^∞ strategy-based robust approach (HSBRA) for adaptive control action

6.3 H^∞ Strategy-Based Robust Approach (HSBRA)

The mathematical conditions, as presented before in Sect. 4.3, are restated with the required additional reinforcements, keeping the H^∞ strategy-based robust approach (HSBRA) in mind, as follows [1, 18, 19]:

- The type of the plant must be like (6.1), having additive disturbances.
- The order of the plant and its relative degree must be equal; i.e., the plant must be an all-pole system.
- The nonlinearity of the plant must be differentiable in an adequate number of times.
- There should exist the first $(n - 1)$ derivatives of the output of the plant for the H^∞ strategy-based learning algorithm.
- The fuzzy basis function $\underline{\xi}(\underline{x})$ must be determinedly exciting for the convergence of the parameter vector $\underline{\theta}$.
- The disturbances and the inherent approximation error must be bounded and the boundary should be known to the designer.
- The control algorithm and adaptation law are required to be carried out in continuous-time mode; otherwise, if it is in discrete-time mode, then the sampling rate must be high enough, so that the plant can be modeled as quasi-continuous.

6.3.1 *The Design Algorithm*

Design methodologies for stable robust direct adaptive fuzzy controllers, utilizing H^∞-based robust control approach [1–4, 10–14], can be divided into two parts, as follows—a) parameter calculations and b) controller adaptation.

(a) ***Parameter Calculations***

(i) Specify $k_1, k_2, \ldots, k_n$ such that $s^n + k_n s^{n-1} + \cdots + k_2 s + k_1 = 0$ be a Hurwitz polynomial.

(ii) Specify a positive definite $n \times n$ matrix Q and solve the Riccati-like equation in (6.14) to obtain $P > 0$ [1].

(iii) Specify the design parameters, i.e., the upper and lower bounds of the closed-loop variables, based on the nature of the plant and control objectives and also specify the values of adaptation gain v, r, and desired attenuation level ρ such that $2\rho^2 \geq r$ and also $\rho^2 \geq r$ [1, 2].

(iv) Define m_i number of fuzzy sets or MFs, F_i^l for fuzzification of input variable x_i, which uniformly covers universe of discourse U_i, where $l = 1, 2, \ldots, m_i$ and $i = 1, 2, \ldots, n$. The $U = U_1 \times U_2 \times \cdots \times U_n$ is usually chosen to be $\{\underline{x} \in R^n : |\underline{x}| \leq M_x\}$, and M_x is the bound of the states.

(v) Construct the fuzzy rule base for fuzzy logic system $u_c(\underline{x}|\theta)$ as:

$$R_c^{(l_1,\ldots l_n)} : \text{F } x_1 \text{ is } F_1^{l_1} \,\&\ldots\&\text{IF } x_n \text{ is } F_n^{l_n} \\ \text{THEN } u_c \text{ is } G^{(l_1,\ldots l_n)} \tag{6.25}$$

where $G^{(l_1,\ldots l_n)}$ are constrained as $\{\underline{\theta} : |\underline{\theta}| \leq M_\theta\}$, M_θ is the bound of the output singletons, and for zero-order T-S method, these are constants of arbitrary values for initial settings. The consequent parts of the rules are utilized to create the singleton vector $\underline{\theta} = [\theta_1\, \theta_2 \ldots \theta_N]^{\mathrm{T}}$, N = total number of rules.

(vi) Define the outputs of fuzzy basis functions

$$\xi_l(\underline{x}) = \frac{\prod_{i=1}^{r} \mu_i^l(x_i)}{\sum_{l=1}^{N} \left(\prod_{i=1}^{r} \mu_i^l(x_i)\right)} \tag{6.26}$$

and create the vector of these basis functions $\underline{\xi}(\underline{x}) = [\xi_1(\underline{x}), \xi_2(\underline{x}), \ldots, \xi_N(\underline{x})]^{\mathrm{T}}$. The $u_c(\underline{x}|\theta)$ is thus constructed as

$$u_c(\underline{x}|\underline{\theta}) = \underline{\theta}^{\mathrm{T}} \underline{\xi}(\underline{x}) \tag{6.27}$$

(b) ***Controller Adaptation***

The feedback control u in (6.7) is applied to the plant (6.1), where u_c is calculated from (6.5) and u_r is calculated from (6.17), and the following adaptation law is applied to adjust the parameter vector $\underline{\theta}$:

$$\dot{\underline{\theta}} = \begin{cases} v\underline{\xi}(\underline{x})bB^{\mathrm{T}}P\underline{e} & \text{if } (|\underline{\theta}| < M_\theta) \text{ or} \\ & (|\underline{\theta}| = M_\theta \text{ and } \underline{e}^{\mathrm{T}}PBb\underline{\xi}^{\mathrm{T}}(\underline{x})\underline{\theta} \leq 0) \\ P\{v\underline{\xi}(\underline{x})bB^{\mathrm{T}}P\underline{e}\} & \text{if } (|\underline{\theta}| = M_\theta \text{ and } \underline{e}^{\mathrm{T}}PBb\underline{\xi}^{\mathrm{T}}(\underline{x})\underline{\theta} > 0) \end{cases} \tag{6.28}$$

where $P\{*\}$ is the projection operator and is given by [1, 2, 10, 18]

$$P\{v\underline{\xi}(\underline{x})bB^{\mathrm{T}}P\underline{e}\} = v\underline{\xi}(\underline{x})bB^{\mathrm{T}}P\underline{e} - v\frac{\underline{e}^{\mathrm{T}}PBb\underline{\xi}^{\mathrm{T}}(\underline{x})\underline{\theta}}{|\underline{\theta}|^2}\underline{\theta}.$$

The flowchart form of presentation of HSBRA algorithm is shown in Fig. 6.2.

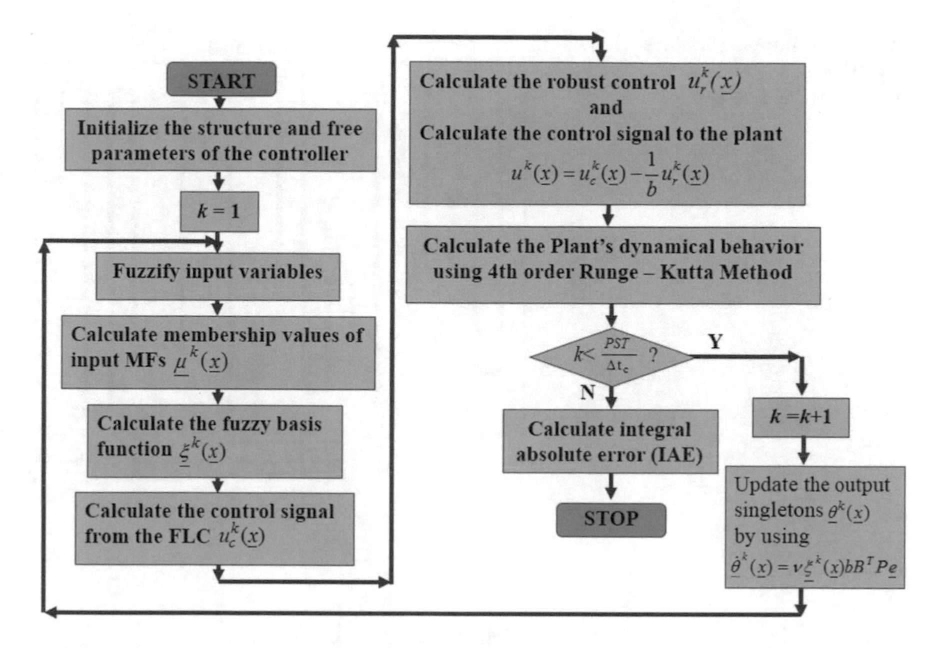

Fig. 6.2 HSBRA algorithm in flowchart form. k plant simulation counter, *PST* plant simulation time, Δt_c controller simulation sampling time

6.4 Case Studies

The effectiveness of the H^{∞}-based AFLC design methodology is evaluated for two benchmark nonlinear systems, i.e., the DC motor system with nonlinear friction characteristics [20, 21] and Duffing's oscillatory system [21–23], as utilized in previous chapters. The nonlinear plant models are simulated each utilizing fixed step fourth-order Runge-Kutta method with sampling time $\Delta t_c = 0.01$ s. Here also, the simulations are carried out for three fixed structure configurations with 3, 5, and 7 input MFs for each input variable. The HSBRA algorithm is simulated for 210 s, where 200 s is utilized for the adaptation purpose and then 10 s of evaluation period with $v = 0$. The disturbance applied to the systems in the following case studies is shown in Fig. 6.3.

6.4.1 *Case Study I: DC Motor Containing Nonlinear Friction Characteristics*

Once again, the controlled plant under consideration is a second-order DC motor containing nonlinear friction characteristics with external disturbance described by the following model [20, 21]:

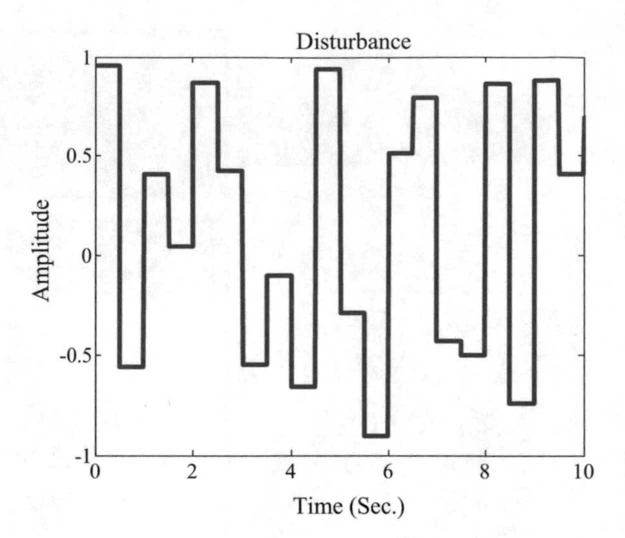

Fig. 6.3 A sample variation of disturbance applied to the systems in the case studies

$$\left.\begin{array}{l} \dot{x}_1 = x_2 \\ \dot{x}_2 = -\frac{f(x_2)}{J} + \frac{C_T}{J}u + d \\ y = x_1 \end{array}\right\} \tag{6.29}$$

where $y = x_1$ is the angular position of the rotor (in rad), x_2 is the angular speed (in rad/s), and u is the current fed to the motor (in A). C_T =10Nm/A, J = 0.1 kg m^2 and $f(x_2) = 5\tan^{-1}(5x_2)$ Nm. The control objective is to track a reference signal y_m = sin(2t).

The control signal obtained as the output from the fuzzy controller, in response to input signals, is given below:

$$u_c = u_c(\underline{x}, w|\underline{\theta}) \tag{6.30}$$

In this HSBRA design methodology, only the positions of the output singletons $\underline{\theta}$ are optimized and the positions of the input MFs and scaling gains for the input and output variables of the fuzzy controller are manually set a priori, similar to the cases of LSBA methodology discussed in Chap. 4. Here, the value of adaptation gain is chosen as v = 1, giving a good tradeoff between adaptation rate and small amplitudes of parameter oscillations which are due to inherent approximation error. The input and output scaling gains of the AFLC are chosen so as to map the ranges of the corresponding variables into the interval [-1, 1]. The other design parameters of the controllers are chosen to be the same as those considered in Sect. 3.6.1 of Chap. 3.

Table 6.1 Comparison of simulation results for case study I (Nonlinear DC motor system): H^{∞} strategy-based robust approach (HSBRA)

Control strategy	No. of input MFs	No. of rules	IAE
HSBRA	3 × 3	9	1.2209
	5 × 5	25	1.2328
	7 × 7	49	1.2333

The results are tabulated in Table 6.1, and it is evident from these results that the performance of the controller does not improve with the increase in the number of input fuzzy MFs as well as number of fuzzy rules. As the performance, in terms of *IAE*, is more or less the same for all three fixed structure configurations with 3, 5, and 7 input MFs for each input variable, a sample representation of performances for 5 input MFs is shown in Fig. 6.4, in a similar manner as the approach that was adopted in Chap. 4.

6.4.2 Case Study II: Duffing's Oscillatory System

In this part of the case study, the Duffing's forced oscillation system with disturbance is considered once again and is given as [21–23]:

$$\left.\begin{aligned} \dot{x}_1 &= x_2 \\ \dot{x}_2 &= -0.1x_2 - x_1^3 + 12\cos(t) + u(t) + d \\ y &= x_1 \end{aligned}\right\} \tag{6.31}$$

The control objective is to track the reference signal y_m, where $y_m = \sin(2t)$. The control signal obtained from the fuzzy controller is given as:

$$u_c = u_c(\underline{x}|\underline{\theta}) \tag{6.32}$$

Here also, the same parameters are chosen for this simulation process as was chosen in case study II in Chap. 4. The design strategy of HSBRA is simulated in a similar fashion as in case study I. Here, the adaptation gain v is set to 0.01 for the entire adaptation period [4, 8, 14].

Figure 6.5 shows temporal variations of system responses for the fixed structure controllers with 5 × 5 input MFs as a sample case, similar to the case study I. Table 6.2 compares the simulation results of case study II for 3 × 3, 5 × 5, and 7 × 7 input MF controller structure configurations. It has been observed from

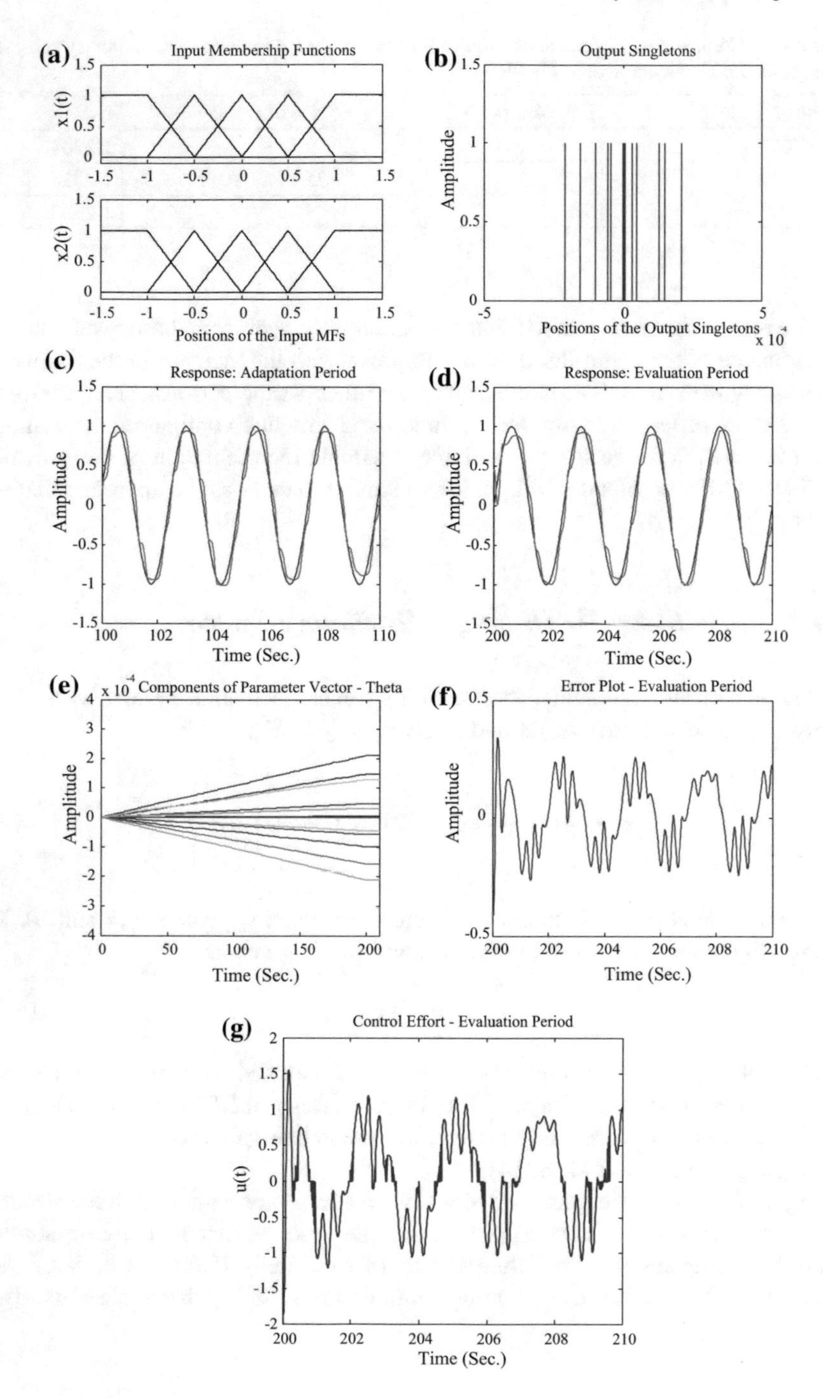
(a)
Input Membership Functions
x1(t)
x2(t)
Positions of the Input MFs
(b)
Output Singletons
Amplitude
Positions of the Output Singletons
x 10^-4
(c)
Response: Adaptation Period
Amplitude
Time (Sec.)
(d)
Response: Evaluation Period
Amplitude
Time (Sec.)
(e)
x 10^-4 Components of Parameter Vector - Theta
Amplitude
Time (Sec.)
(f)
Error Plot - Evaluation Period
Amplitude
Time (Sec.)
(g)
Control Effort - Evaluation Period
u(t)
Time (Sec.)

◂**Fig. 6.4** Responses of the adaptive fuzzy controller with HSBRA, for 5 × 5 configuration of input MFs, of case study I (Nonlinear DC motor system): **a** positions of input MFs, **b** positions of output singletons, **c** adaptation period response for 100–110 s, **d** evaluation period response for 200–210 s, **e** variation of component vector θ, **f** evaluation period error variation, and **g** evaluation period control effort variation

Table 6.2 that the performance of AFLCs, designed according to HSBRA algorithm, in terms of *IAE*, does not improve with the increase in number of input MFs. The best performance is obtained, in this case study, with 3 × 3 input MF configuration, but it is a very marginal improvement over the 5 × 5 input MF configuration. The result achieved in case of 7 × 7 input MF configuration is much less convincing than those of the other cases. Overall, it can be concluded that, due to the chaotic nature of the Duffing's system, the performance of the HSBRA control strategy does not provide a good enough solution, when subjected to random disturbance variations.

6.5 Summary

In this chapter, H^{∞}-based robust adaptive control technique has been utilized for designing the AFLCs. The stability of closed-loop system and the convergence of the plant output to a desired reference signal, in the presence of external disturbances, are precisely guaranteed in this design methodology. The H^{∞}-based robust adaptive control design strategy has been implemented for two benchmark nonlinear systems for different controller configurations, and the drawbacks of this design method are summarized.

The next chapter will discuss the contemporary concept of hybridization of H^{∞}-based robust adaptive control-based local adaptation and stochastic optimization-based global optimization for designing the direct robust stable adaptive fuzzy controllers and how the performances can be enhanced, using these state-of-the-art controller design philosophies.

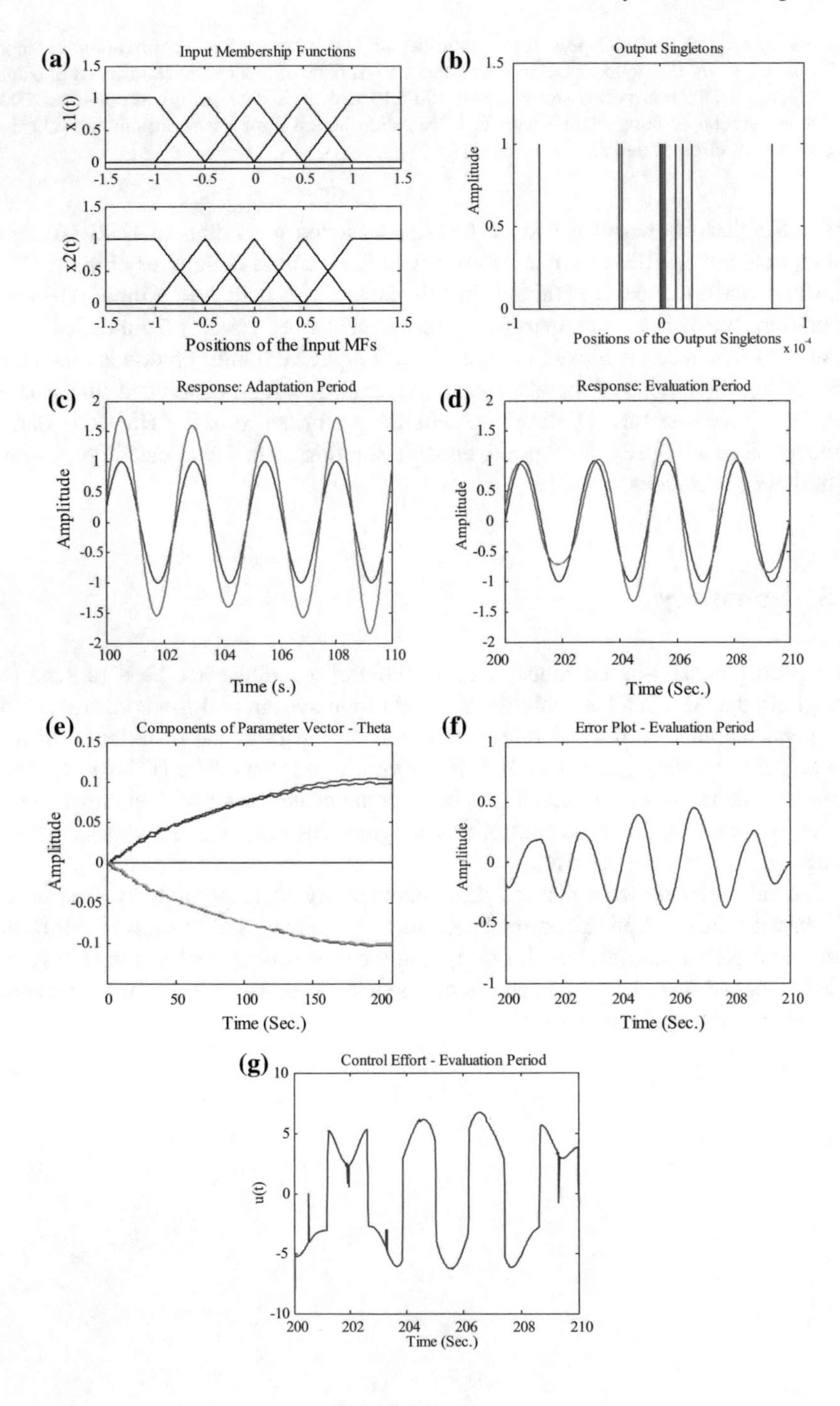
(a)
Input Membership Functions
x1(t)
x2(t)
Positions of the Input MFs
(b)
Output Singletons
Amplitude
Positions of the Output Singletons
x 10^-4
(c)
Response: Adaptation Period
Amplitude
Time (s.)
(d)
Response: Evaluation Period
Amplitude
Time (Sec.)
(e)
Components of Parameter Vector - Theta
Amplitude
Time (Sec.)
(f)
Error Plot - Evaluation Period
Amplitude
Time (Sec.)
(g)
Control Effort - Evaluation Period
u(t)
Time (Sec.)

◀**Fig. 6.5** Responses of the adaptive fuzzy controller with HSBRA, for 5 × 5 configuration of input MFs, of case study II (Duffing's system): **a** positions of input MFs, **b** positions of output singletons, **c** adaptation period response for 100–110 s, **d** evaluation period response for 200–210 s, **e** variation of component vector θ, **f** evaluation period error variation, and **g** evaluation period control effort variation

Table 6.2 Comparison of simulation results for case study II (Duffing's system): H^{∞} strategy-based robust approach (HSBRA)

Control strategy	No. of input MFs	No. of rules	IAE
HSBRA	3 × 3	9	1.9484
	5 × 5	25	1.9543
	7 × 7	49	2.2557

References

1. B.S. Chen, C.H. Lee, Y.C. Chang, H^{∞} tracking design of uncertain nonlinear SISO systems: adaptive fuzzy approach. IEEE Trans. Fuzzy Syst. **4**(1), 32–43 (1996)
2. B.S. Chen, C.L. Tsai, D.S. Chen, Robust H-infinity and mixed H^2/H^{∞} filters for equalization designs of nonlinear communication systems: Fuzzy interpolation approach. IEEE Trans. Fuzzy Syst. **11**(3), 384–398 (June 2003)
3. H.-X. Li, S. Tong, A hybrid adaptive fuzzy control for a class of nonlinear MIMO system. IEEE Trans. Fuzzy Syst. **11**(1), 24–34 (2003)
4. X. Zhengrogn, C. Qingwei, H. Weili, Robust H^{∞} control for fuzzy descriptor systems with state delay. IEEE J. Syst. Eng. Electron. **16**(1), 98–101 (2005)
5. H. Yishao, Z. Dequn, C. Xiaoxui, D. Lin, H^{∞} tracking design for a class of decentralized nonlinear systems via fuzzy adaptive observer. IEEE J. Syst. Eng. Electron. **20**(4), 790–799 (2009)
6. W. Miaoxin, L. Jizhen, L. Xianqjie, L. Juncheng, Robust H^{∞} controller design for a class of nonlinear fuzzy time-delay systems. IEEE J. Syst. Eng. Electron. **20**(6), 1286–1289 (2009)
7. G. Chao, L. Xiao-Geng, W. Jun-Wei, W. Fei, Robust H^{∞} decentralized fuzzy tracking control for bank-to-turn missiles, in *Proceedings of the 32nd Chinese Control Conference*, China, July 26–28, 2013, pp. 3498–3503
8. S. Park, H. Ahn, Robust T-S fuzzy H^{∞} control of interior permanent magnet motor using weighted integral action, in *Proceedings of 2014 14th International Conference on Control, Automation and Systems (ICCAS 2014)*, Korea, Oct 22–25, 2014, pp. 1504–1507
9. J. Xiong, X-H. Chang, Observer-based robust non-fragile H^{∞} control for uncertain T-S fuzzy singular systems, in *Proceedings of the 36th Chinese Control Conference*, China, July 26–28, 2017, pp. 4233–4238
10. W. Assawinchaichote, S.K. Nguang, Fuzzy H^{∞} output feedback control design for singularly perturbed systems: An LMI approach, in *Proceedings of 42nd IEEE Conference on Decision and Control*, Hawaii, USA, Dec 2003, pp. 863–868
11. S.K. Nguang, W. Assawinchaichote, P. Shi, Y. Shi, Robust H^{∞} control design for uncertain fuzzy systems with Markovian jumps: an LMI approach, in *Proceedings of the American Control Conference*, Portland USA, June 2005, pp. 1805–1810
12. T. Agustinah, A. Jazidie, Md. Nuh, Fuzzy tracking control based on H^{∞} performance for nonlinear systems, *WSEAS Trans. Syst. Control* **6**(11), 393–403 (Nov 2011)
13. S.-M. Wu, C.-C. Sun, H.-Y. Chung, W.-J. Chang, Mixed H_2/H_{∞} region-based fuzzy controller design for continuous-time fuzzy systems. J. Intell. Fuzzy Syst. **18**, 19–30 (2007)

14. H.-N. Wu, H.-X Li, H^{∞} fuzzy observer-based control for a class of nonlinear distributed parameter systems with control constraints. IEEE Trans. Fuzzy Syst. **16**(2), 502–516 (Apr 2008)
15. Y. Guan, S. Zhou, W. X. Zheng, H^{∞} dynamic output feedback control for fuzzy systems with quantized measurements, in *Proceedings of Joint 48th IEEE Conference on Decision and Control and 28th Chinese Control Conference*, Shanghai, P.R. China, Dec 2009, pp. 7765–7770
16. J. Pan, S. Fei, Y. Ni, M. Xue, New approaches to relaxed stabilization conditions and H^{∞} control designs for T-S fuzzy systems. J Control Theory Appl. **10**(1), 82–91 (2012)
17. K. Das Sharma, A. Chatterjee, Patrick Siarry, A. Rakshit, CMA—H^{∞} hybrid design of robust stable adaptive fuzzy controllers for non-linear systems, in *Proceedings of 1st International Conference Frontiers in Optimization: Theory and Applications (FOTA-2016)*, Nov 2016, India
18. K. Das Sharma, Hybrid methodologies for stable adaptive fuzzy control, Ph.D. dissertation, Jadavpur University, 2012
19. L.-X. Wang, *A Course in Fuzzy Systems and Control* (Englewood Cliffs, Prentice Hall, NJ, 1997)
20. K. Das Sharma, A. Chatterjee, F. Matsuno, A Lyapunov theory and stochastic optimization based stable adaptive fuzzy control methodology, in *Proceedings of SICE Annual Conference 2008*, Japan, Aug 20–22, pp. 1839–1844
21. K. Das Sharma, A. Chatterjee, A. Rakshit, A hybrid approach for design of stable adaptive fuzzy controllers employing Lyapunov theory and particle swarm optimization. IEEE Trans. Fuzzy Syst. **17**(2), 329–342 (Apr 2009)
22. K. Das Sharma, A. Chatterjee, A. Rakshit, Design of a hybrid stable adaptive fuzzy controller employing Lyapunov theory and harmony search algorithm. IEEE Trans. Contr. Syst. Tech. **18**(6), 1440–1447 (Nov 2010)
23. K. Fischle, D. Schroder, An improved stable adaptive fuzzy control method. IEEE Trans. Fuzzy Syst. 7(1), 27–40 (Feb 1999)

Chapter 7
Fuzzy Controller Design V: Robust Hybrid Adaptive Approaches

7.1 Introduction

Optimization of design variables in different form is an age-old challenge encountered by the engineering community. Many a times the fitness functions of such optimizations are mostly non-differentiable, or even if differentiable, their derivatives may not be calculated explicitly. Furthermore, the problems are non-convex type in general, and so, finding the global optimum is quite difficult. To get rid of these kinds of problems, Hansen et al. proposed the covariance matrix adaptation evolution strategy (CMA-ES), popularly known as the CMA strategy [1]. It is an evolutionary algorithm that creates a number of probable solution points utilizing the normal distribution and it is a derivative-free optimization. Although the original version is applicable only to the unconstrained problems and can handle small population size during operation, the further variants developed over a period of time show that the CMA strategy can be applied for the constrained optimization problems with large population sizes too [2, 3].

This chapter presents a systematic design procedure for stable adaptive fuzzy logic controllers (AFLCs) using hybridizations of H^{∞} strategy-based robust approach (HSBRA) [4–8], and CMA algorithm-based stochastic optimization technique [1, 9]. This hybrid design strategy of fuzzy controllers tries to combine the advantages of both H^{∞} control-based robust local adaptation and stochastic optimization-based global search method, in a bid to develop a superior method [9–13]. The aims of this design scheme are to execute simultaneous adaptations of both the structural features of FLC and its free parameters, such that it will attempt: (i) to guarantee the stability of the closed-loop system and (ii) to accomplish high degrees of automation in the design process, utilizing CMA algorithm, by getting rid of many manually tuned parameters [9]. Hence, the objective of this chapter is to design a stable adaptive fuzzy controller which can deliver a high degree of automation in the design process, guarantee asymptotic stability in the sense of Lyapunov, and also achieve satisfactory transient performance. The scheme is

K. Das Sharma et al., *Intelligent Control*, Cognitive Intelligence and Robotics,
https://doi.org/10.1007/978-981-13-1298-4_7

implemented for two benchmark nonlinear case studies, and the results are analyzed to show the usefulness of the scheme.

7.2 Hybrid Design Methodology Using H^∞ Strategy

The control objectives and fuzzy controller formulation in hybrid design methodology are quite similar to that of the Lyapunov-based strategy. The basics are described in this chapter; for details, please see Chap. 5.

Control Objectives

Let us consider that the objective is to design an adaptive strategy for an *n*th-order SISO nonlinear plant given as [5, 6]:

$$\left.\begin{aligned} x^{(n)} &= f(\underline{x}) + bu + d \\ y &= x \end{aligned}\right\} \tag{7.1}$$

where $f(\cdot)$ is an unknown continuous function, $u \in R$ and $y \in R$ are the input and output of the plant, b is an unknown positive constant, and d is unknown but bounded external disturbances. Let us assume that the state vector is given as $\underline{x} = (x_1, x_2, \ldots, x_n)^{\mathrm{T}} = (x, \dot{x}, \ldots, x^{(n-1)})^{\mathrm{T}} \in R^n$.

The control objective is to force the plant output $y(t)$ to follow a given bounded reference signal $y_m(t)$ under the constraints that all closed-loop variables involved must be bounded to guarantee the closed-loop stability of the system [9–13]. So, the objective is to design a stable adaptive fuzzy controller for the system described in (7.1) and where the tracking error is $e = y_m - y$.

Now, the ideal control law with H^∞ control action for the system in (7.1) is given as [4, 5, 12–14]:

$$u^* = \frac{1}{b}\left[-f(\underline{x}) + y_m^{(n)} + \underline{k}^T \underline{e} - u_r\right] \tag{7.2}$$

where $y_m^{(n)}$ is the *n*th derivative of the output of the reference model, the control signal u_r is applied to attenuate the external disturbance and the error due to fuzzy approximation of the AFLC [4–6], $\underline{e} = (e, \dot{e}, \ldots, e^{(n-1)})^{\mathrm{T}}$ is the error vector, and $\underline{k} = (k_1, k_2, \ldots, k_n)^{\mathrm{T}} \in R^n$ is the vector describing the desired closed-loop dynamics for the error.

The ideal control law u^* guarantees perfect tracking, i.e., $y(t) \equiv y_m(t)$ if $\underset{t\to\infty}{\mathrm{Lt}}\ e(t) = 0$. As f and b are not known precisely, the ideal u^* of (7.2) cannot be implemented in practice. The design objective of asymptotically stable tracking may be achieved if ω, the summation of the fuzzy approximation error (ε) and the external disturbance (*d*), is zero; otherwise, the H$^\infty$ tracking performance will come into play, given as [4, 5]:

$$\int_0^T \underline{e}^{\mathrm{T}} Q \underline{e} \mathrm{d}t \leq \underline{e}^{\mathrm{T}}(0) P \underline{e}(0) + \frac{1}{v} \underline{\tilde{\theta}}^{\mathrm{T}}(0) \underline{\tilde{\theta}}(0) + \rho^2 \int_0^T \omega^{\mathrm{T}} \omega \mathrm{d}t \quad \forall T \in [0, \infty) \omega \in L_2[0, T] \tag{7.3}$$

for given weighting matrices $Q = Q^{\mathrm{T}} \geq 0$, $P = P^{\mathrm{T}} \geq 0$, an adaptation gain $v > 0$, and prescribed attenuation level ρ.

Now, to ensure stability it is assumed that the control law *u*(*t*) is actually the summation of the fuzzy control $u_{\mathrm{c}}(\underline{x}|\underline{\theta})$ and robust control term, given as [4]:

$$u(t) = u_{\mathrm{c}}(\underline{x}|\underline{\theta}) - \frac{1}{b} u_r(\underline{x}) \tag{7.4}$$

Here, it has been assumed like all other earlier cases discussed in this book that the AFLC is constructed using a zero-order Takagi–Sugeno (T-S) fuzzy system. Then, $u_{\mathrm{c}}(\underline{x}|\underline{\theta})$ for the AFLC is given in the form [4, 10, 12, 14]:

$$u_{\mathrm{c}}(\underline{x}|\underline{\theta}) = \sum_{l=1}^{N} \theta_l * \alpha_l(\underline{x}) \Big/ \left(\sum_{l=1}^{N} \alpha_l(\underline{x}) \right) = \underline{\theta}^{\mathrm{T}} * \underline{\xi}(\underline{x}) \tag{7.5}$$

where $\underline{\theta} = [\theta_1\ \theta_2 \ldots \theta_N]^{\mathrm{T}}$ = the vector of the output singletons, $\alpha_l(\underline{x})$ = = the firing degree of rule "*l*", *N* = the total number of rules, $\mu_i^l(x_i)$ = the membership value of the *i*th input membership function (MF) in the activated *l*th rule, $\underline{\xi}(\underline{x})$ = vector containing normalized firing strengths of all fuzzy IF-THEN rules = $(\xi_1(\underline{x}), \xi_2(\underline{x}), \ldots, \xi_N(\underline{x}))^{\mathrm{T}}$ and

$$\xi_l(\underline{x}) = \alpha_l(\underline{x}) \Big/ \left(\sum_{l=1}^{N} \alpha_l(\underline{x}) \right) \tag{7.6}$$

The H$^\infty$ control action can be formed as described in Chap. 6 and is given as [4, 5]:

$$u_r = -\frac{1}{r} B^{\mathrm{T}} P \underline{e} \tag{7.7}$$

Let us define a quadratic form of tracking error as $V_e = \frac{1}{2} \underline{e}^{\mathrm{T}} P \underline{e} + \frac{1}{2v} \underline{\tilde{\theta}}^{\mathrm{T}} \underline{\tilde{\theta}}$ where *P* is a symmetric positive definite matrix satisfying the Lyapunov equation [10]. With the use of u_r, it can be shown that $\dot{V}_e \leq -\frac{1}{2} \underline{e}^T Q \underline{e} + \frac{1}{2} \rho^2 \omega^T \omega$, which in turn satisfied the condition for the H^{∞} tracking control $\frac{1}{2}\int_{0}^{T} \underline{e}^{\mathrm{T}} Q \underline{e} \mathrm{d}t \leq \frac{1}{2}\underline{e}^{\mathrm{T}}(0) P \underline{e}(0) + \frac{1}{2v}\underline{\tilde{\theta}}^{\mathrm{T}}(0)\underline{\tilde{\theta}}(0) + \frac{1}{2}\rho^2 \int_{0}^{T} \omega^{\mathrm{T}}\omega$, where Q is a positive definite matrix [4, 5, 9]. Thus, using the control law $u_r(\underline{x})$, one can guarantee the robustness in $\underline{x}$ with the parameter adaptation law expressed as [4, 9]:

$$\underline{\dot{\theta}} = v\underline{\xi}(\underline{x}) b B^{\mathrm{T}} P \underline{e} \tag{7.8}$$

where $v > 0$ is the adaptation gain or learning rate.

Stochastic Optimization Algorithm-Based Design of Fuzzy Controller

A systematic design strategy for fuzzy controllers has already been detailed in Chap. 3 of this book. Different SOAs have so far been utilized in this book to design the fuzzy controllers, where the structure and the other free parameters, e.g., supports of the input MFs, input and output scaling gains of the controller are stochastically adapted. To determine the optimal structure of the FLC [10, 12] in SOA-based generic design, a candidate solution vector (CSV) in solution space can be formed in a similar manner as described in previous chapters, given as [10, 12]:

$$\begin{aligned}\underline{Z} = & [\text{structural flags for MFs} \mid \text{center locations of the MFs} \mid \ldots \\ & \mid \text{scaling gains} \mid \text{positions of the output singletons}]\end{aligned} \tag{7.9}$$

The structural flags are particularly important to determine the structure of the FLCs (see Sect. 3.3.1 for details) [10].

Figure 3.4 shows the flowchart representation of SOA-based approach (SOABA) algorithm for the FLC with supervisory control action. Candidate controller simulation (CCS) algorithm, as shown in Fig. 3.2, has been utilized to implement the SOABA algorithm with an initial pool of CSVs. In this case also, the fitness function is chosen as the integral absolute error (*IAE*) where $IAE = \sum_{n=0}^{\mathrm{PST}} e(n)\Delta t_c$, and PST = plant simulation time, Δt_c = step size or sampling time.

Based on the fitness value, CSVs are updated according to the improvisation rules for the particular SOA under consideration [10].

Hybrid Design Methodology

In these design methodologies, the complementary features of H^{∞} tracking control-based robust approach and the SOA-based approach are combined to obtain a superior systematic design procedure for direct AFLCs, for a class of SISO systems. A generic schematic architecture of robust hybrid adaptive fuzzy control action is shown in Fig. 7.1.

The optimization process of the controller is derived in three different ways to explore the possibility of obtaining superior results [12] as discussed in Chap. 5. These three proposed robust hybrid strategies are described now.

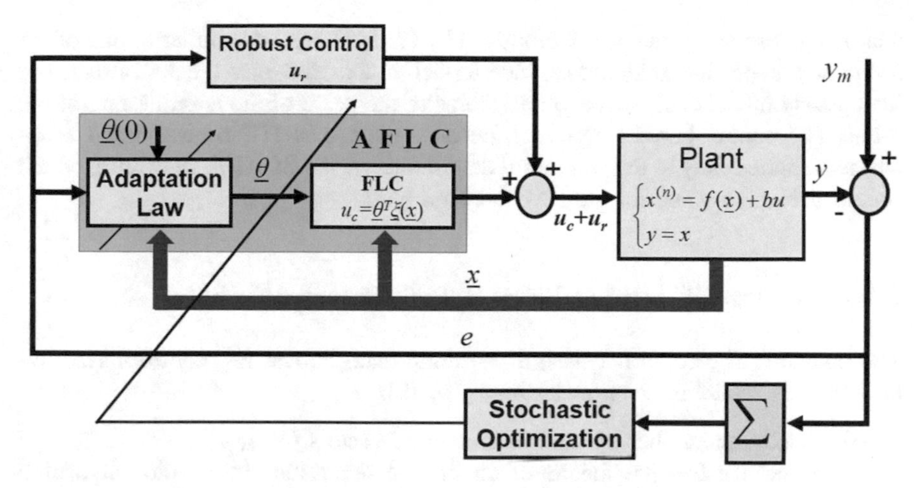

Fig. 7.1 A schematic architecture of robust hybrid adaptive fuzzy control action

7.2.1 H$^\infty$ Strategy-Based Cascade Model

Very similar to the cascade model as discussed in Chap. 5, in this model also a stochastic optimization algorithm and robust H$^\infty$ control-based optimization are used in cascade fashion. Thus, here also, the structure and the free parameters of the fuzzy controller are optimized by the stochastic optimization-based approach as well as the positions of the output singletons are primarily tuned in a global search space by the SOA method [10, 12]. Once the initial global tuning of $\underline{\theta}$ by the SOA technique is accomplished, the robust H$^\infty$ control-based method optimizes the $\underline{\theta}$ locally to further improve the performance of the controller as shown in Fig. 7.2. Thus, in hybrid cascade model, the performance of the robust H$^\infty$ control-based method is enhanced by the SOA method. In this cascade model, the learning rate is adjusted in such a fashion that robust H$^\infty$ control adaptation (7.8) searches only the

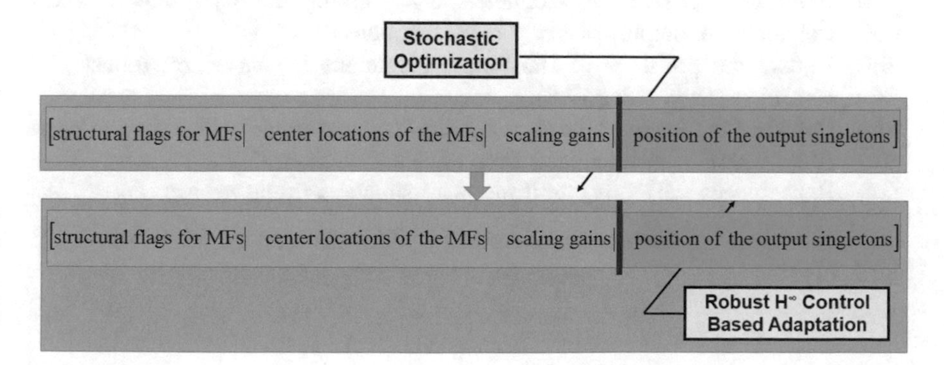

Fig. 7.2 Robust H$^\infty$ control-based hybrid cascade adaptation scheme

near neighborhood of the best $\underline{\theta}$ produced by the SOA method similar to that of the Lyapunov theory-based hybrid cascade model. In this case, also the performance of the cascade model is sensitive to the choice of the value of the learning rate of the robust H^{∞} control-based adaptation, because the robust H^{∞} control-based adaptation is applied only to the solution $\underline{\theta}$ determined by the SOABA algorithm, which acts as the initial value for the HSBRA algorithm.

7.2.1.1 Robust H^{∞} Strategy-Based Hybrid Cascade Algorithm

The outline of the controller design algorithm using robust H^{∞} control-based hybrid cascade model is given as follows [10, 12]:

(i) Select the number of candidate solution vectors (N_{CSV}).
(ii) Select the free parameters of the chosen stochastic optimization algorithm (SOA).
(iii) The stopping criterion can be set in two ways, either by choosing the maximum number of iterations ($\mathrm{iter}_{\max}$) or by using minimum error criteria, i.e., continue the SOA adaptation until a maximum pre-specified allowable error is attained by the controller.
(iv) Initialize N_{CSV} number of randomly generated CSVs.
(v) Determine the structure of the candidate controller based on structural flags as incorporated in the candidate solution vector $\underline{Z}$ (see (7.9)).
(vi) Calculate the fitness value, e.g., *IAE*, of the controller for each CSV that is used as a candidate controller here, by using CCS algorithm, as shown in flowchart form, in Fig. 3.2.
(vii) Improvise a new CSV based on the updating rules of the SOA.
(viii) Repeat steps (v)–(vii), until the termination criterion is fulfilled.
(ix) Obtain an optimal/near-optimal CSV after the termination of the SOABA method.
(x) Determine the optimal structure of the candidate controller based on structural flags as incorporated in $\underline{Z}$ (see (7.9)).
(xi) Initialize other HSBRA parameters, e.g., learning rate, Lyapunov matrix.
(xii) Calculate the outputs of fuzzy basis functions.
(xiii) Update the positions of the output singletons ($\underline{\theta}$) based on robust H^{∞} control strategy, using (7.8).
xiv) Calculate the feedback control u as in (7.4) and implement it for the plant (7.1), where u_c is calculated from (7.5) and u_r from (7.7).
(xv) Repeat steps (xii)–(xiv) until the plant simulation time expires.

The flowchart representation of robust hybrid cascade model is shown in Fig. 7.3 [4].

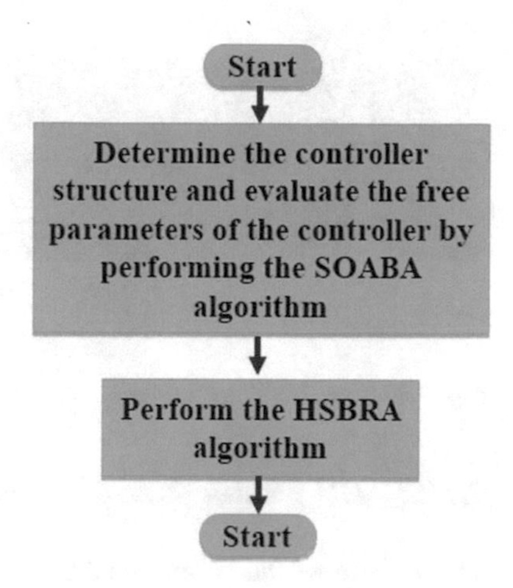

Fig. 7.3 Flowchart representation of robust H^∞ control-based hybrid cascade algorithm

7.2.2 *H^∞ Strategy-Based Concurrent Model*

In this stable adaptive fuzzy controller design process, similar to the previously discussed model in Chap. 5, the robust H^∞ control approach and the SOA-based approach run concurrently or parallel to optimize the (i) structure of the controller, (ii) scaling gains, and (iii) positions of the output singletons [2, 4, 5, 8].

In this method too, a particle $\underline{Z}$ is divided into two sub-groups as [2, 4]:

$$\underline{Z} = [\underline{\psi} \mid \underline{\theta}] \tag{7.10}$$

where

$$\left.\begin{aligned} \underline{\psi} &= [\text{structural flags for MFs} \mid \text{center locations of the MFs} \mid \ldots \\ &\qquad \mid \text{scaling gains}] \\ \underline{\theta} &= [\text{position of the output singletons}] \end{aligned}\right\} \tag{7.11}$$

Like the earlier hybrid designs, the partition of $\underline{Z}$ vector separates the parameters of the AFLC in such a fashion that $\underline{\psi}$ has nonlinear influence and $\underline{\theta}$ has linear influence on u_c, respectively [14]. In HSBRA, the values of the free parameters in the vector $\underline{\psi}$ are defined a priori and the value of the corresponding $\underline{\theta}$ minimizing the tracking error emphasizing the robust H^∞ control strategy can be obtained by the adaptation law as given in (7.8). So, in HSBRA, similar to the methodology followed in LSBA, the $\underline{\psi}$ is set by hand tuned trial and error method. In this robust hybrid concurrent method, SOA is applied to optimize $\underline{\psi}$ and $\underline{\theta}$ in tandem by a method of CSV improvisation in optimization iteration. In addition to that, the adaptation law, as in (7.8), is applied for every updated candidate controller to adjust the value of $\underline{\theta}$ only. So, in this robust design method also, $\underline{\theta}$ explores both the local search space and global search space, simultaneously, in a twofold manner, as shown in Fig. 7.4.

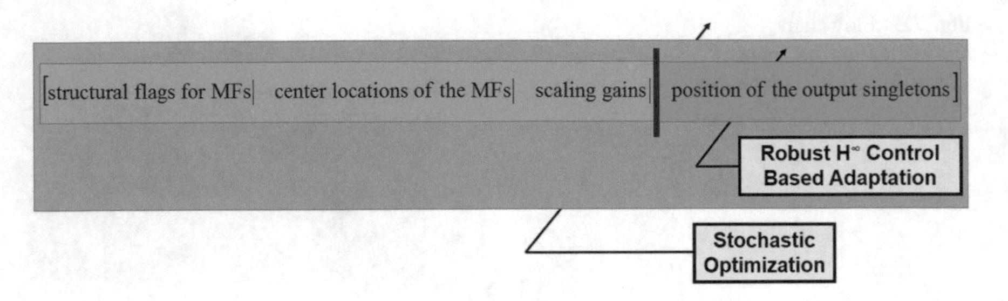

Fig. 7.4 Robust H$^\infty$ control-based hybrid concurrent adaptation scheme

7.2.2.1 H$^\infty$ Strategy-Based Hybrid Concurrent Algorithm

The outline of the controller design algorithm using robust H$^\infty$ control-based hybrid concurrent model is given as follows [10–12]:

(i) Select the number of candidate solution vectors (N_{CSV}).
(ii) Select the free parameters of the chosen stochastic optimization algorithm (SOA).
(iii) Initialize other HSBRA parameters, e.g., learning rate, Lyapunov matrix.
(iv) The stopping criterion can be set in two ways, either by choosing the maximum number of iterations (iter$_{\max}$) or by using a minimum error criterion, i.e., continue the SOA adaptation until a maximum pre-specified allowable error is attained by the controller.
(v) Initialize N_{CSV} number of randomly generated candidate solution vectors (CSV).
(vi) Determine the structure of the candidate controller based on structural flags as incorporated in the candidate solution vector $\underline{Z}$ (see (7.9)).
(vii) Calculate the outputs of fuzzy basis functions.
(viii) Update the positions of the output singletons ($\underline{\theta}$) based on robust H$^\infty$ control theory using (7.8).
(ix) Calculate the feedback control u as in (7.4) and implement it for the plant (7.1), where u_{c} is calculated from (7.5) and u_r from (7.7).
(x) Repeat steps (vii)–(ix) until the plant simulation time expires.
(xi) Calculate the fitness value, e.g., *IAE*, of the controller for each CSV that acts as a candidate controller.
(xii) Replace $\underline{\theta}$ of $\underline{Z}$ by $\underline{\theta}_{\mathrm{HSBRA}}$ obtained after step (x).
(xiii) Improvise a new CSV based on the updating rules of the SOA.
(xiv) Repeat steps (vi)–(xiii) until the termination criterion is fulfilled.

The flowchart representation of robust H$^\infty$ control-based hybrid concurrent model is shown in Fig. 7.5.

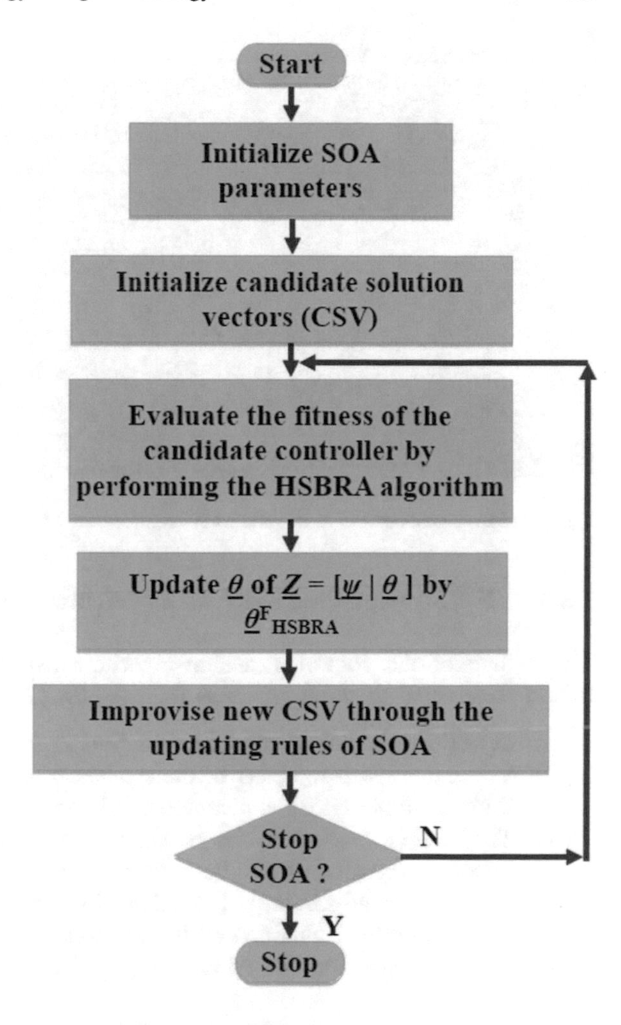

Fig. 7.5 Flowchart representation of robust H^{∞} control-based hybrid concurrent algorithm

7.2.3 H^{∞} Strategy-Based Preferential Model

In this model, the SOABA method and the robust hybrid concurrent method both are employed to keep the tracking error as minimum as possible [10, 12], similar to that of the preferential model discussed in Chap. 5. Here also, each CSV in the population, i.e., each candidate controller, evaluates the CCS algorithm and the HSBRA algorithm separately during the same iteration. The structure of this hybrid preferential model has been discussed in detail in Sect. 5.2.3 in Chap. 5. The CSV may follow different path of optimization in different iterations depending on the fitness value achieved in two comparative paths, namely CCS algorithm and HSBRA algorithm, as described in Fig. 7.6. This method is proposed with an objective of avoiding getting stuck in local minima [10].

Candidate Solution Vectors →

Iterations ↓

HSBRA	CCS	HSBRA	CCS	CCS	HSBRA	HSBRA	CCS	CCS	HSBRA
CCS	HSBRA	CCS	HSBRA	HSBRA	CCS	HSBRA	HSBRA	CCS	HSBRA
CCS	HSBRA	HSBRA	CCS	CCS	CCS	CCS	HSBRA	HSBRA	CCS
HSBRA	CCS	CCS	CCS	HSBRA	HSBRA	HSBRA	CCS	CCS	HSBRA
HSBRA	CCS	HSBRA	HSBRA	CCS	HSBRA	CCS	HSBRA	HSBRA	CCS
CCS	HSBRA	HSBRA	CCS	CCS	HSBRA	CCS	CCS	HSBRA	CCS
HSBRA	HSBRA	CCS	HSBRA	HSBRA	CCS	CCS	HSBRA	CCS	HSBRA
HSBRA	CCS	HSBRA	HSBRA	CCS	HSBRA	HSBRA	CCS	HSBRA	CCS
CCS	CCS	CCS	HSBRA	HSBRA	CCS	HSBRA	HSBRA	CCS	HSBRA
HSBRA	CCS	HSBRA	CCS	HSBRA	CCS	CCS	CCS	HSBRA	CCS

Fig. 7.6 Movements of CSVs in robust H^{∞} control-based hybrid preferential algorithm

7.2.3.1 H^{∞} Strategy-Based Hybrid Preferential Algorithm

The outline of the controller design algorithm using robust H^{∞} control-based hybrid preferential model is given as follows [10, 12]:

(i) Select the number of candidate solution vectors (N_{CSV}).
(ii) Select the free parameters of the stochastic optimization algorithm (SOA).
(iii) Initialize other HSBRA parameters, like learning rate, Lyapunov matrix.
(iv) The stopping criterion can be set in two ways, either by choosing the maximum number of iterations ($iter_{max}$) or by using a minimum error criteria, i.e., continue the SOA adaptation until a maximum pre-specified allowable error is attained by the controller.
(v) Initialize N_{CSV} number of randomly generated candidate solution vectors (CSV).
(vi) Determine the structure of the candidate controller based on structural flags as incorporated in the candidate solution vector $\underline{Z}$ (see (7.9)).
(vii) Independently evaluate:

   a. CSS algorithm and calculate the IAE_{CCS} as shown in flowchart in Fig. 3.2.
   b. HSBRA algorithm and calculate the IAE_{HSBRA} as shown in flowchart in Fig. 6.2.

(viii) If ($IAE_{HSBRA} < IAE_{CCS}$), then update the values of the output singletons $\underline{\theta}$ by the final adapted value from the HSBRA algorithm and set $IAE = IAE_{HSBRA}$; otherwise, keep the $\underline{\theta}$ unchanged and set $IAE = IAE_{CCS}$.
(ix) Improvise a new CSV based on the updating rules of the SOA.
(x) Repeat steps (vi)–(ix) until the termination criterion is fulfilled.

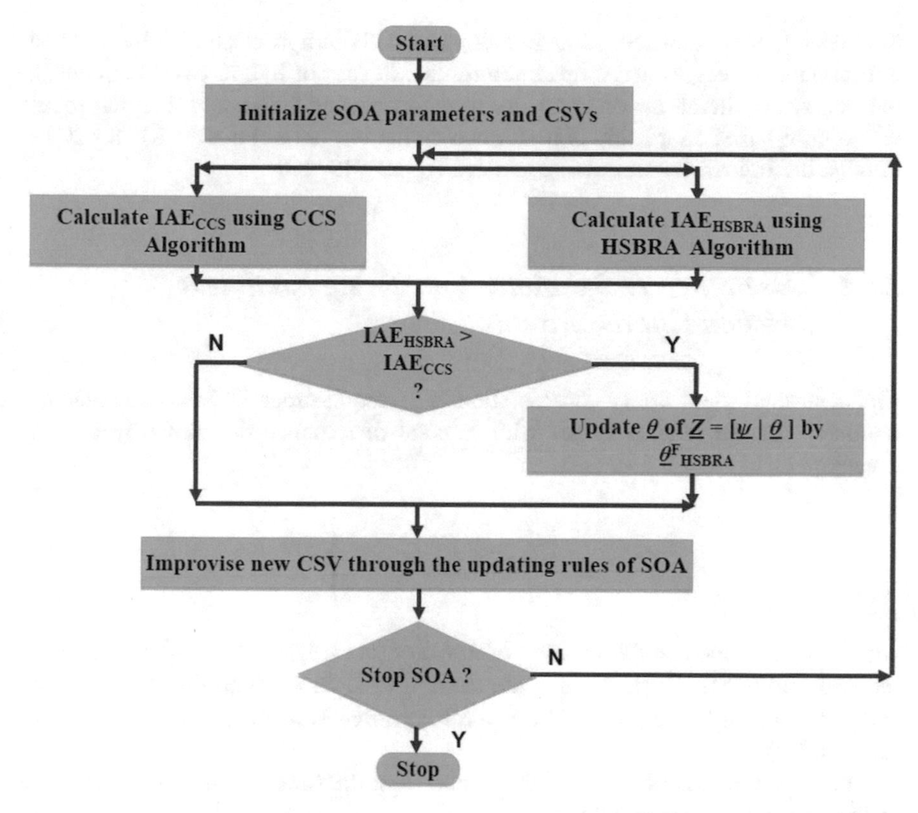

Fig. 7.7 Flowchart representation of robust H^{∞} control-based hybrid preferential algorithm

The flowchart representation of robust H^{∞} control-based hybrid preferential model is shown in Fig. 7.7.

7.3 Case Studies

To demonstrate the effectiveness of the robust hybrid adaptive control schemes, in this section also the same two benchmark case studies are considered, which were previously utilized in other sections of this book. To study the performances in terms of *IAE*, nonlinear systems with fixed structure configurations and a variable structure configuration, with minimum 2 and maximum 7 numbers of MFs for each input, are simulated, respectively. The process models are simulated each using a fixed step fourth-order Runge-Kutta method with sampling time $\Delta t_c = 0.01$ sec. CMA is one of the potential contemporary SOAs employed in this case study to hybridize with robust H^{∞} control-based adaptation strategy. In all simulation case studies, except the hybrid cascade model, the plant is evaluated for 10 s, after the training is completed (200 iterations of CSV improvisations) by different optimization strategies, to compare the results. In case of hybrid cascade model, at the outset, 200 iterations of CMA are used to train the CSVs and then the robust H^{∞} control-based adaptation is performed on the resultant optimal CSV for 200 s. Finally, the trained controller is evaluated for 10 s [9, 10].

7.3.1 *Case Study I: DC Motor Containing Nonlinear Friction Characteristics*

The controlled plant under consideration is a second-order DC motor containing nonlinear friction characteristics with external disturbance described by the following model [10, 12, 14]:

$$\left.\begin{aligned} \dot{x}_1 &= x_2 \\ \dot{x}_2 &= -\tfrac{f(x_2)}{J} + \tfrac{C_T}{J}u + d \\ y &= x_1 \end{aligned}\right\} \tag{7.12}$$

where $y = x_1$ is the angular position of the rotor (in rad), x_2 is the angular speed (in rad/sec), and u is the current fed to the motor (in A). C_T = 10 Nm/A, J = 0.1 kgm^2 and $f(x_2) = 5\tan^{-1}(5x_2)$ Nm. The control objective is to track a reference signal y_m = sin (2*t*).

The control signal obtained as the output from the fuzzy controller, in response to input signals, is given below:

$$u_c = u_c(\underline{x}, w|\underline{\theta}) \tag{7.13}$$

To perform this simulation case study, a population size of ten potential candidate solutions is considered. As in earlier case studies, in a similar fashion, a fixed number of evenly distributed input MFs is used for creating an initial population of CMA, in fixed structure controller design. In variable structure controller design, the structural flags are utilized to automatically vary the structure of the candidate controller in each iteration, as detailed in Sect. 3.3 of this book. Further, the free parameters of the controller and the positions of the output singletons are tuned by the CMA algorithm to minimize the tracking error. Here also, twenty simulations are performed to maintain and assess the stochastic behavior of the optimization process [1, 2].

The cascade model, the concurrent model, and the preferential model of hybrid adaptive fuzzy controllers are implemented in a similar fashion as discussed earlier in Sect. 5.2 of this book. The difference with the previous discussions is that, here, the H^{∞}-based robust adaptation strategy and the CMA-based global optimization technique have been hybridized to achieve the desired control objectives.

The results of simulation studies are tabulated in Table 7.1. The simulations are carried out for three fixed structures of 3 × 3, 5 × 5, and 7 × 7 input MFs and a

Table 7.1 Comparison of simulation results for case study I (Nonlinear DC motor system): CMA-based hybrid design strategies

Control strategy	No. of MFs	No. of rules	Avg. IAE	Best IAE	Std. dev.
CMABA	3 × 3	9	0.9484	0.8241	0.1715
CMA-Hybd-Cas			1.3230	1.2140	0.1733
CMA-Hybd-Con			1.2569	1.1917	0.0948
CMA-Hybd-Pref			0.8252	0.7275	0.0792
CMABA	5 × 5	25	0.6587	0.5857	0.0901
CMA-Hybd-Cas			0.9425	0.8551	0.0639
CMA-Hybd-Con			0.7566	0.6744	0.0616
CMA-Hybd-Pref			0.6826	0.5491	0.1164
CMABA	7 × 7	49	0.7280	0.5716	0.1714
CMA-Hybd-Cas			1.2125	1.1217	0.0836
CMA-Hybd-Con			1.1860	0.9728	0.2508
CMA-Hybd-Pref			0.7079	0.6015	0.0980
CMABA-V	4 × 4	16	–	0.6031	–
CMA-Hybd-Cas-V	4 × 4	16	–	0.7032	–
CMA-Hybd-Con-V	5 × 2	10	–	0.7193	–
CMA-Hybd-Pref-V	3 × 5	15	–	0.8089	–

variable structure with a potentially biggest possible structure of 7 × 7 MFs. The temporal variations of system responses with the fixed structure controllers of 5 × 5 input MFs for CMA-based approach (CMABA), CMA-Hybd-Cas model, CMA-Hybd-Con model, and CMA-Hybd-Pref model are shown in Figs. 7.8, 7.9, 7.10, and 7.11, respectively. On the other hand, Figs. 7.12, 7.13, 7.14, and 7.15 depict the system performances of the same control schemes, with the variable structure input MFs, in a very similar fashion as the cases presented in Chap. 5 of this book [10].

The noteworthy issue in this case study is that the performance of robust adaptation strategy-based cascade hybrid model is quite reasonable and very much comparable with its other counter parts like concurrent model and preferential model as evident from the results presented in Table 7.1. From the evaluation period transient responses, as depicted in figures, it is observed that the tracking performance is much smoother and less oscillatory in case of cascade hybrid model. But, on the whole, the performance of preferential hybrid model is superior in different configurations of hybrid AFLCs.

7.3.2 *Case Study II: Duffing's Oscillatory System*

In this case study, the Duffing's forced oscillation system with disturbance is considered and is given as [10, 12, 15]:

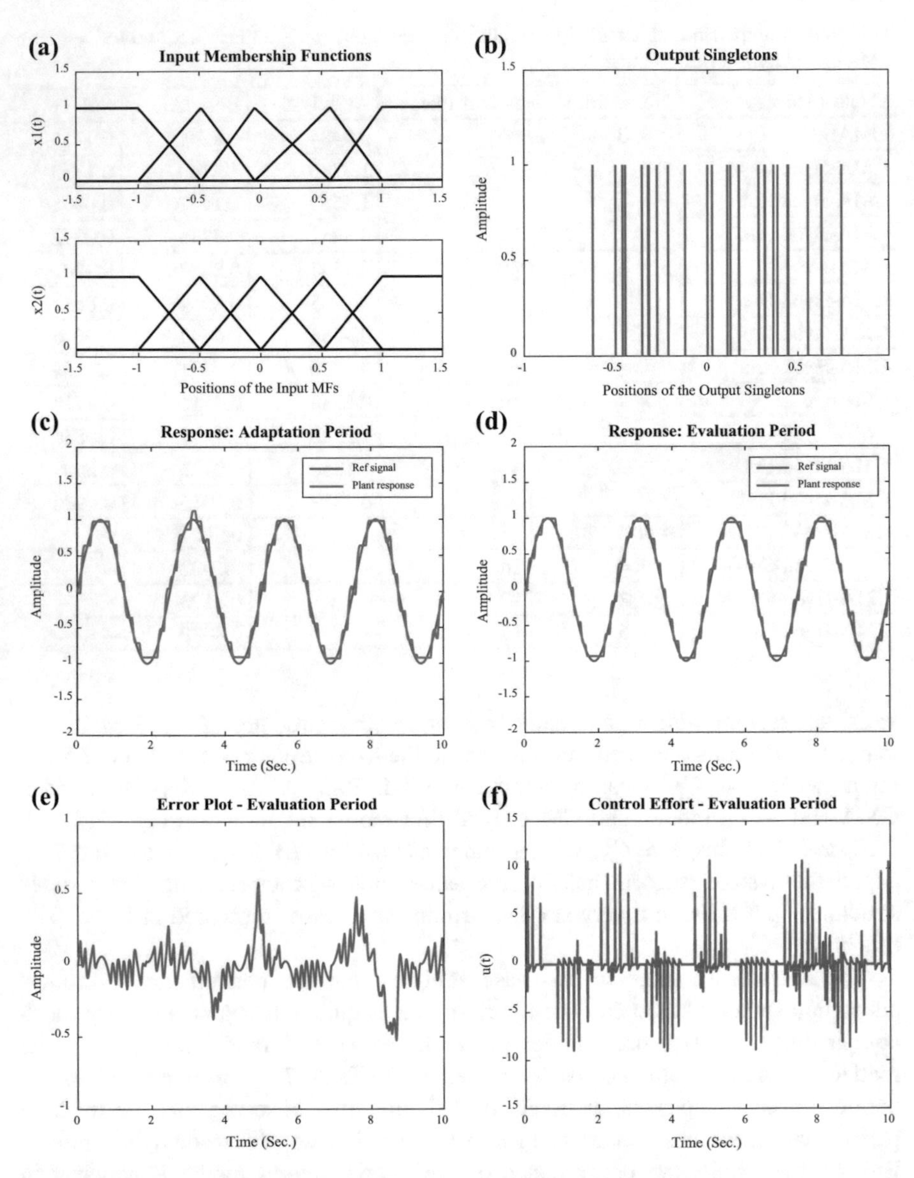

Fig. 7.8 Responses of the adaptive fuzzy controller with CMABA model, for 5 × 5 configuration of input MFs, of case study I (Nonlinear DC motor system): **a** positions of input MFs, **b** positions of output singletons, **c** adaptation period response after 100 CMA generations, **d** evaluation period response for 0–10 s after 200 CMA generations, **e** evaluation period error variation, and **f** evaluation period control effort variation

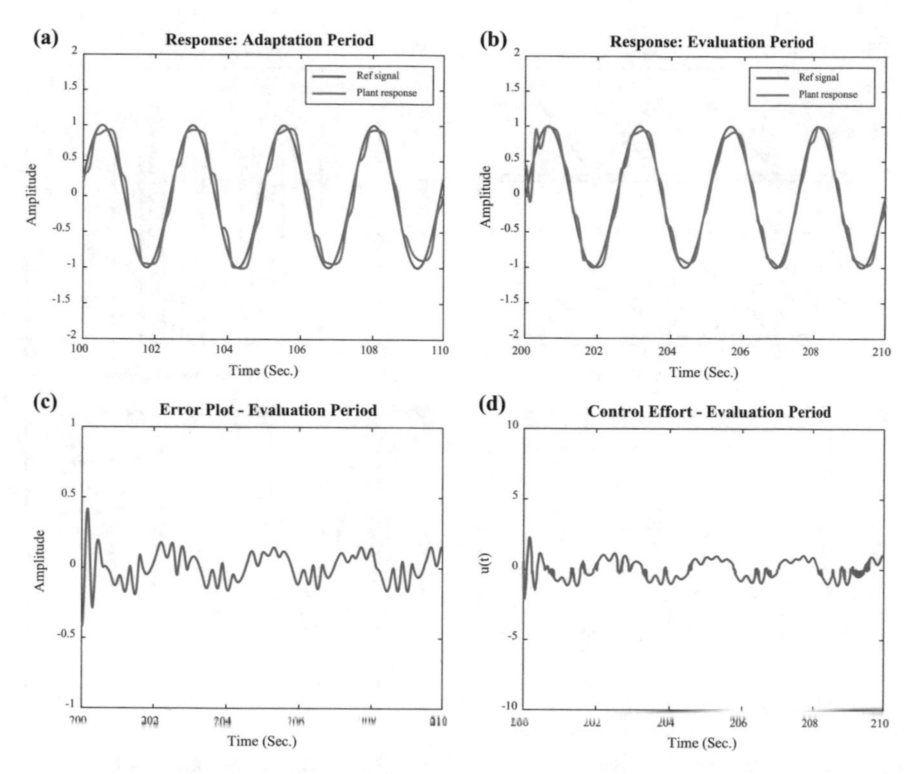

Fig. 7.9 Responses of the adaptive fuzzy controller with CMA-Hybd-Cas model, for 5 × 5 configuration of input MFs, of case study I (Nonlinear DC motor system): **a** adaptation period response after 100 s, **b** evaluation period response for 200–210 s, **c** evaluation period error variation, **d** evaluation period control effort variation

$$\left.\begin{array}{l}\dot{x}_1 = x_2 \\ \dot{x}_2 = -0.1x_2 - x_1^3 + 12\cos(t) + u(t) + d \\ y = x_1\end{array}\right\} \tag{7.14}$$

The control objective is to track the reference signal y_m, where $y_m = \sin(2t)$. The control signal obtained from the fuzzy controller is given as:

$$u_c = u_c(\underline{x}|\underline{\theta}) \tag{7.15}$$

The design parameters for this simulation case study are chosen in a similar fashion as was chosen in case study II in Chaps. 4 and 6. The hybrid design strategies, as well as the CMA-based AFLC design approach, are simulated for different fixed MF configurations and a variable MF configuration in a similar manner as in case study I. The Duffing's system is a highly nonlinear system and it shows chaotic behavior if left unforced. Now, in this case study a randomly variable disturbance signal is applied to the Duffing's system and that makes the situation

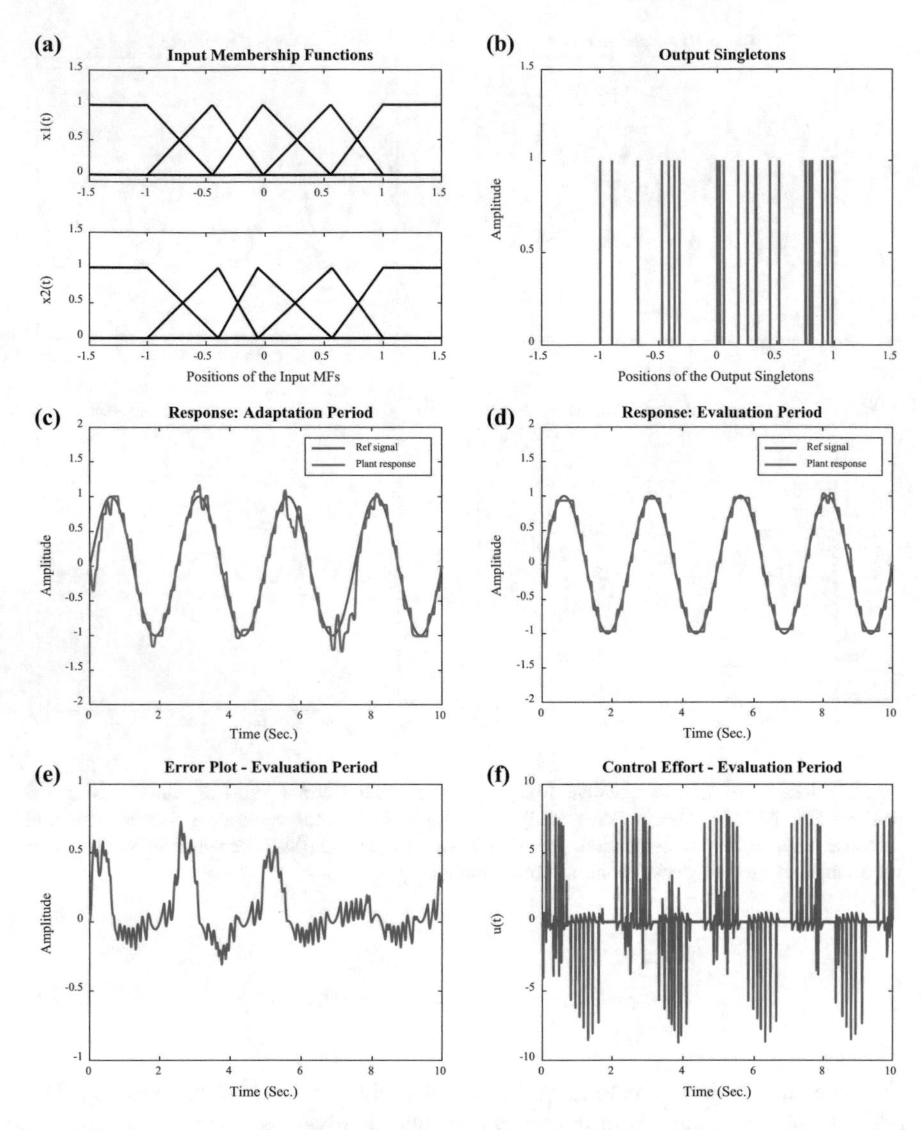

Fig. 7.10 Responses of the adaptive fuzzy controller with CMA-Hybd-Con model, for 5×5 configuration of input MFs, of case study I (Nonlinear DC motor system): **a** positions of input MFs, **b** positions of output singletons, **c** adaptation period response after 100 CMA generations, **d** evaluation period response for 0–10 s after 200 CMA generations, **e** evaluation period error variation, **f** evaluation period control effort variation

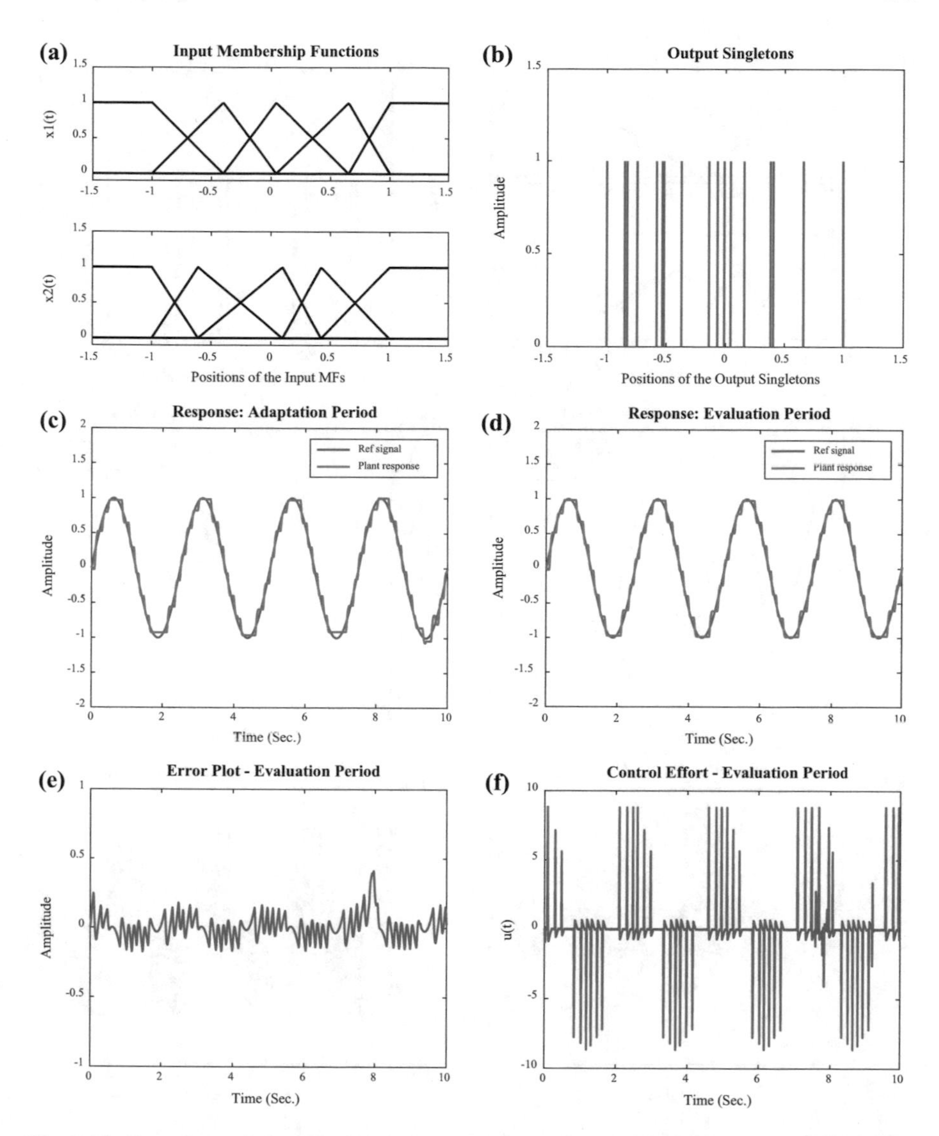

Fig. 7.11 Responses of the adaptive fuzzy controller with CMA-Hybd-Pref model, for 5×5 configuration of input MFs, of case study I (Nonlinear DC motor system): **a** positions of input MFs, **b** positions of output singletons, **c** adaptation period response after 100 CMA generations, **d** evaluation period response for 0–10 s after 200 CMA generations, **e** evaluation period error variation, **f** evaluation period control effort variation

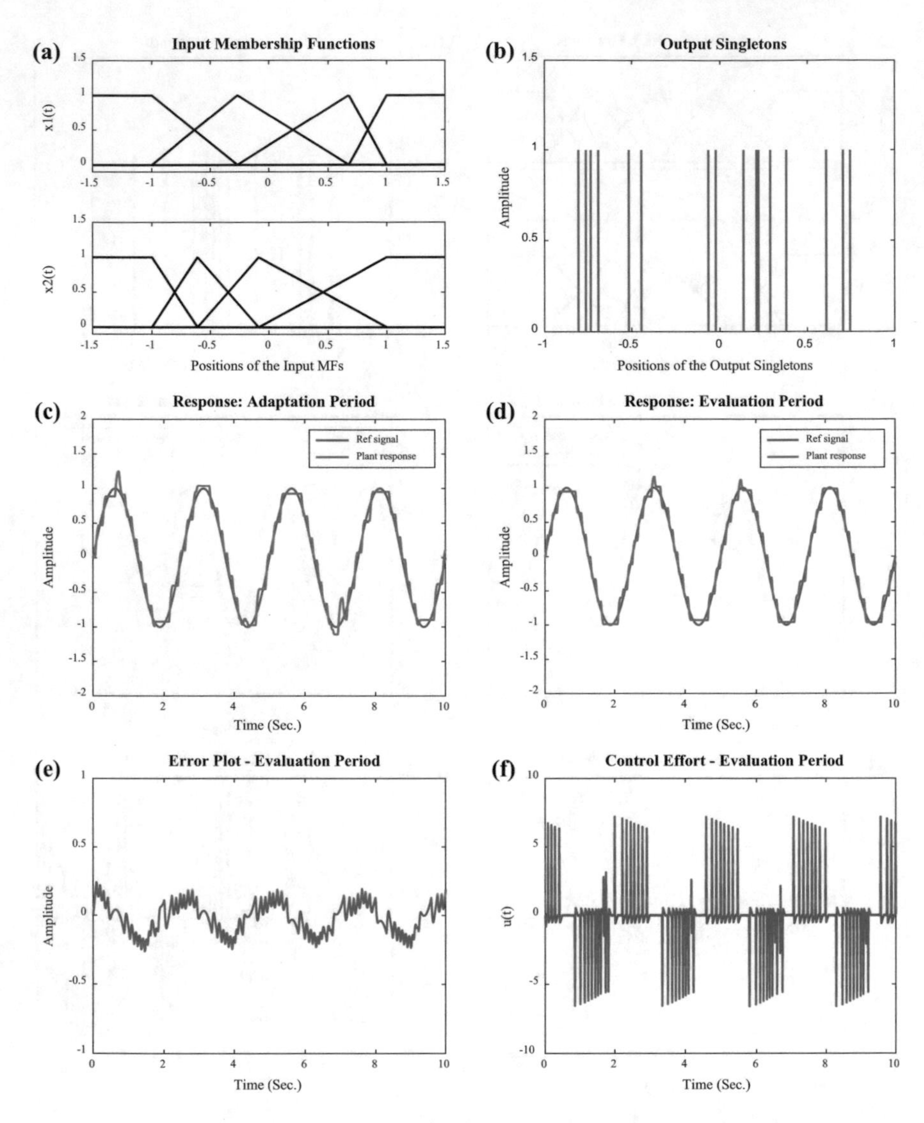

Fig. 7.12 Responses of the adaptive fuzzy controller with CMABA model, for variable structure configuration of input MFs, of case study I (Nonlinear DC motor system): **a** positions of input MFs, **b** positions of output singletons, **c** adaptation period response after 100 CMA generations, **d** evaluation period response for 0–10 s after 200 CMA generations, **e** evaluation period error variation, **f** evaluation period control effort variation

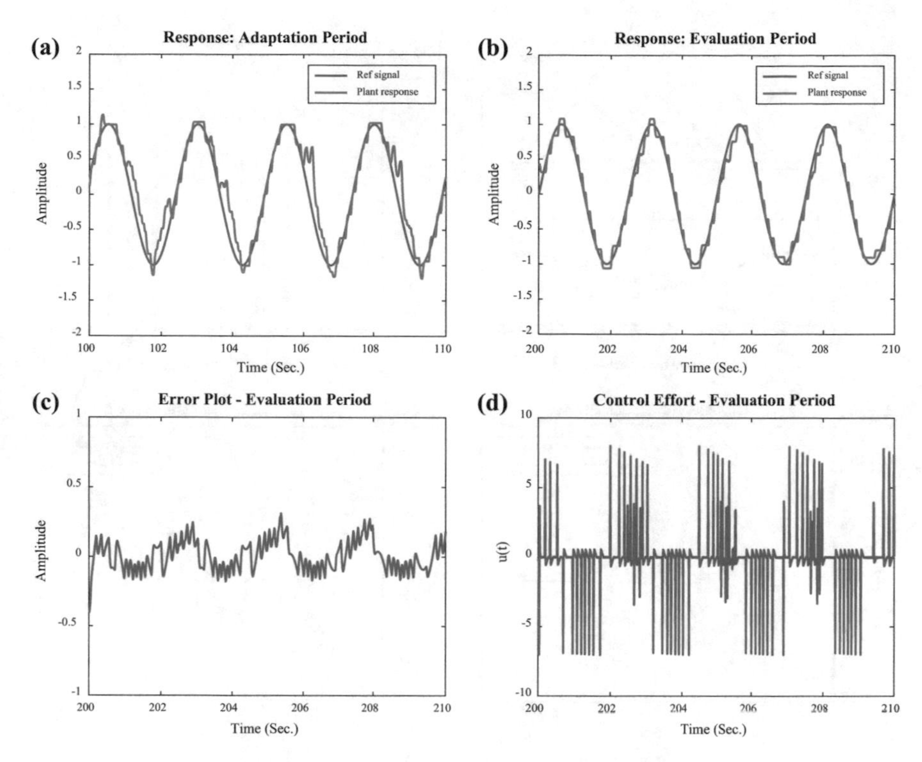

Fig. 7.13 Responses of the adaptive fuzzy controller with CMA-Hybd-Cas model, for variable structure configuration of input MFs, of case study I (Nonlinear DC motor system): **a** adaptation period response after 100 s, **b** evaluation period response for 200–210 s, **c** evaluation period error variation, **d** evaluation period control effort variation

critical. The performances of the hybrid controllers are tabulated in Table 7.2 and corresponding graphical representations are shown in Figs. 7.16, 7.17, 7.18, 7.19, 7.20, 7.21, 7.22, and 7.23.

Although the adaptation gain ν is set to a small value of 0.0001 for adaptation period, in the case of the cascade hybrid model, the performance becomes worst compared to CMABA control strategy. Thus, it can be inferred that further tuning of the output singletons after having an optimal/near-optimal solutions from CMABA strategy, will probably lead the solution vector away from the optimal

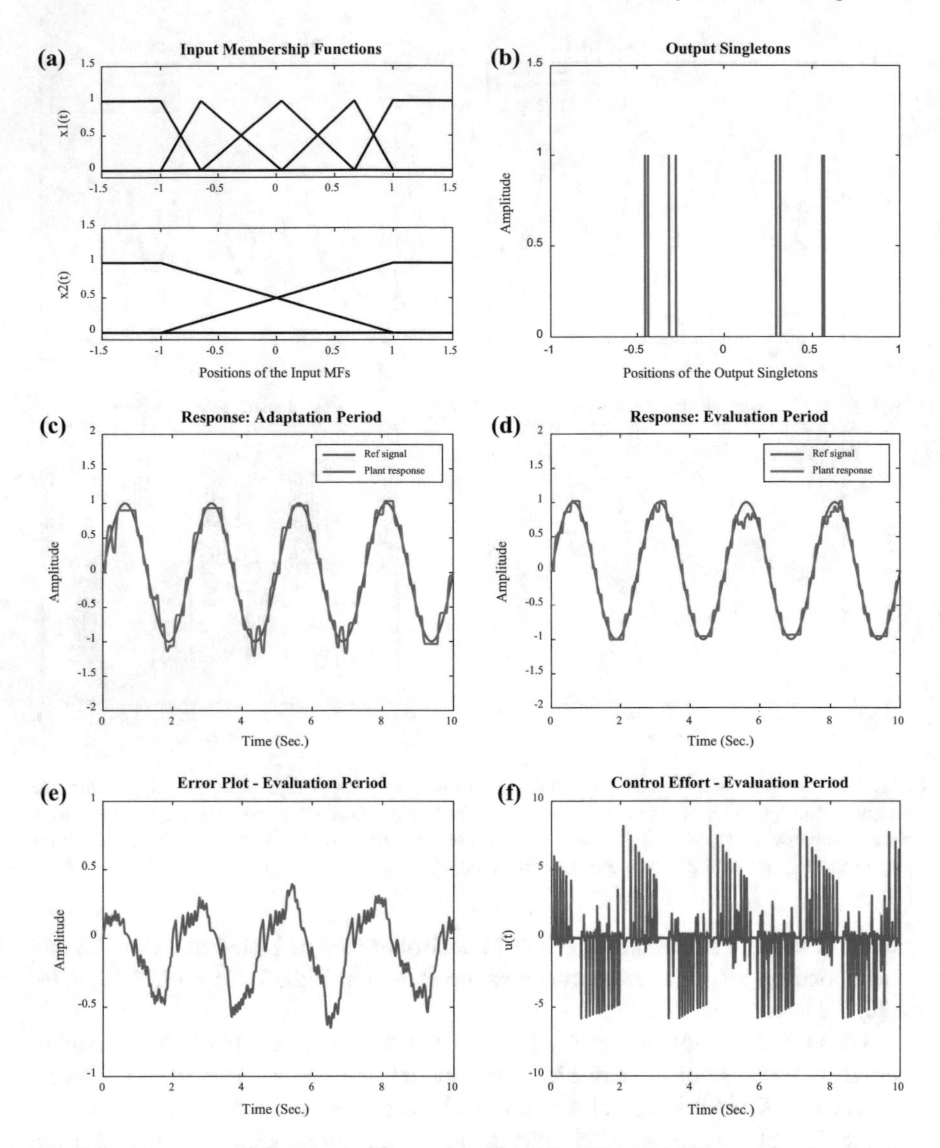

Fig. 7.14 Responses of the adaptive fuzzy controller with CMA-Hybd-Con model, for variable structure configuration of input MFs, of case study I (Nonlinear DC motor system): **a** positions of input MFs, **b** positions of output singletons, **c** adaptation period response after 100 CMA generations, **d** evaluation period response for 0–10 s after 200 CMA generations, **e** evaluation period error variation, **f** evaluation period control effort variation

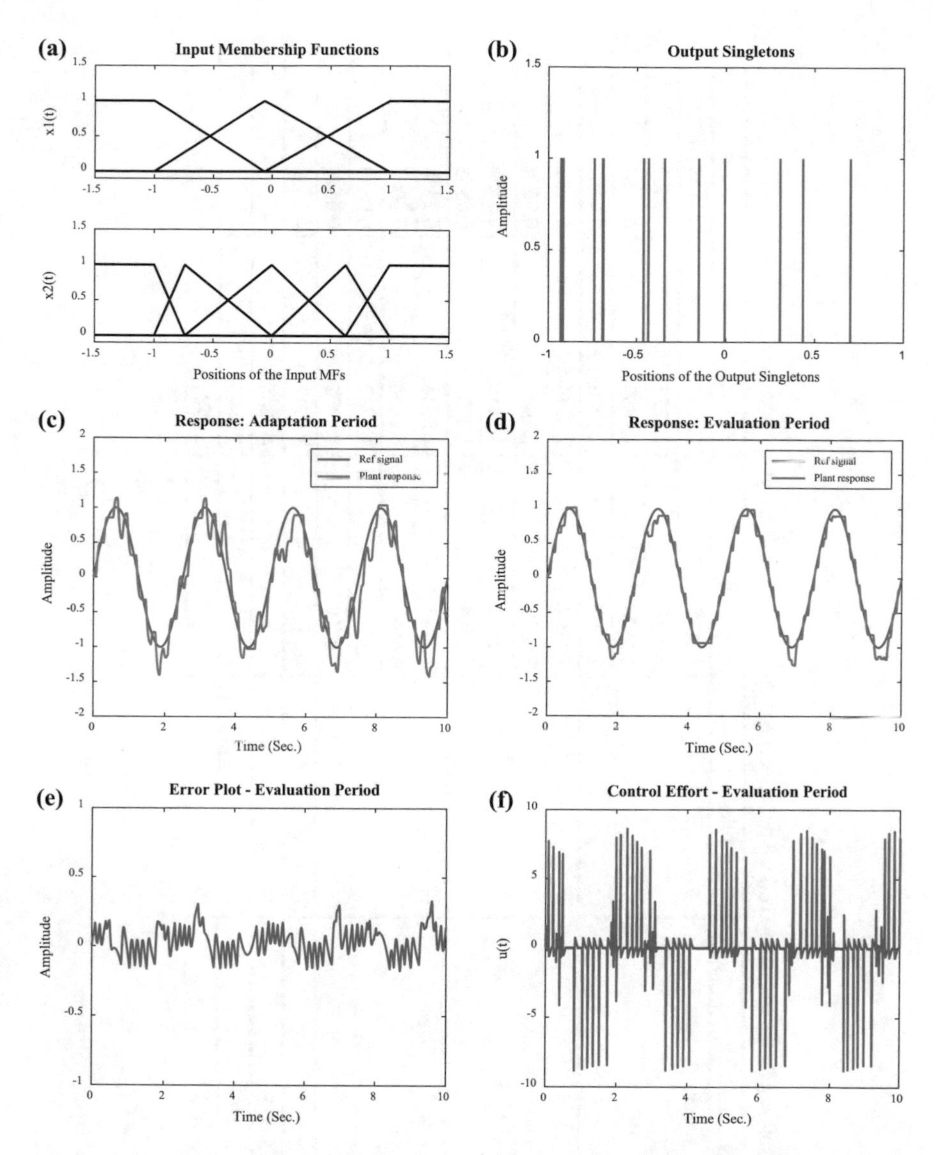

Fig. 7.15 Responses of the adaptive fuzzy controller with CMA-Hybd-Pref model, for variable structure configuration of input MFs, of case study I (Nonlinear DC motor system): **a** positions of input MFs, **b** positions of output singletons, **c** adaptation period response after 100 CMA generations, **d** evaluation period response for 0–10 s after 200 CMA generations, **e** evaluation period error variation, **f** evaluation period control effort variation

Table 7.2 Comparison of simulation results for case study II (Duffing's system): CMA-based hybrid design strategies

Control strategy	No. of MFs	No. of rules	Avg. IAE	Best IAE	Std. dev.
CMABA	3 × 3	9	1.3179	1.1291	0.1743
CMA-Hybd-Cas			6.6088	6.2223	0.3604
CMA-Hybd-Con			1.2526	1.1430	0.1557
CMA-Hybd-Pref			1.2002	1.1278	0.0687
CMABA	5 × 5	25	1.4102	1.1164	0.1297
CMA-Hybd-Cas			6.6037	6.1967	0.3674
CMA-Hybd-Con			1.2931	1.1236	0.1356
CMA-Hybd-Pref			1.1296	1.0727	0.0856
CMABA	7 × 7	49	1.3295	1.2238	0.1448
CMA-Hybd-Cas			6.6682	6.5190	0.3003
CMA-Hybd-Con			1.3345	1.2417	0.0882
CMA-Hybd-Pref			1.2514	1.1972	0.0644
CMABD-V	5 × 7	35	–	1.0737	–
CMA-Hybd-Cas-V	5 × 7	35	–	4.3245	–
CMA-Hybd-Con-V	3 × 4	12	–	2.4814	–
CMA-Hybd-Pref-V	3 × 6	18	–	1.5783	–

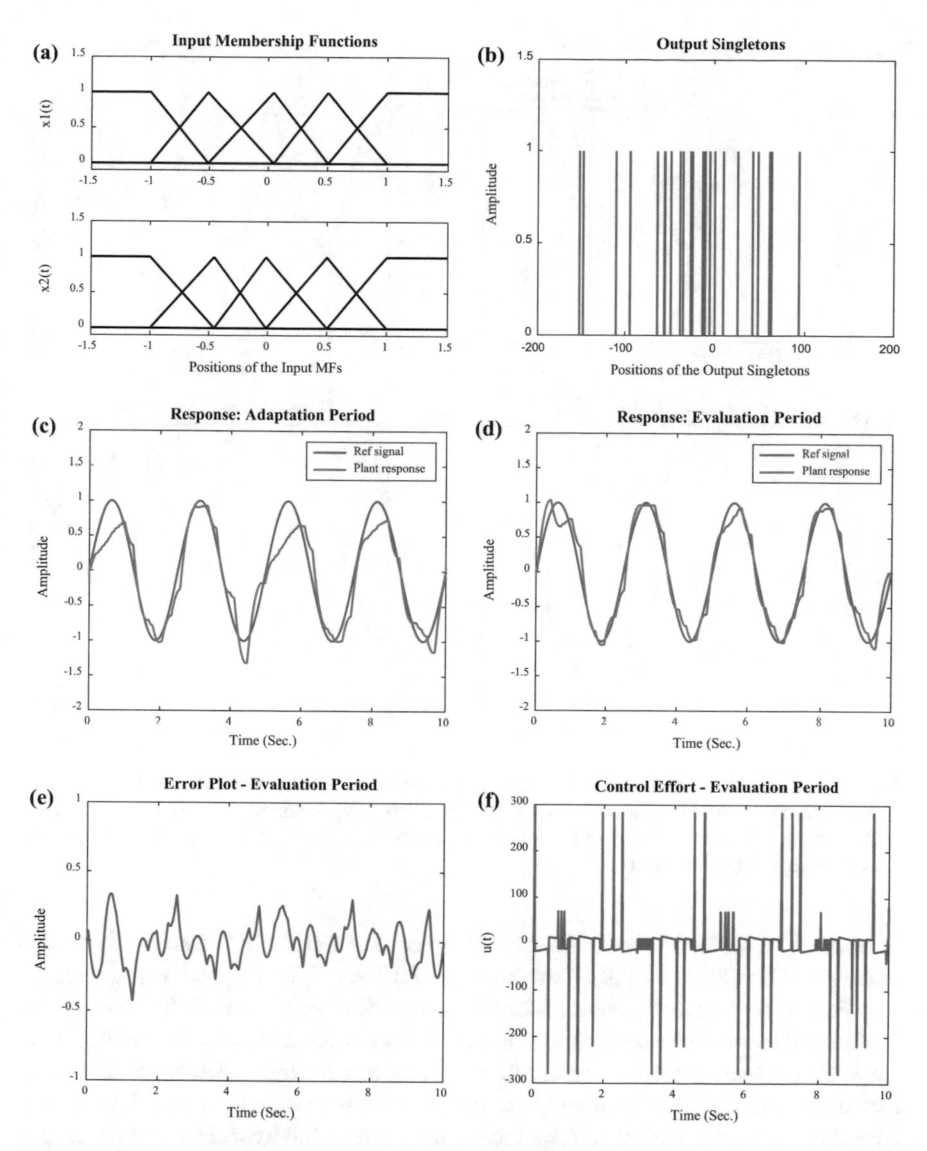

Fig. 7.16 Responses of the adaptive fuzzy controller with CMABA model, for 5×5 configuration of input MFs, of case study II (Duffing's system): **a** positions of input MFs, **b** positions of output singletons, **c** adaptation period response after 100 CMA generations, **d** evaluation period response for 0–10 s after 200 CMA generations, **e** evaluation period error variation, **f** evaluation period control effort variation

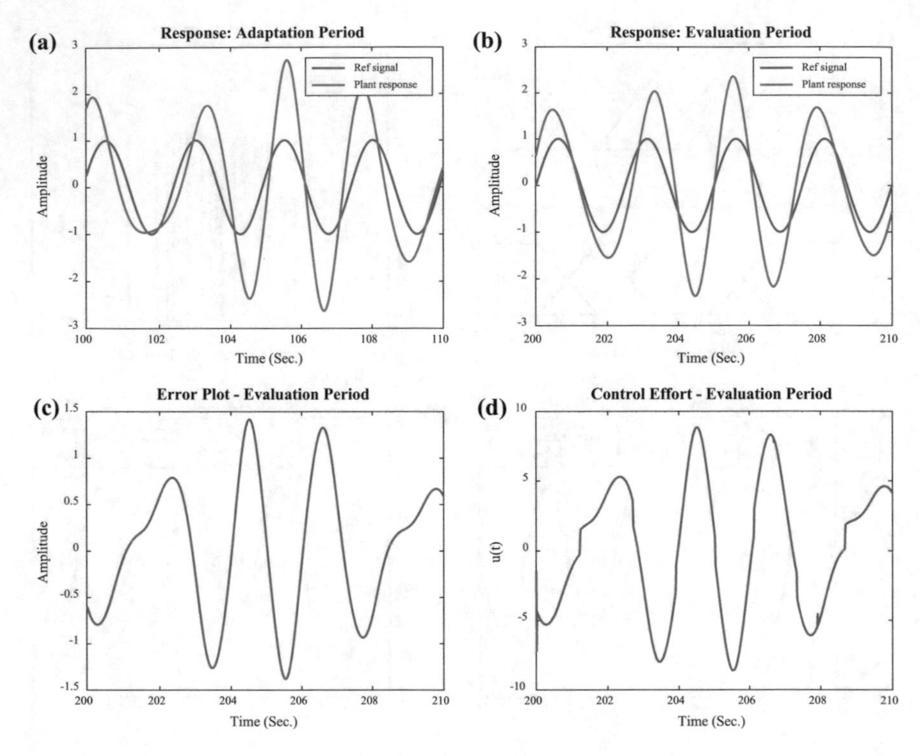

Fig. 7.17 Responses of the adaptive fuzzy controller with CMA-Hybd-Cas model, for 5 × 5 configuration of input MFs, of case study II (Duffing's system): **a** adaptation period response after 100 s, **b** evaluation period response for 200–210 s, **c** evaluation period error variation, **d** evaluation period control effort variation

solution. The performances of the cascade hybrid model for both fixed and variable structure MF configurations, shown in Figs. 7.17 and 7.21, respectively, depict the situation in a graphical manner. Moreover, it is observed from the figures that the performances of the other hybrid controllers like concurrent and preferential one, may be improved further if the number of iterations for the CMA improvisation is increased. But that will incur added computational burden at the same time. In this case study, also the overall performance of the preferential hybrid control strategy is observed to be superior, on the whole.

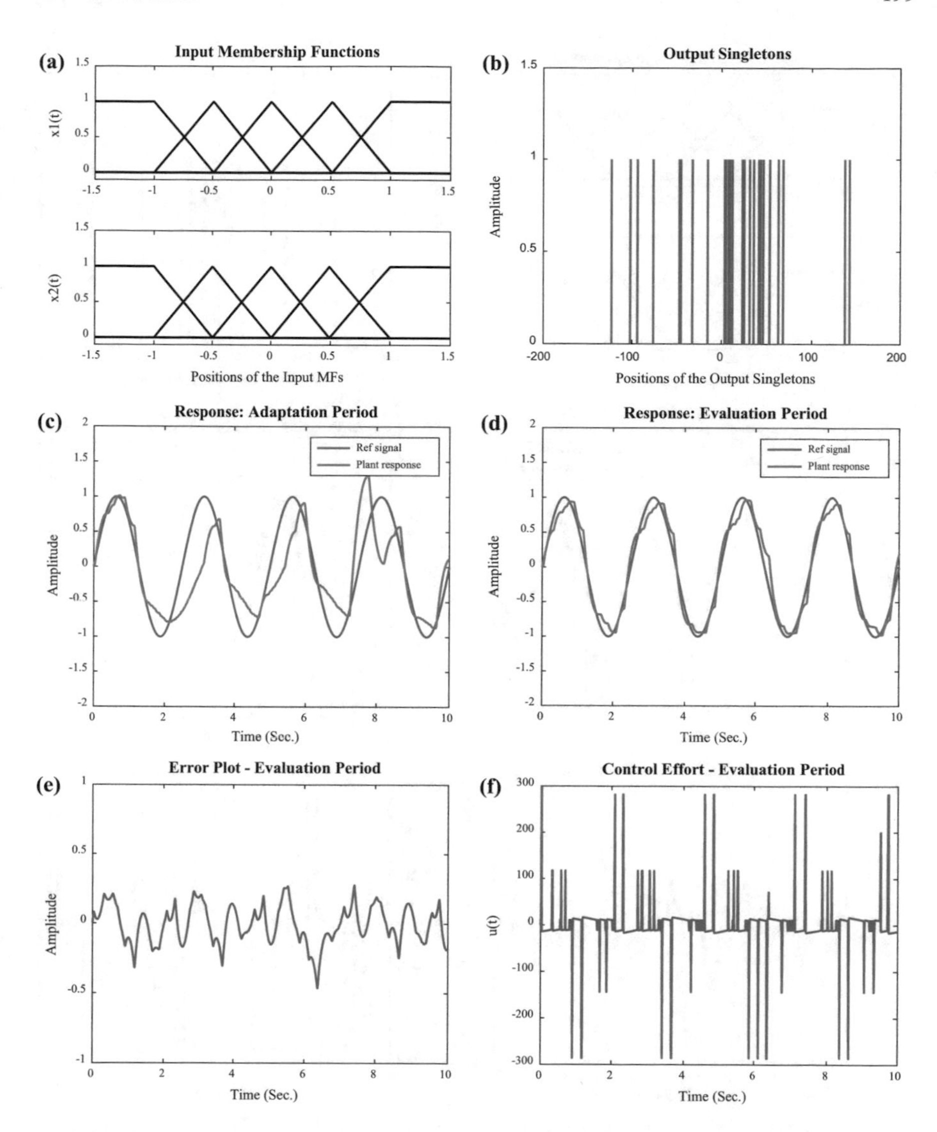

Fig. 7.18 Responses of the adaptive fuzzy controller with CMA-Hybd-Con model, for 5×5 configuration of input MFs, of case study II (Duffing's system): **a** positions of input MFs, **b** positions of output singletons, **c** adaptation period response after 100 CMA generations, **d** evaluation period response for 0–10 s after 200 CMA generations, **e** evaluation period error variation, **f** evaluation period control effort variation

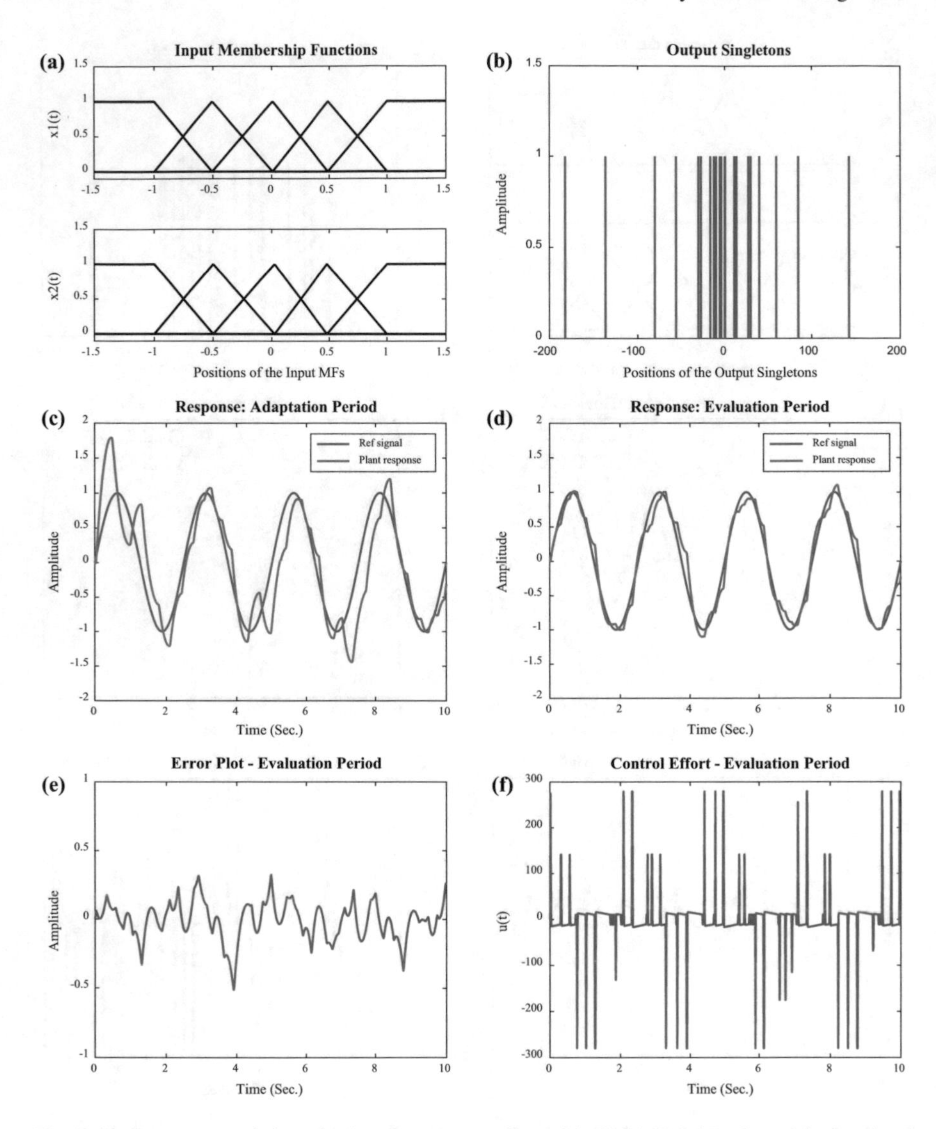

Fig. 7.19 Responses of the adaptive fuzzy controller with CMA-Hybd-Pref model, for 5×5 configuration of input MFs, of case study II (Duffing's system): **a** positions of input MFs, **b** positions of output singletons, **c** adaptation period response after 100 CMA generations, **d** evaluation period response for 0–10 s after 200 CMA generations, **e** evaluation period error variation, **f** evaluation period control effort variation

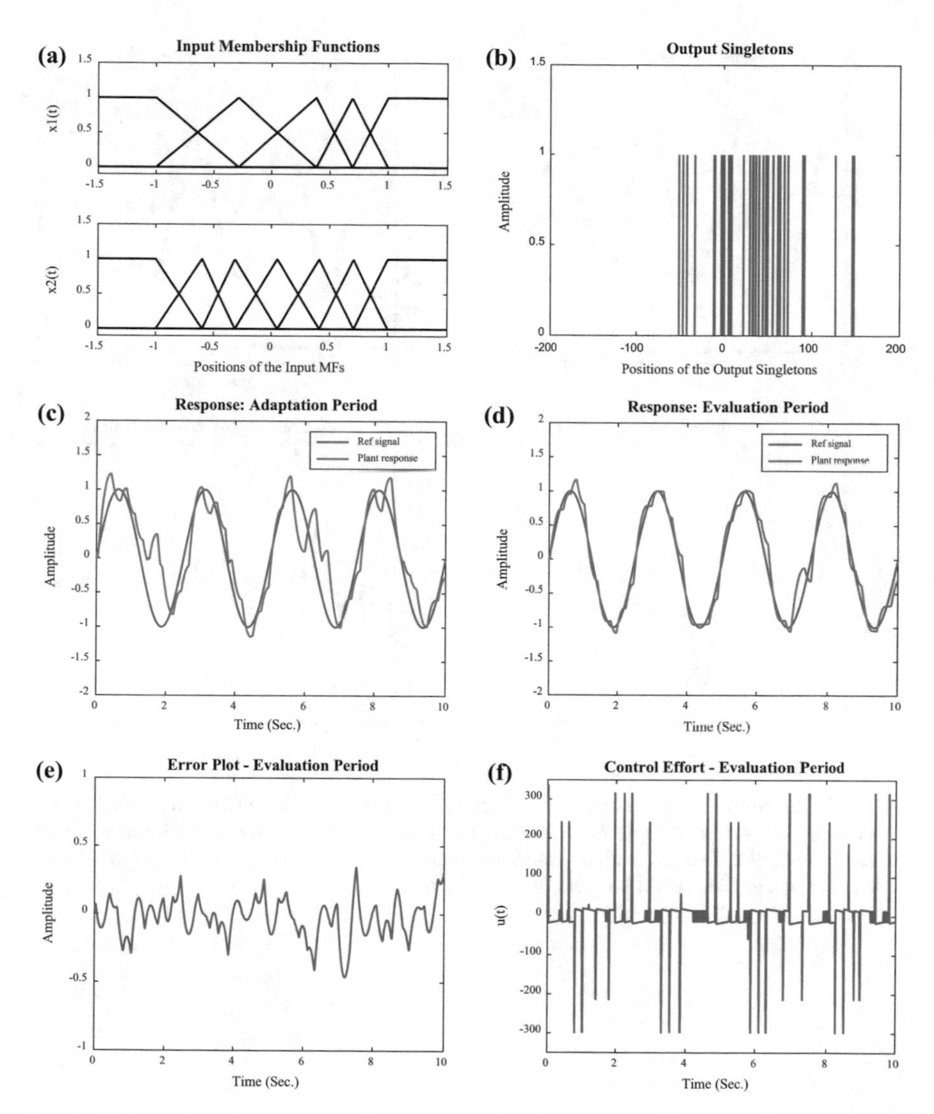

Fig. 7.20 Responses of the adaptive fuzzy controller with CMABA model, for variable structure configuration of input MFs, of case study II (Duffing's system): **a** positions of input MFs, **b** positions of output singletons, **c** adaptation period response after 100 CMA generations, **d** evaluation period response for 0–10 s after 200 CMA generations, **e** evaluation period error variation, **f** evaluation period control effort variation

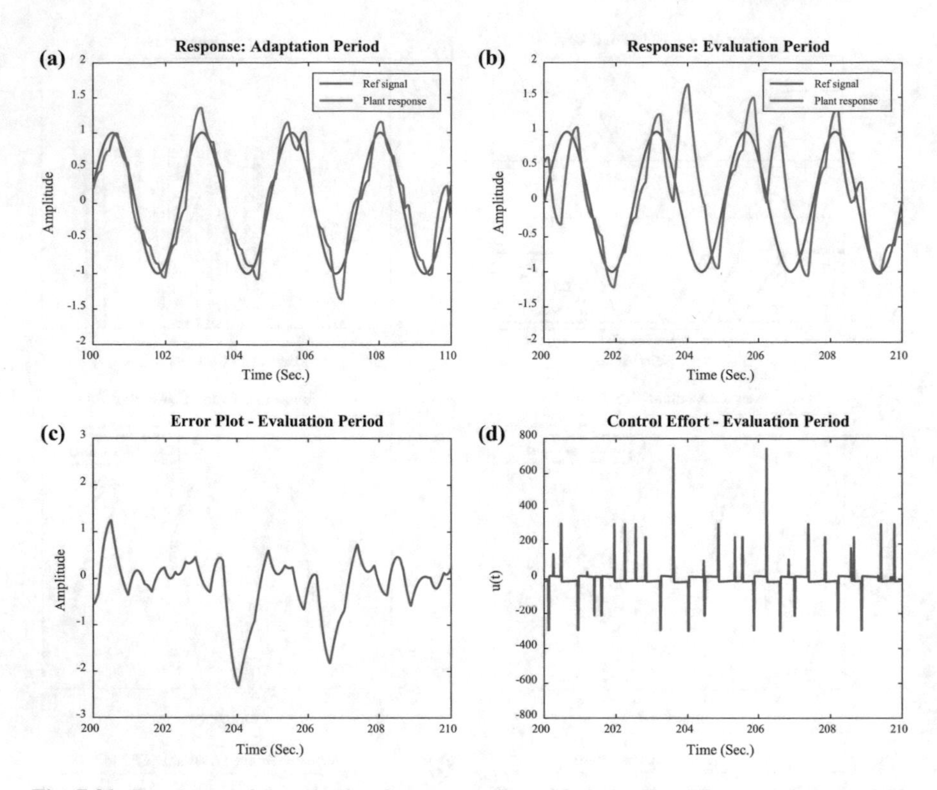

Fig. 7.21 Responses of the adaptive fuzzy controller with CMA-Hybd-Cas model, for variable structure configuration of input MFs, of case study II (Duffing's system): **a** adaptation period response after 100 s, **b** evaluation period response for 200–210 s, **c** evaluation period error variation, **d** evaluation period control effort variation

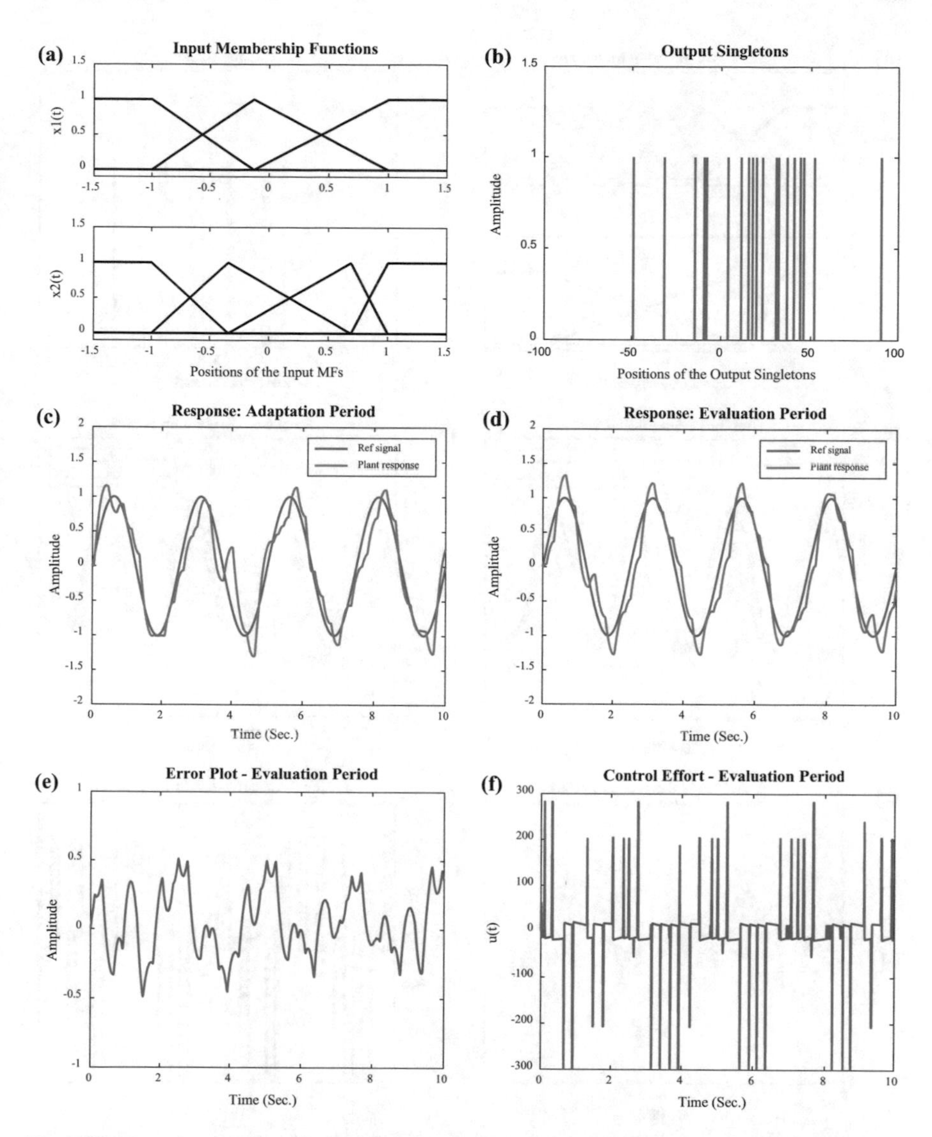

Fig. 7.22 Responses of the adaptive fuzzy controller with CMA-Hybd-Con model, for variable structure configuration of input MFs, of case study II (Duffing's system): **a** positions of input MFs, **b** positions of output singletons, **c** adaptation period response after 100 CMA generations, **d** evaluation period response for 0–10 s after 200 CMA generations, **e** evaluation period error variation, **f** evaluation period control effort variation

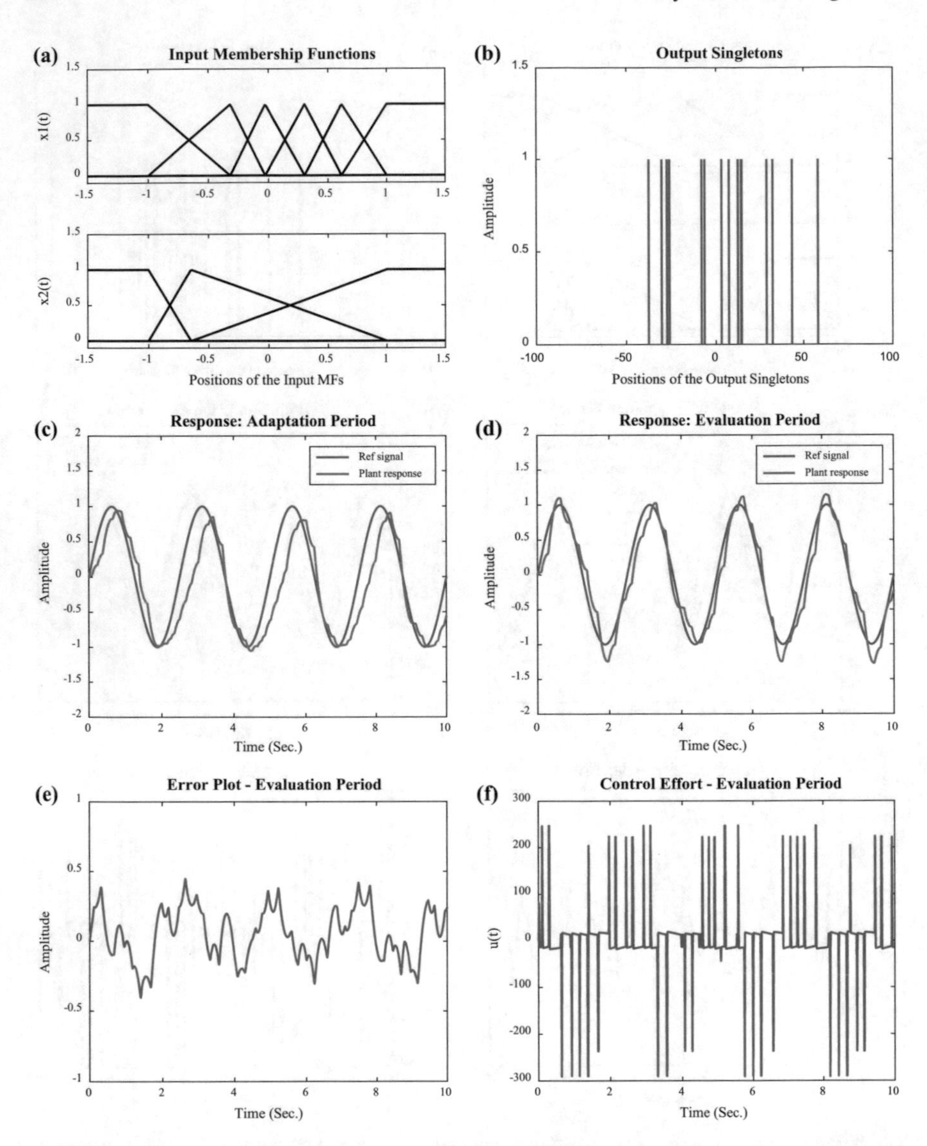

Fig. 7.23 Responses of the adaptive fuzzy controller with CMA-Hybd-Pref model, for variable structure configuration of input MFs, of case study II (Duffing's system): **a** positions of input MFs, **b** positions of output singletons, **c** adaptation period response after 100 CMA generations, **d** evaluation period response for 0–10 s after 200 CMA generations, **e** evaluation period error variation, **f** evaluation period control effort variation

7.4 Summary

In this chapter, H^∞-based robust adaptive control technique has been hybridized with a contemporary stochastic optimization algorithm, namely covariance matrix adaptation, which is popularly called the CMA strategy. Similar to the hybridization schemes described in Chap. 5, here also cascade model, concurrent model, and preferential model of hybrid controllers are designed and implemented for the benchmark case studies with different fixed structure and one variable structure MF configurations. The critical analyses of the controller performances are described in detail.

To further demonstrate the usefulness of the hybrid controllers discussed so far in this book, several real implementations are presented in next few chapters. In the next chapter, a real implementation experiment of controlling the temperature of an air heater system will be described in detail.

References

1. N. Hansen, A. Ostermeier, A. Gawelczyk, On the adaptation of arbitrary normal mutation distributions in evolution strategies: the generating set adaptation, in *Proceedings 6th International Conference on Genetic Algorithms* (San Francisco, USA, 1995), pp. 57–64
2. A. Auger, N. Hansen, A restart CMA evolution strategy with increasing population size, in *Proceedings of the Congress on Evolutionary Computation* (Edinburgh, UK, 2005), pp. 1769–1776
3. P.N. Suganthan, N. Hansen, J.J. Liang, K. Deb, Y.-P.Chen, A. Auger, S. Tiwari, *Problem definitions and evaluation criteria for the cec 2005 special session on real-parameter optimization*. Technical Report (Nanyang Tech. University, Singapore, 2005)
4. B.S. Chen, C.H. Lee, Y.C. Chang, H^∞ tracking design of uncertain nonlinear SISO systems: adaptive fuzzy approach. IEEE Trans. Fuzzy Syst. **4**(1), 32–43 (1996)
5. B.S. Chen, C.L. Tsai, D.S. Chen, Robust H-infinity and mixed H^2/H^∞ filters for equalization designs of nonlinear communication systems: fuzzy interpolation approach. IEEE Trans. Fuzzy Syst. **11**(3), 384–398 (2003)
6. H.-X. Li, S. Tong, A hybrid adaptive fuzzy control for a class of nonlinear MIMO system. IEEE Trans. Fuzzy Syst. **11**(1), 24–34 (2003)
7. X. Zhengrogn, C. Qingwei, H. Weili, Robust H^∞ control for fuzzy descriptor systems with state delay. IEEE J. Syst. Eng. Electron. **16**(1), 98–101 (2005)
8. H. Yishao, Z. Dequn, C. Xiaoxui, D. Lin, H^∞ tracking design for a class of decentralized nonlinear systems via fuzzy adaptive observer. IEEE J. Syst. Eng. Electron. **20**(4), 790–799 (2009)
9. K. Das Sharma, A. Chatterjee, P. Siarry, A. Rakshit, CMA—H^∞ hybrid design of robust stable adaptive fuzzy controllers for non-linear systems, in *Proceedings of 1st International Conference on Frontiers in Optimization: Theory and Applications (FOTA-2016)*, (Kolkata, India, 2016 in press)
10. K. Das Sharma, *Hybrid Methodologies for Stable Adaptive Fuzzy Control*. Ph.D. dissertation, Jadavpur University, 2012
11. K. Das Sharma, A. Chatterjee, F. Matsuno, A Lyapunov theory and stochastic optimization based stable adaptive fuzzy control methodology, in *Proceedings of SICE Annual Conference 2008* (Japan, 2008), Aug 20–22, pp. 1839–1844

12. K. Das Sharma, A. Chatterjee, A. Rakshit, A hybrid approach for design of stable adaptive fuzzy controllers employing Lyapunov theory and particle swarm optimization. IEEE Trans. Fuzzy Syst. **17**(2), 329–342 (2009)
13. K. Das Sharma, A. Chatterjee, A. Rakshit, Design of a hybrid stable adaptive fuzzy controller employing Lyapunov theory and harmony search algorithm. IEEE Trans. Contr. Syst. Tech. **18**(6), 1440–1447 (2010)
14. K. Fischle, D. Schroder, An improved stable adaptive fuzzy control method. IEEE Trans. Fuzzy Syst. **7**(1), 27–40 (1999)
15. L.-X. Wang, *A Course in Fuzzy Systems and Control* (Englewood Cliffs, Prentice Hall, NJ, 1997)

Part IV
Applications

Chapter 8
Experimental Study I: Temperature Control of an Air Heater System with Transportation Delay

8.1 Introduction

In industrial thermal processes, sometimes it is a difficult task to place the temperature sensor at a location close to the heating element, such as in different furnaces [1–5]. Thus, due to the physical separation between the sensors and the heating element, a transport delay is introduced in the thermal process. Along with this difficulty, the system may suffer from significant external disturbances which may primarily arise because of the presence of one or more blower(s) in the process, which can effectively change the hot air mass flow rate. The temperature controller, employed to control the temperature of the process, at different locations, should be able to cope with such variations and unwanted or unforeseeable situations. Researchers, during the last two decades, have continuously demonstrated that the conventional PID-type controllers are not suitable for these types of applications [1, 6–8] and several research works have corroborated the fact that fuzzy logic controllers (FLCs) possess tremendous potential to find their applications in these fields [9–11]. Several researchers have also been motivated by the hybrid design methodologies combining fuzzy control theory with genetic algorithm to accommodate the complexities of the system to be controlled [12–17]. In this chapter, discussions are presented about a small scale, real version of a typical industrial thermal process which is developed in our Electrical Measurement and Instrumentation Laboratory, Electrical Engineering Department, Jadavpur University, Kolkata, India, to emulate an air heater thermal process. This laboratory scale, experimental setup has been developed to test the suitability of our designed stable adaptive fuzzy control methodologies, when implemented in real practice. Typically, such a process can be modeled as a first-order linear system, but the situation gets complicated because of the variable heat transfer dynamics which arises due to variations in speed of the blower fan, fitted at one end of the air flow duct [18]. The air mass flow rate is primarily dependent on the blower speed; thus, a change in that speed effectively acts as a load disturbance, influencing the heat transfer dynamics of the air heater process. In this chapter, elaborate discussions are presented about our

K. Das Sharma et al., *Intelligent Control*, Cognitive Intelligence and Robotics,
https://doi.org/10.1007/978-981-13-1298-4_8

designed hybrid stable adaptive fuzzy controllers (AFLCs), which combine the good features of Lyapunov theory [13] and stochastic optimization algorithm-based strategies to evolve superior control alternatives [14, 19–21], when applied to effectively control the temperature of this air heater process at different cross sections of the air flow duct. These proposed hybrid controllers can effectively control the temperature of a process with such variable heat transfer dynamics and the same controllers are also able to cope with the transportation delay inherently present in the process due to the physical separation between the heating element and temperature sensors.

As in previous chapters, in this chapter too, representative contemporary stochastic optimization techniques, i.e., particle swarm optimization (PSO) and harmony search (HS) algorithms [22–25], are individually employed and hybridized with Lyapunov theory to design our proposed stable AFLCs. The Lyapunov theory provides a local adaptation feature to the fuzzy controllers and ensures the stability of the controller. On the other hand, a multi-agent stochastic optimization algorithm is adept in providing good global-search capability, required to provide high degree of automation in the controller design strategies. Two variants of these contemporary hybrid design strategies, namely concurrent and preferential models, as discussed in previous chapters, are utilized for controlling the temperature of the air heater process. The motivation of this research was to design a stable fuzzy controller in off-line manner, considering the real-life operational environment of the system, and then apply the same for real-life operation. In this present work, at the onset, the process parameters are estimated with real-time input–output data. Then, utilizing this information, the fuzzy controller's structure and free parameters are obtained using stochastic optimization algorithm and LSBA approach, which, in turn, guaranteed the stability of the designed controller. The performances of the controllers are first validated in simulation. Finally, the designed controllers have been implemented in real-life laboratory experiments for our indigenously built laboratory-scale air heater system and the satisfactory performances are demonstrated, for both with transport-delay and without transport delay situations, and, also, in the presence of fixed and randomly varying load disturbances.

8.2 Hardware Description of the Experimental Platform

The experimental platform for this practical experiment for controlling the temperature of an air heater system and employing stable AFLC consists of a laboratory-scale hardware system, indigenously developed in our laboratory, as mentioned before, interfaced to a PC through USB port. Figure 8.1 shows the experimental setup under consideration where the front view, right side view and left side view of the setup are shown in (a), (b), and (c) of Fig. 8.1, respectively. The experimental arrangement draws air from the atmosphere using a blower fan, the air is then heated as it passes over a heater coil, and then this air is released into the atmosphere through a duct. The control objective is to maintain the temperature of the air at a desired level. Temperature control is achieved by varying the applied

Fig. 8.1 Air heater system experimental setup **a** front view, **b** right side view, and **c** left side view

(a)

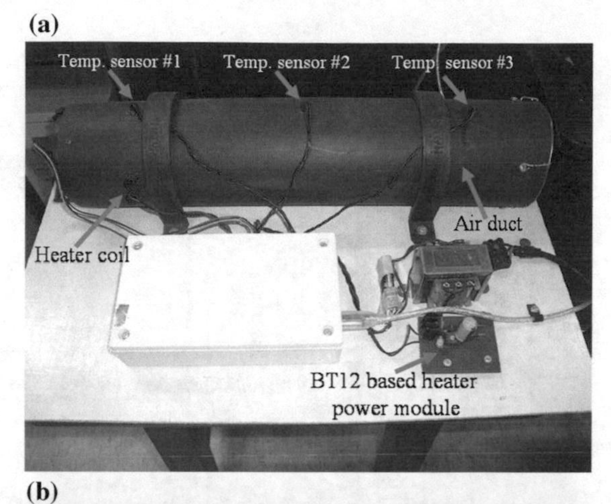

(b)

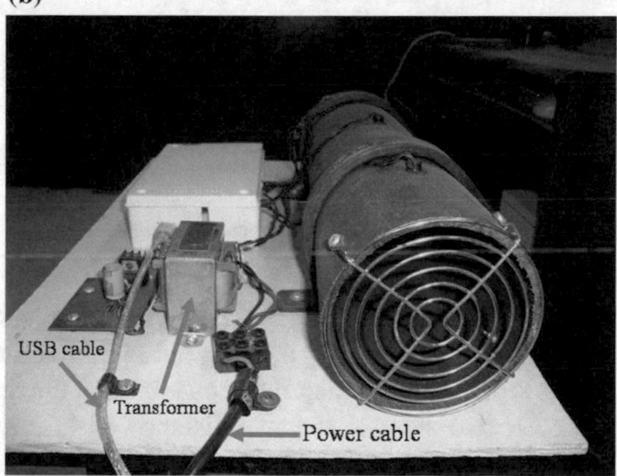

(c)

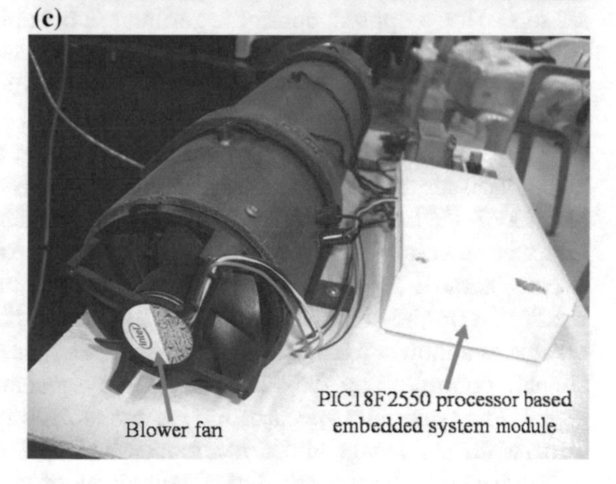

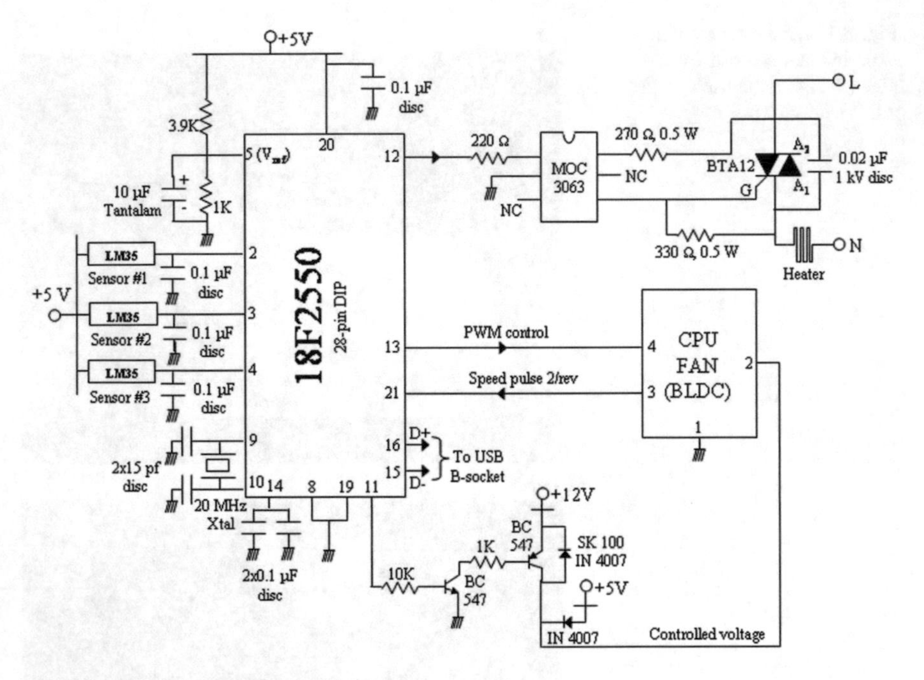

Fig. 8.2 Circuit layout of the air heater system

voltage to the heater coil. The air temperature is sensed by using LM35 temperature sensors [26], placed along the path of the flow at different positions of the duct. The spatial separation between the sensor #1 and the heater coil is negligibly small but the other two sensors are at a distance of 15 and 30 cm, respectively, from the heater coil, and that introduces a transport delay into the system for those two sensors. The proposed control algorithms are implemented in Visual Basic on standard PC and the PC communicates with the designed air heater system which comprises an embedded system with 28-pin DIP PIC18F2550 processor, through USB link [27]. The power module of the heater coil consists of a BTA12 triac [28]. The gate terminal of the BTA12 is triggered by the control signal output from the PIC processor according to the controller decision through an opto-isolator MOC3063 [29]. An interrupt-driven software is developed for this sophisticated temperature control scheme which has a sampling rate of 10 ms for each PWM cycle generated for control action. The controller output is generated to suitably modify the on-time of the duty cycle of PWM. A 4-pin BLDC-type CPU fan [30] is utilized as a blower for the air heater system, and the speed of the blower fan is also suitably controlled by the PIC processor by generating a suitable PWM control signal. The maximum speed of the blower is 2800 rpm and that can be varied, in terms of the percentage of the maximum speed, during the experiment which acts as a disturbance to the process. The circuit layout of the air heater system is shown in detail in Fig. 8.2.

8.3 System Identification of the Air Heater System

The design technique of a stable adaptive fuzzy controller, for controlling the temperature of air at different points along the cross section of the duct in the air heater system, requires the knowledge of the nominal parameters of the process for tuning the controller. Thus, before designing the controllers, the parameters of the air heating process are estimated, i.e., the process to be controlled is identified, using a stochastic optimization algorithm [14, 15].

The system identification problem can be viewed as the construction of a mathematical model for a dynamic process using the knowledge and observations of that process. The objectives of system identification are to (i) select a feasible model structure describing the process to be controlled and (ii) estimate the model parameters with a minimum estimation error. Figure 8.3 gives a block diagram representation of the air heater system.

Let the voltage input to the process be V_{in}. The temperature sensor produces an output voltage V_{op}, where "p" represents the sensor number. The output of each temperature sensor is proportional to ΔT, the rise in air temperature in the duct from the ambient air temperature. So, the sensor output voltages are V_{o1}, $V_{o2,}$ and V_{o3} for sensor #1, sensor #2, and sensor #3, respectively, and they are expressed in general as:

$$V_{\text{op}} = Sk_p \Delta T \tag{8.1}$$

where Sk_p, p = 1, 2, and 3, are the gains of the individual temperature sensors.

Let τ_{d2} be the transportation delay for sensor #2 and this denotes the time taken by the hot air to flow from the heater coil to the sensor #2. Similarly, τ_{d3} denotes the transportation delay for sensor #3. However, as mentioned before, the sensor #1 is placed very near to the heater coil, and thus, it can be assumed that there is no

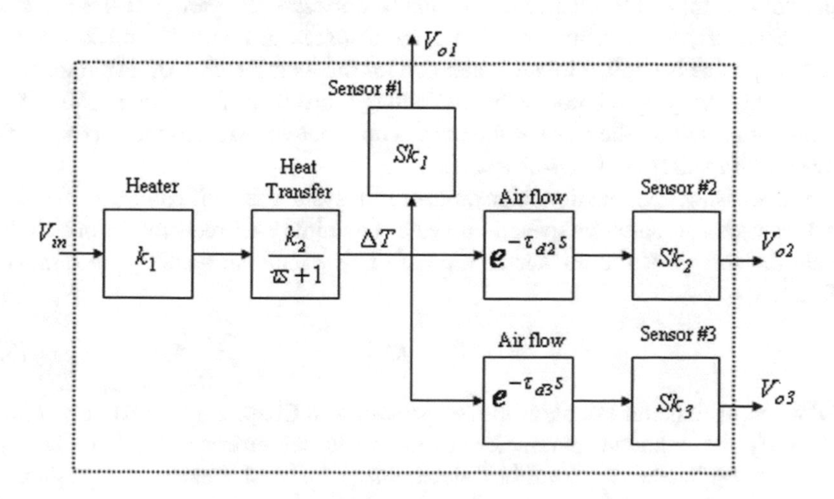

Fig. 8.3 Block diagram of air heater process system

transport delay τ_{d1} involved for sensor #1. Thus, the transfer functions between the heater input voltage and the sensor output voltages of sensor #1, sensor #2, and sensor #3, respectively, can be expressed as:

$$\begin{aligned} G_1(s) &= \frac{V_{o1}(s)}{V_{\text{in}}(s)} = \frac{K_1}{\tau s + 1}, \\ G_2(s) &= \frac{V_{o2}(s)}{V_{\text{in}}(s)} = \frac{K_2 e^{-\tau_{d2} s}}{\tau s + 1}, \\ G_3(s) &= \frac{V_{o3}(s)}{V_{\text{in}}(s)} = \frac{K_3 e^{-\tau_{d3} s}}{\tau s + 1} \end{aligned} \tag{8.2}$$

where $K_1 = k_1k_2Sk_1$, $K_2 = k_1k_2Sk_2$, and $K_3 = k_1k_2Sk_3$ are the dc gains of that process for different sensor positions along the duct, and τ is the RC time constant of the heat transfer dynamics. Since the temperature sensors used to measure the temperature of the process at different cross sections of the duct are identical in nature, so $Sk_1 = Sk_2 = Sk_3$ and, thus, the dc gains K_1, $K_2$and K_3 are practically equal, and therefore, we can write, $K_1 = K_2 = K_3 = K$.

In this chapter, the state model of the air heater process considering temperature measurement at sensor #1 can be viewed as:

$$\begin{aligned} \dot{x}_1 &= -\left(\frac{1}{\tau}\right)x_1 + \left(\frac{K}{\tau}\right)u(t) \\ y &= x_1 \end{aligned} \tag{8.3}$$

where $y = V_{o1}$, the output voltage of sensor #1, in Volt (V), and $u = V_{\text{in}}$ in Volt (V).

The unknown parameters of this process are the dc gain of the process (K) and the time constant of the heat transfer dynamics (τ). The open-loop test data of the input voltage applied to the heater coil and its corresponding temperature measured from sensor #1, for duration of 1500 s, are acquired at intervals of 100 ms each. The voltage that is applied to the heater coil for the identification of its parameters is a randomly varying signal with its voltage level in the range 150–200 V. Figure 8.4a, b shows the plots of the open-loop input voltage and the corresponding temperature measured at sensor #1.

In stochastic optimization algorithm-based system identification, a candidate solution vector in solution space is a vector containing all required information to obtain the model of the air heater process. This candidate vector $\underline{Z}$ is formed as [19, 20]:

$$\underline{Z} = [K|\tau] \tag{8.4}$$

PSO algorithm and HS algorithm as presented in Chap. 2 are separately utilized to identify the unknown parameters of the air heater process [14, 15]. Here, the stochastic optimization algorithm determines the best vector comprising the unknown parameters as shown in (8.4) for which discrepancy between the model

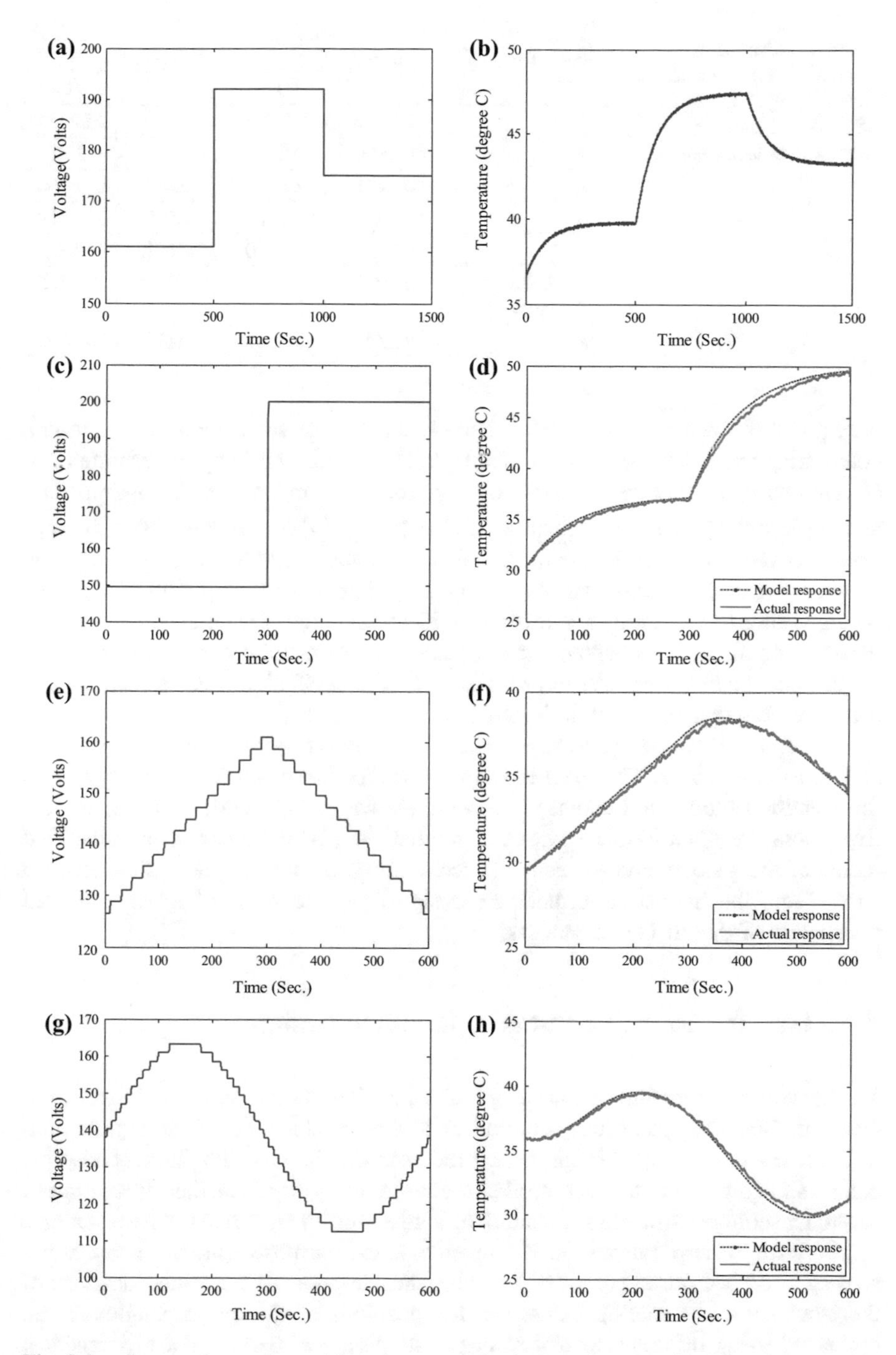

Fig. 8.4 **a** Input voltage plot and **b** temperature plot in open-loop configuration for parameter estimation, **c**, **e**, and **g** are different test input voltages, and **d**, **f**, and **g** are corresponding open-loop temperature responses, respectively

Table 8.1 Estimated parameters of the air heater system using PSO- and HSA-based system identification technique

No. of test	PSO-based		HSA-based	
	K	τ (s)	K	τ (s)
1	0.2500	89.7758	0.2415	90.9467
2	0.2500	90.2467	0.2489	91.0673
3	0.2487	92.8731	0.2546	89.8700
4	0.2526	89.1973	0.2505	89.3419
5	0.2511	91.6730	0.2461	90.5032
Standard deviation	0.0015	1.4977	0.0049	0.7315
Average	0.2505	90.7532	0.2483	90.3458

output and the actual experimental output data, for the same input, is minimum, considering the entire set of input–output real-life data. The integral absolute error (*IAE*) is used as the fitness function (or performance index) for the optimization technique and *IAE* can be defined as $IAE = \sum_{n=0}^{N} d(n)\Delta t_c$, where $d(n)$ = discrepancy between the model output and the actual experimental output data, N = number of data points, and Δt_c = sampling time for collected data. The results obtained after five test runs are utilized for identification of the parameters of the air heater system. In this chapter, those parameters are estimated as average of the parameters obtained over five test runs using PSO and HS algorithms separately and these are obtained as shown in Table 8.1.

Further, the identified process is tested for different test input voltages as shown in Fig. 8.4c, e, and g. The open-loop responses for those test input voltages from the identified model and actual process are shown in Fig. 8.4d, f, and h, respectively, and the open-loop responses show that the performances of the identified model of the process and the actual process are quite close to each other, in each case. Thus, the hybrid controllers are designed on the basis of these estimated parameters of the air heater process.

8.4 Hybrid Adaptive Fuzzy Controller Design

The Lyapunov theory and stochastic optimization algorithm are employed together to design the stable adaptive fuzzy controllers, as discussed in Chap. 5, at first, for controlling the temperature of the air heater process in simulations with different adaptive control strategies. Once these controllers are successfully designed, then the optimized controller settings are utilized to control the temperature of air in real implementations.

For both design purpose and implementation purpose, the air temperature sensing is carried out at sensor #1 of the air heater system. The center locations of the input MFs, the scaling gains, and the positions of the output singletons are optimized using different control strategies as discussed earlier. For this practical system, the AFLCs are designed with fixed structure, having 5 input MFs, and with variable structure, with potentially biggest 7 input MFs, for the simulation case study as well as for the experimental case study. Global version PSO and HS algorithms are utilized to design the AFLCs in both individual and hybrid configurations as detailed in Sects. 3.3 and 3.4 of Chap. 3 for PSOBA and HSABA, respectively, and Sect. 5.2 of Chap. 5 for hybrid concurrent and hybrid preferential control strategies. It should be noted here that, during tabulation of results, *-V denotes the variable structure configurations of the controllers; i.e., PSOBA-V means PSOBA algorithm in variable structure configuration.

8.4.1 Simulation Case Study

The air temperature sensing is carried out at sensor #1 of the air heater system, for both design and implementation purposes. For this practical system, the AFLCs are designed with potentially biggest 7 input MFs, for the simulation case study as well as for the experimental case study. During the simulation, the process models are simulated each using a fixed step fourth-order Runge-Kutta method. To perform the simulation, a population size of ten candidate solution vectors (CSV) is chosen. For each simulation, 200 iterations of CSV update are performed. For the stable adaptive control strategies, as mentioned before, tuning of the center locations of the input MFs, free parameters of the controller, and the positions of the output singletons are carried out, to minimize the integral absolute error (*IAE*) between the reference signal and the actual temperature measured from the air heater process. The sampling time is taken as 100 ms in conformation with the hardware driver circuit requirement. The other parameters are chosen as: $P = 0.1$, $f^U = 50$ and $b = \frac{K}{\tau} = 0.0028$, in case of PSO-based design and $b = \frac{K}{\tau} = 0.0027$, in case of HS algorithm-based design. The value of the adaptation gain is chosen as $v = 0.0001$, considering both exploration capability of the design algorithm and inherent approximation error present in the process. The simulation is performed for 600 s. In variable structure controller design, the structural flags are utilized to optimize the structure of the controller and the input MFs and the corresponding positions of the output singletons are also set according to these structural flags.

8.4.1.1 PSO-Based Design

In this simulation case study, PSO-Hybd-Con model produced best control configuration in fixed structure mode. In variable structure configuration, the PSOBA evolved as the superior strategy, although PSO-Hybd-Con model produces an IAE slightly higher but with least number of rules (here it is only 3), as indicated in Table 8.2. It is also noted from Table 8.2 that, in fixed structure configuration, the IAE value of the PSO-Hybd-Pref model is slightly higher than that of the PSO-Hybd-Con model, but less than the IAE value obtained in PSOBA design strategy. In variable structure configuration of the AFLC design, PSO-Hybd-Pref model produces second best result, both in terms of IAE value and number of rules.

Table 8.2 Comparison of simulation results for the air heater system: PSO-based design strategies

Control strategy	No. of MFs	IAE
PSOBA	5	275.04
PSO-Hybd-Con		268.39
PSO-Hybd-Pref		273.28
PSOBA-V	5	276.42
PSO-Hybd-Con-V	3	287.70
PSO-Hybd-Pref-V	4	282.80

Evaluation time = 600 s

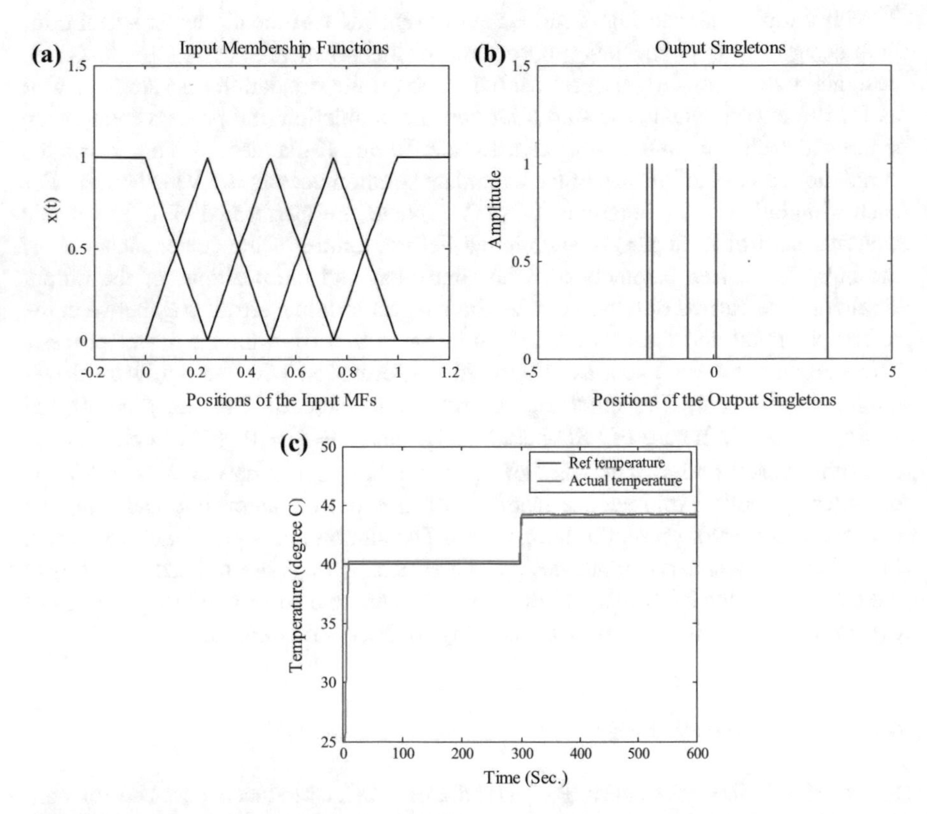

Fig. 8.5 Responses of the adaptive fuzzy controller with PSOBA, for fixed structure configuration, of the air heater system during simulation study: **a** positions of input MFs, **b** positions of output singletons, and **c** evaluation period response after 200 PSO generations during simulation

The locations of the center positions of input MFs and the positions of output singletons with PSOBA, PSO-Hybd-Con, and PSO-Hybd-Pref control strategies, for both fixed and variable structure configurations, are shown in (a) and (b) of Figs. 8.5, 8.6, 8.7, 8.8, 8.9, and 8.10, respectively. In (c) of Figs. 8.5, 8.6, 8.7, 8.8,

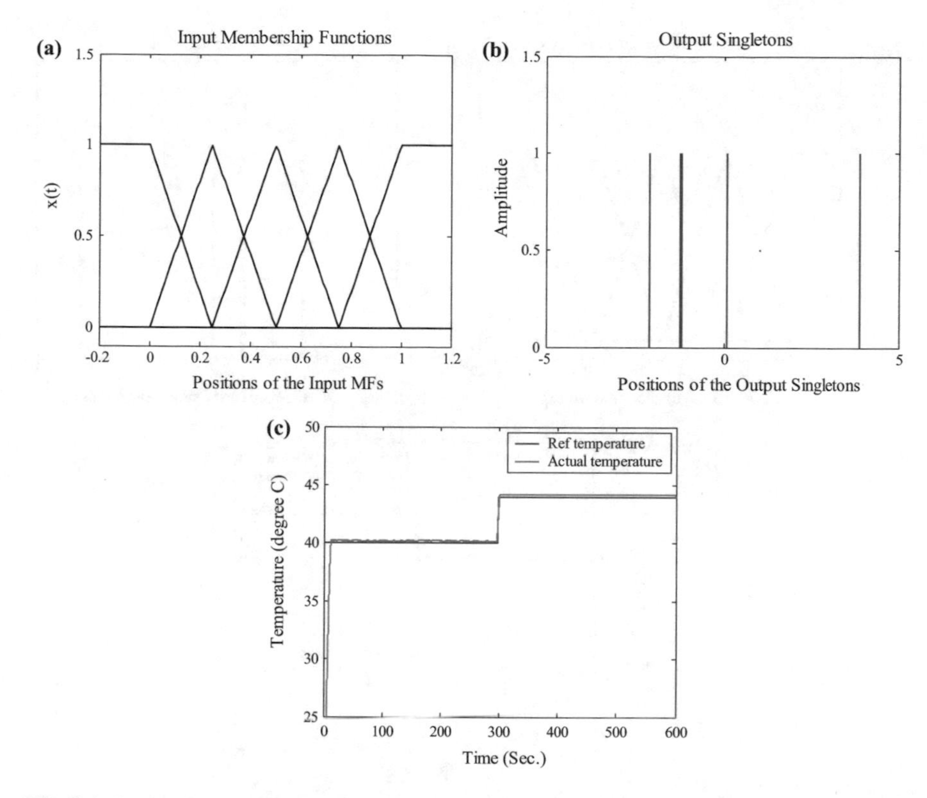

Fig. 8.6 Responses of the adaptive fuzzy controller with PSO-Hybd-Con model, for fixed structure configuration, of the air heater system during simulation study: **a** positions of input MFs, **b** positions of output singletons, and **c** evaluation period response after 200 PSO generations during simulation

8.9, and 8.10, the evaluation period responses are shown after 200 PSO generations in simulation, for different control strategies.

8.4.1.2 HS Algorithm-Based Design

In this case study, HS-Hybd-Con model-based control strategy produces best results, in both fixed and variable structure configurations, and HS-Hybd-Pref model-based controllers are the second best solution as pointed out by the *IAE* values and number of rules in Table 8.3. The locations of the center positions of input MFs and the positions of output singletons with HSBA, HS-Hybd-Con model, and HS-Hybd-Pref model control strategies are shown in (a) and (b) of Figs. 8.11, 8.12, 8.13, 8.14, 8.15, and 8.16, respectively, and in (c) option of the above figures, the evaluation period responses during simulation case studies are shown.

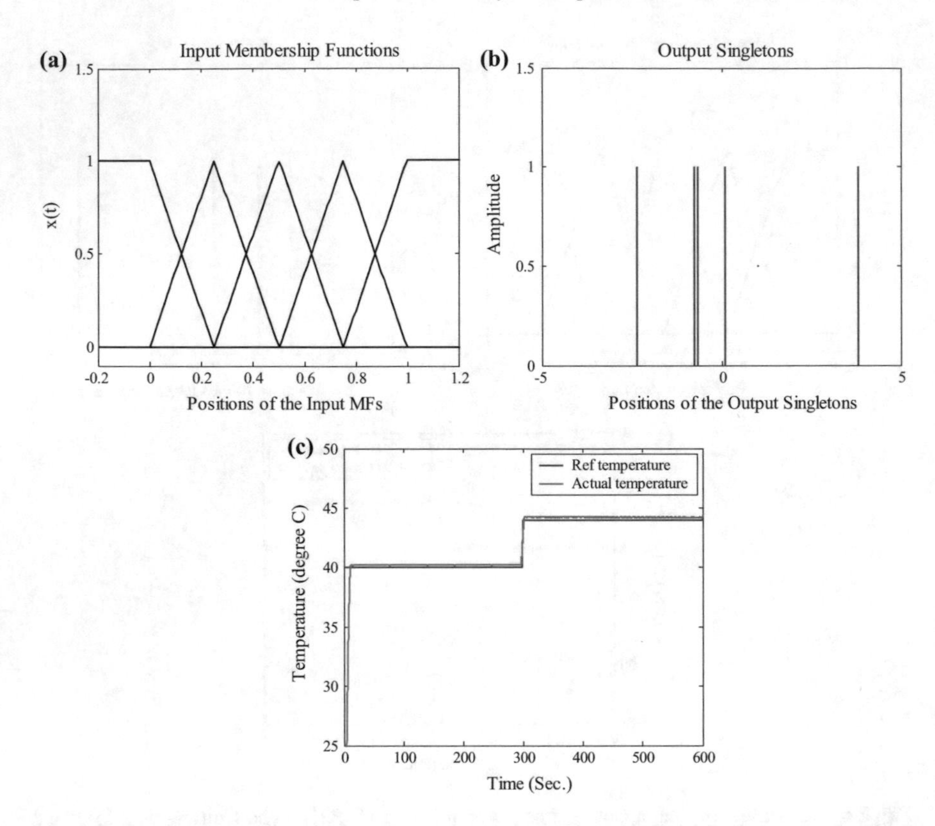

Fig. 8.7 Responses of the adaptive fuzzy controller with PSO-Hybd-Pref model, for fixed structure configuration, of the air heater system during simulation study: **a** positions of input membership functions, **b** positions of output singletons, and **c** evaluation period response after 200 PSO generations during simulation

In this simulation case study, it is revealed from the responses of different control strategies that the center locations of MFs for the input variable of each AFLC are much randomly optimized over its universe of discourse than that of the case in PSO-based design. The results obtained, in terms of *IAE* values and temperature control responses, depict that the performance of HSA-based design and PSO-based design during simulation are quite similar.

The performance of the AFLCs designed with different control strategies as depicted from the evaluation period response of the air heater process in simulation shows that the rise time in all the cases is significantly small, which may not be achieved in practical experimentation as the thermal process is generally quite sluggish in nature.

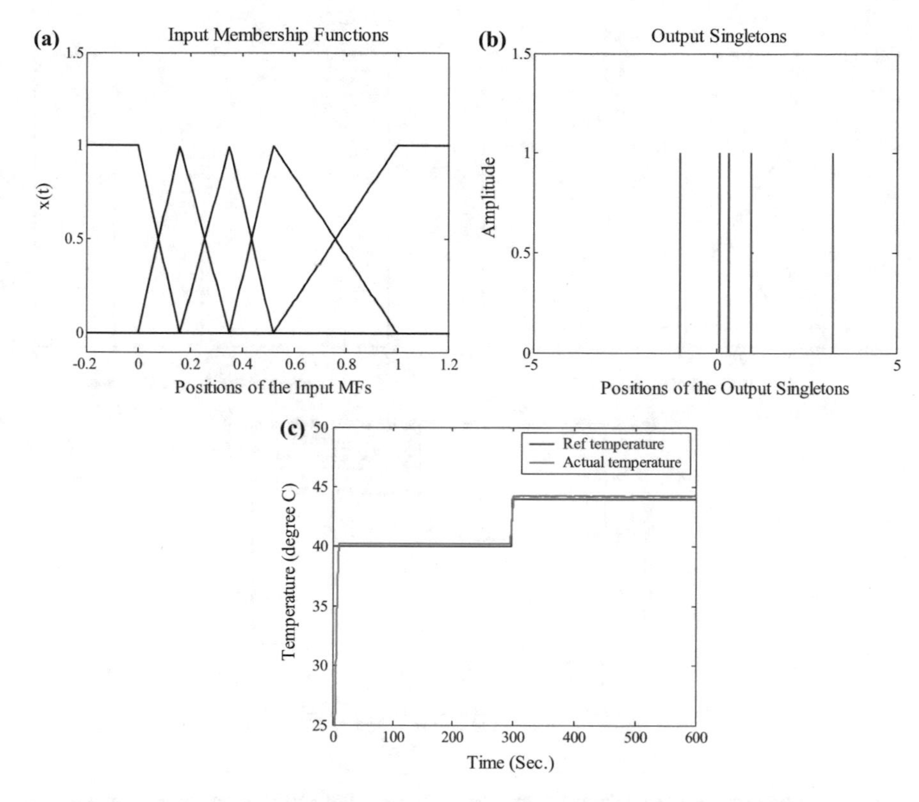

Fig. 8.8 Responses of the adaptive fuzzy controller with PSOBA, for variable structure configuration, of the air heater system during simulation study: **a** positions of input MFs, **b** positions of output singletons and **c** evaluation period response after 200 PSO generations during simulation

8.4.2 Experimental Case Study I

To demonstrate the utility of the stochastic optimization-based AFLC design strategies in real implementation situation, the temperature control of the air heater process is next performed in practical condition. The AFLCs trained during the simulation case studies are used to perform the temperature control in real implementation. As discussed about the air heater process before, there are three temperature sensors and, among these, sensor #1 is placed just above the heater coil. Thus, the heat transfer dynamics measured at sensor #1 does not involve or negligible, if any, transport delay. The heat transfer dynamics is significantly sensitive to the blower speed as this can change the air mass flow rate. Therefore, the blower speed can be treated as a disturbance for the air heater process. A steady value of the blower speed provides a fixed disturbance, and here, the blower speed is kept

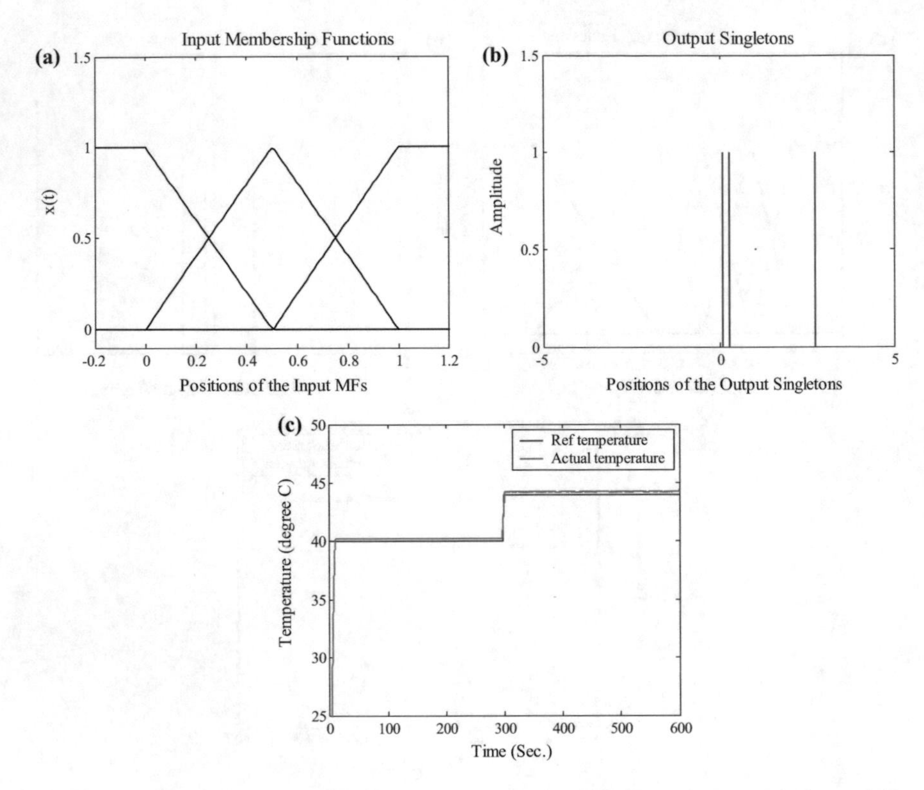

Fig. 8.9 Responses of the adaptive fuzzy controller with PSO-Hybd-Con model, for variable structure configuration, of the air heater system during simulation study: **a** positions of input MFs, **b** positions of output singletons, and **c** evaluation period response after 200 PSO generations during simulation

fixed at 50% of the maximum speed for fixed type of disturbance. On the other hand, a randomly variable blower speed provides variable disturbances, and here, it is varied randomly between 60 and 80% of the maximum blower speed, as shown in Fig. 8.17.

8.4.2.1 PSO-Based Design

The AFLCs trained during the simulation case studies are employed for real implementations. Figures 8.18 and 8.19 show the performances of the temperature controller for controlling the temperature at sensor #1, utilizing fixed and variable structure versions of the PSOBA, PSO-Hybd-Con model, and PSO-Hybd-Pref model, for fixed blower speed disturbance and randomly variable blower speed disturbances, respectively.

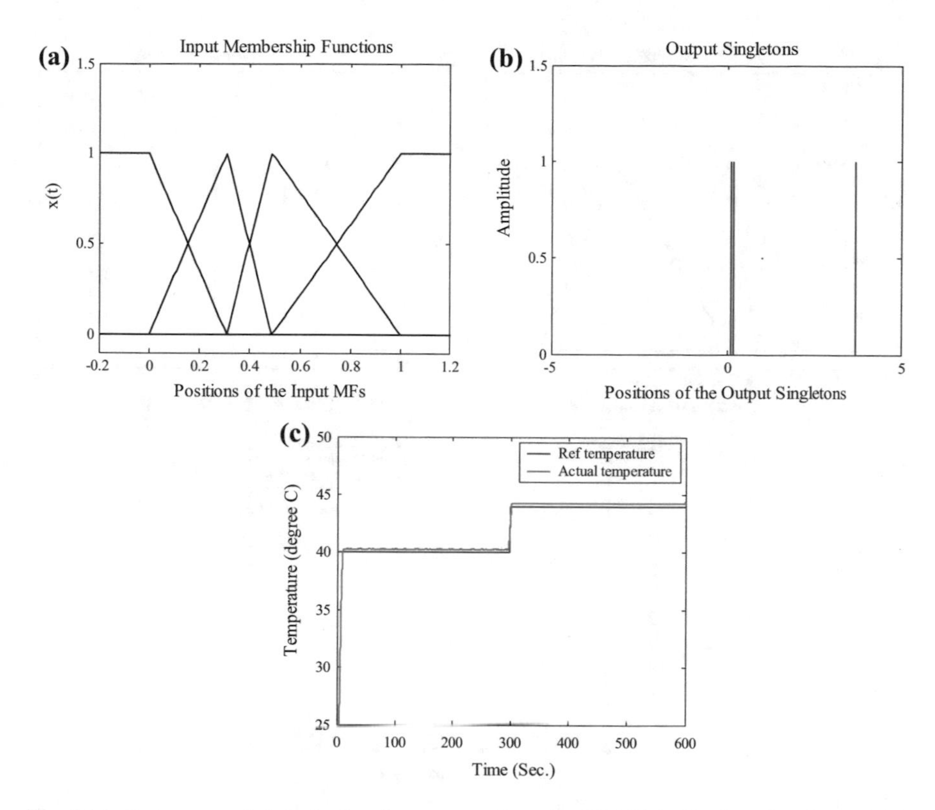

Fig. 8.10 Responses of the adaptive fuzzy controller with PSO-Hybd-Pref model, for variable structure configuration, of the air heater system during simulation study: **a** positions of input MFs, **b** positions of output singletons and **c** evaluation period response after 200 PSO generations during simulation

Table 8.3 Comparison of simulation results for the air heater system: HSA-based design strategies

Control strategy	No. of MFs	IAE
HSABA	5	276.70
HSA-Hybd-Con		273.88
HSA-Hybd-Pref		276.66
HSABA-V	7	278.22
HSA-Hybd-Con-V	4	276.92
HSA-Hybd-Pref-V	5	296.86

Evaluation time = 600 s

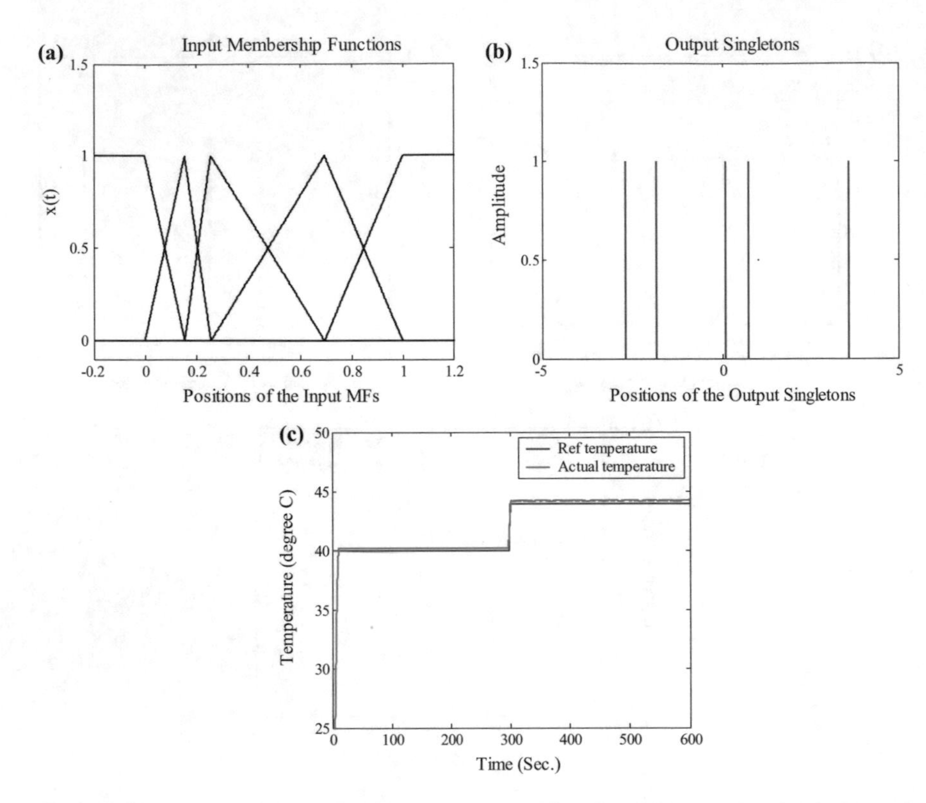

Fig. 8.11 Responses of the adaptive fuzzy controller with HSABA, for fixed structure configuration, of the air heater system during simulation study: **a** positions of input MFs, **b** positions of output singletons, and **c** evaluation period response after 200 HS generations during simulation

The performances of the different control strategies are evaluated by calculating the *IAE* values for 600 s evaluation period. Table 8.4 presents a comparison of competing controllers, in terms of *IAE*, in evaluation phase, for the temperature measured at sensor #1. The results reported in Table 8.4 demonstrate that the proposed PSO-Hybd-Pref controller evolved as the best option in fixed structure version, whereas variable structure PSO-Hybd-Con model out performs all other competing controllers, implemented for fixed disturbance case study. The variable structure versions of the two hybrid controllers were successful in reducing the total number fuzzy rules implemented, which are useful from the point of view of reducing computational burden in real implementations. Table 8.4 also demonstrates that the performance of the fixed structure PSO-Hydb-Pref model controller evolved as the superior control option, compared to other options in randomly varying disturbance environment too. In this case too, the variable structure PSO-Hybd-Con model becomes the best control solution, in terms of both *IAE* and number of fuzzy rules.

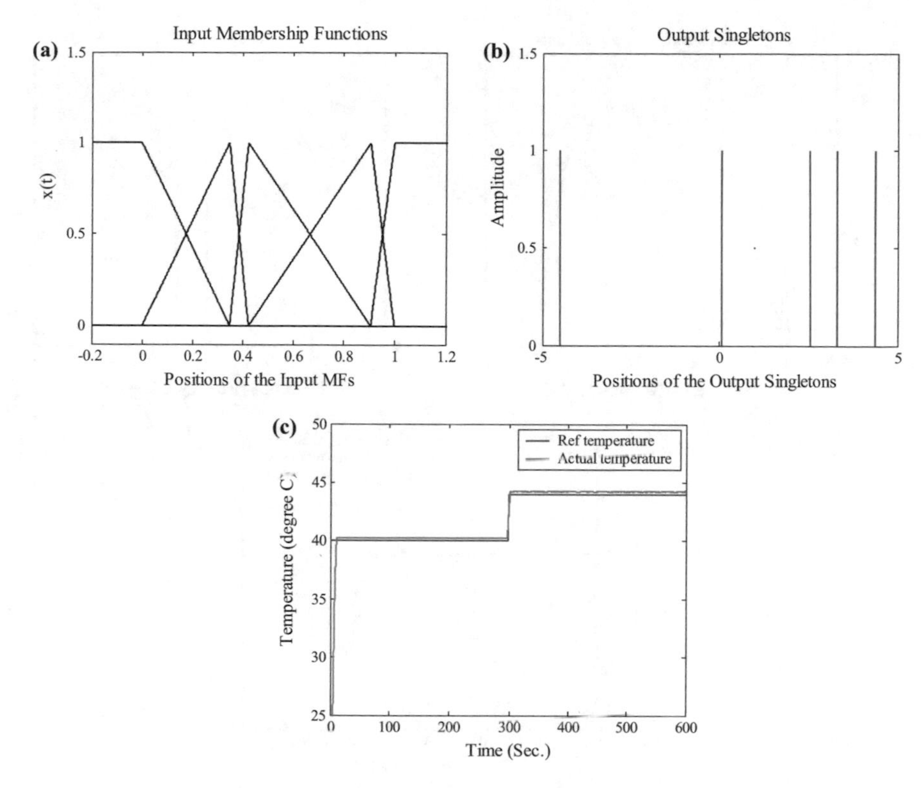

Fig. 8.12 Responses of the adaptive fuzzy controller with HSA-Hybd-Con model, for fixed structure configuration, of the air heater system during simulation study: **a** positions of input MFs, **b** positions of output singletons, and **c** evaluation period response after 200 HS generations during simulation

8.4.2.2 HS Algorithm-Based Design

The HS algorithm-based AFLCs, which are tuned in an off-line manner during simulation case studies, are implemented in practical situation to control the temperature of the air heater process. The temperature measurement is carried out at sensor #1. The other design considerations, like calculation of *IAE*, disturbances, are same as the experimental case study I of PSO-based design technique. Table 8.5 indicates the performances of the different control strategies and here also, for fixed blower speed disturbance, HSA-Hybd-Pref model evolved as the best or second best control option in fixed and variable structure controller configurations. Again, for randomly variable blower speed disturbance, for temperature measured at sensor #1, hybrid preferential model produces best result in both fixed structure and

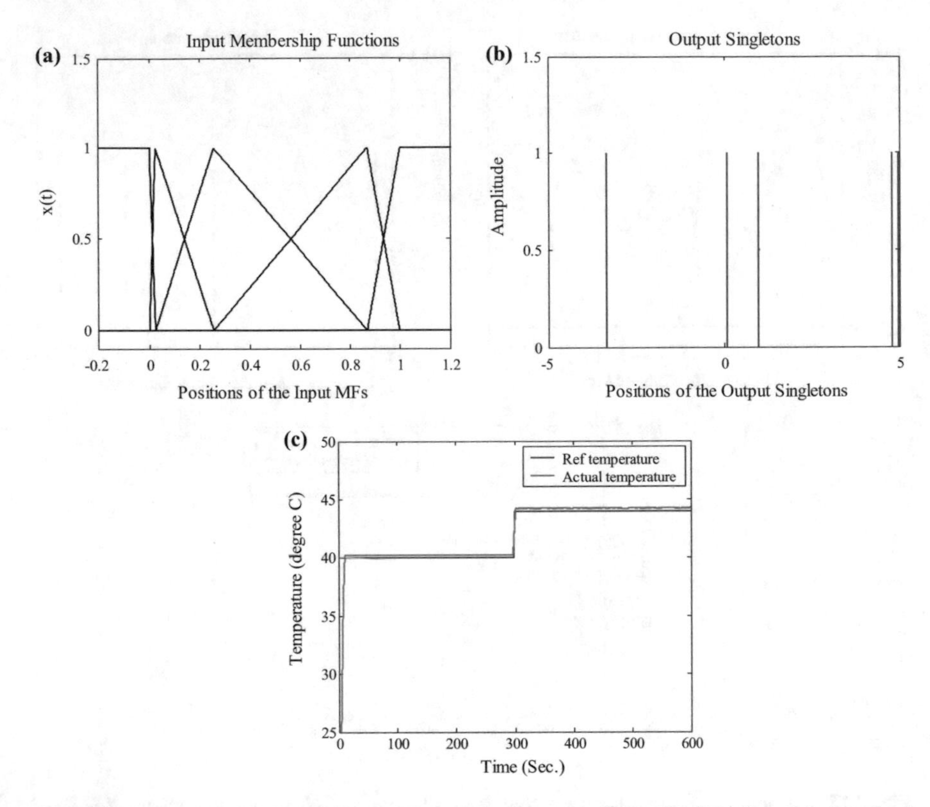

Fig. 8.13 Responses of the adaptive fuzzy controller with HSA-Hybd-Pref model, for fixed structure configuration, of the air heater system during simulation study: **a** positions of input MFs, **b** positions of output singletons and **c** evaluation period response after 200 HS generations during simulation

variable structure configuration of the controllers. Figures 8.20 and 8.21 show the performances of HSABA, HSA-Hybd-Con model, and HSA-Hybd-Pref model of control strategy, for fixed structure and variable structure configurations, with fixed blower speed disturbance and randomly variable blower speed disturbances, respectively.

8.4.3 Experimental Case Study II

To demonstrate the robustness of the designed hybrid and non-hybrid controllers, in both fixed and variable structure configuration, they are employed to control the temperature of the air heater process, where the temperature is now measured at the

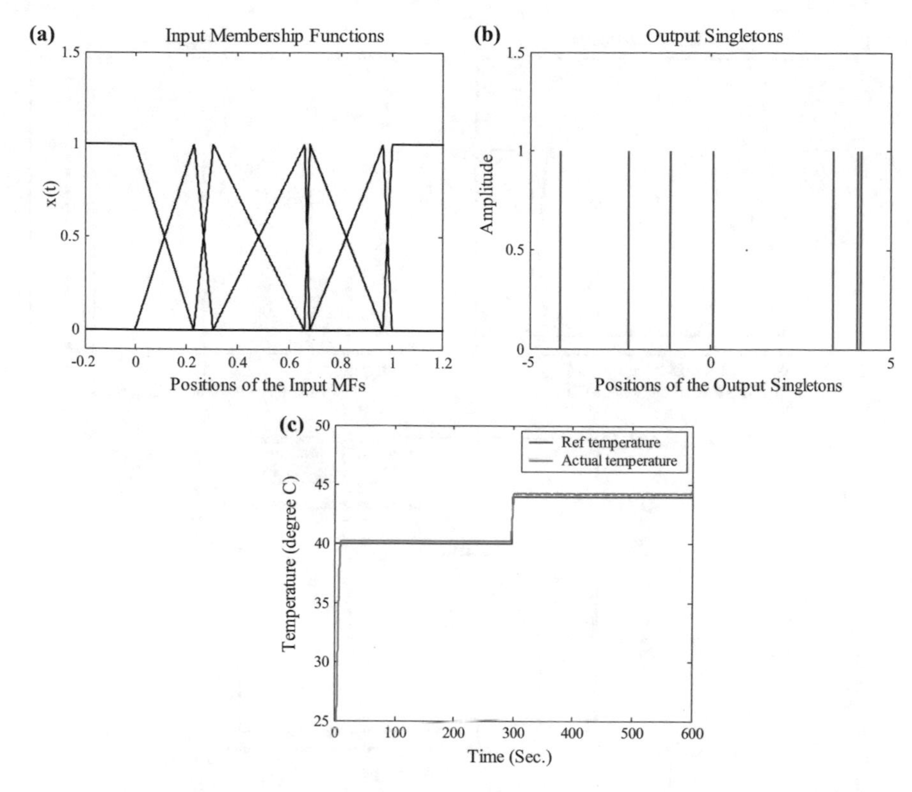

Fig. 8.14 Responses of the adaptive fuzzy controller with HSABA, for variable structure configuration, of the air heater system during simulation study: **a** positions of input MFs, **b** positions of output singletons, and **c** evaluation period response after 200 HS generations during simulation

location of the sensor #3 and feedback for performing control action. Here, the same controllers are implemented that were designed on the basis of temperature sensed at sensor #1. The objective of this case study is to demonstrate that the controller designed for the "no transport delay" situation is still robust enough for those situations where it is not possible to place the temperature sensor right at the ideal place where the temperature should be monitored but has to be placed at a distant place, because of the practical constraints.

8.4.3.1 PSO-Based Design

The experimental results for the PSO-based AFLCs, as depicted in Table 8.6, illustrate that although the *IAE* values are higher than the corresponding values of the previous case study but the basic requirement of temperature control is achieved

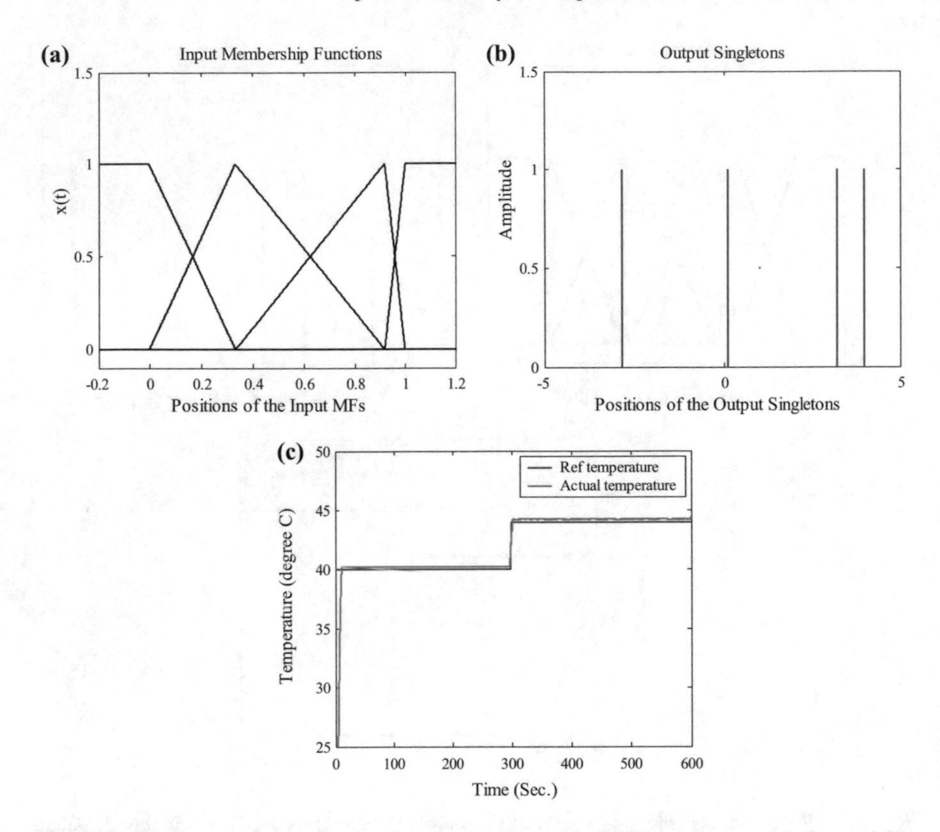

Fig. 8.15 Responses of the adaptive fuzzy controller with HS-Hybd-Con model, for variable structure configuration, of thc air heater system during simulation study: **a** positions of input MFs, **b** positions of output singletons, and **c** evaluation period response after 200 HS generations during simulation

in both fixed and randomly variable disturbance environments, as shown in Figs. 8.22 and 8.23, respectively. In this case study also, PSO-Hybd-Pref model mostly evolves as the second best control solution, in both fixed and variable structure controller configurations, for both fixed and randomly variable disturbance situations. However, PSO-Hybd-Con configuration also provides encouraging performances in more than one occasion. Most surprisingly, in variable structure configuration, with randomly variable disturbance, PSOBA gives the best transient performance in terms of *IAE* value.

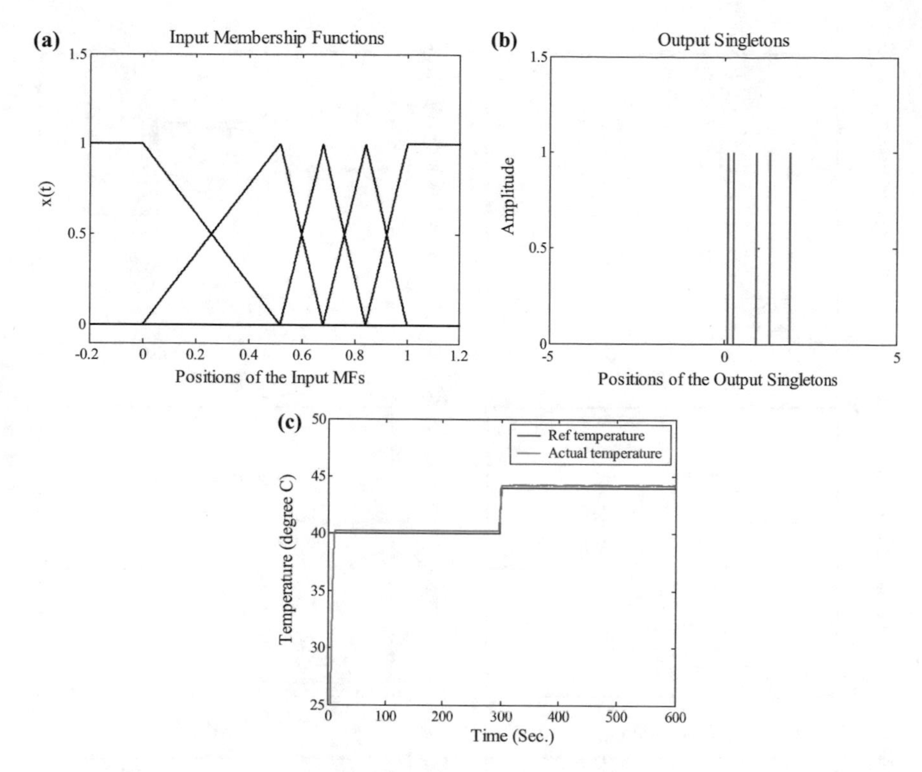

Fig. 8.16 Responses of the adaptive fuzzy controller with HS-Hybd-Pref model, for variable structure configuration, of the air heater system during simulation study: **a** positions of input MFs, **b** positions of output singletons, and **c** evaluation period response after 200 HS generations during simulation

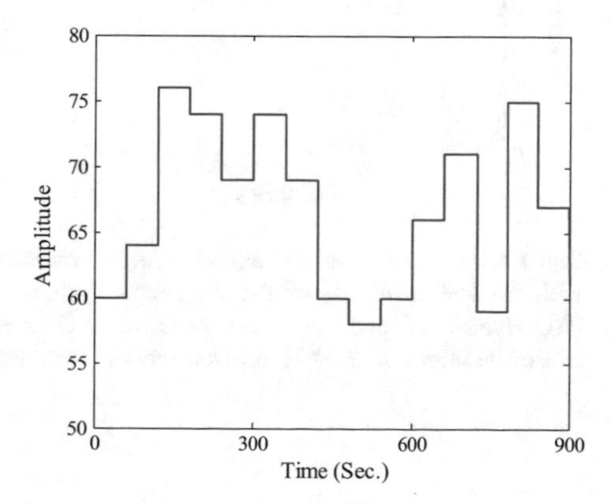

Fig. 8.17 Plot of randomly variable blower speed applied as disturbance to the air heater system

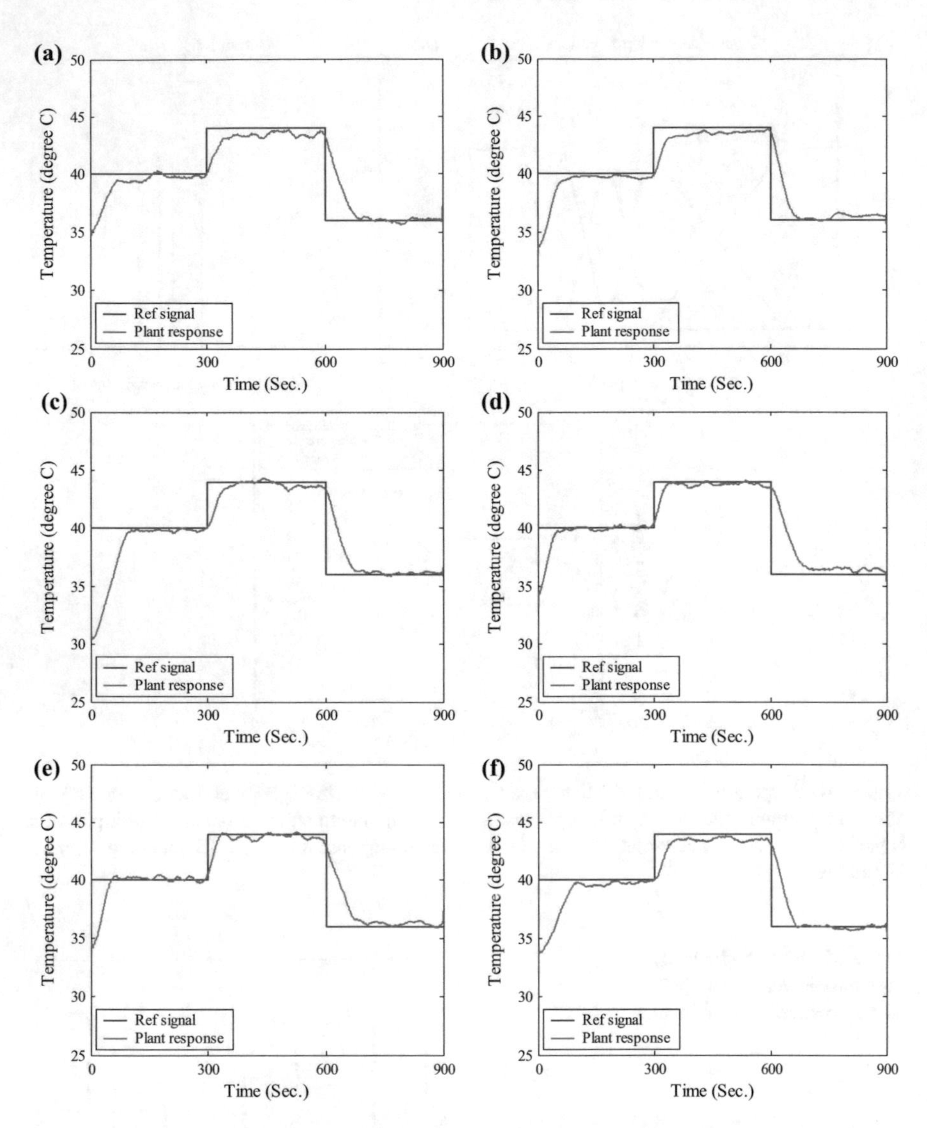

Fig. 8.18 Responses of the adaptive fuzzy controller in practical experimentation, for sensor #1 with fixed disturbance, of the air heater system: **a** fixed structure PSOBA, **b** fixed structure PSO-Hybd-Con model, **c** fixed structure PSO-Hybd-Pref model, **d** variable structure PSOBA, **e** variable structure PSO-Hybd-Con model, **f** variable structure PSO-Hybd-Pref model

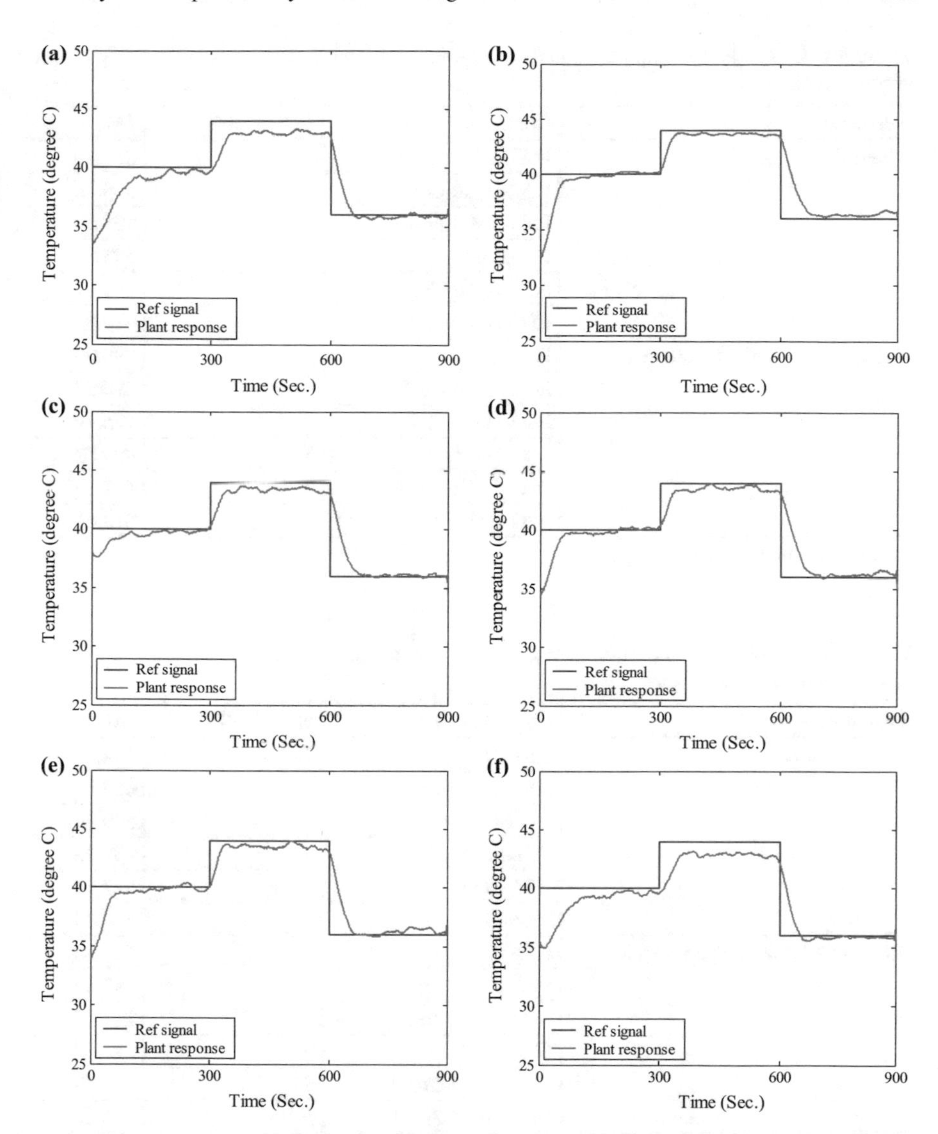

Fig. 8.19 Responses of the adaptive fuzzy controller in practical experimentation, for sensor #1 with randomly variable disturbance, of the air heater system: **a** fixed structure PSOBA, **b** fixed structure PSO-Hybd-Con model, **c** fixed structure PSO-Hybd-Pref model, **d** variable structure PSOBA, **e** variable structure PSO-Hybd-Con model, **f** variable structure PSO-Hybd-Pref model

Table 8.4 Comparison of experimental results for the air heater system for sensor #1: PSO-based design strategies

Disturbance	Control strategy	No. of MFs	IAE
Fixed disturbance	PSOBA	5	551.29
	PSO-Hybd-Con		528.06
	PSO-Hybd-Pref		458.34
	PSOBA-V	5	545.82
	PSO-Hybd-Con-V	3	445.91
	PSO-Hybd-Pref-V	4	479.35
Randomly variable disturbance	PSOBA	5	590.79
	PSO-Hybd-Con		499.57
	PSO-Hybd-Pref		475.49
	PSOBA-V	5	499.82
	PSO-Hybd-Con-V	3	459.13
	PSO-Hybd-Pref-V	4	615.61

Evaluation time = 600 s

Table 8.5 Comparison of experimental results for the air heater system for sensor #1: M-HS-based design strategies

Disturbance	Control strategy	No. of MFs	IAE
Fixed disturbance	HSABA	5	544.69
	HSA-Hybd-Con		657.41
	HSA-Hybd-Pref		549.90
	HSABA-V	7	489.32
	HSA-Hybd-Con-V	4	527.34
	HSA-Hybd-Pref-V	5	437.29
Randomly variable disturbance	HSABA	5	623.82
	HSA-Hybd-Con		693.12
	HSA-Hybd-Pref		577.48
	HSABA-V	7	516.62
	HSA-Hybd-Con-V	4	516.39
	HSA-Hybd-Pref-V	5	456.53

Evaluation time = 600 s

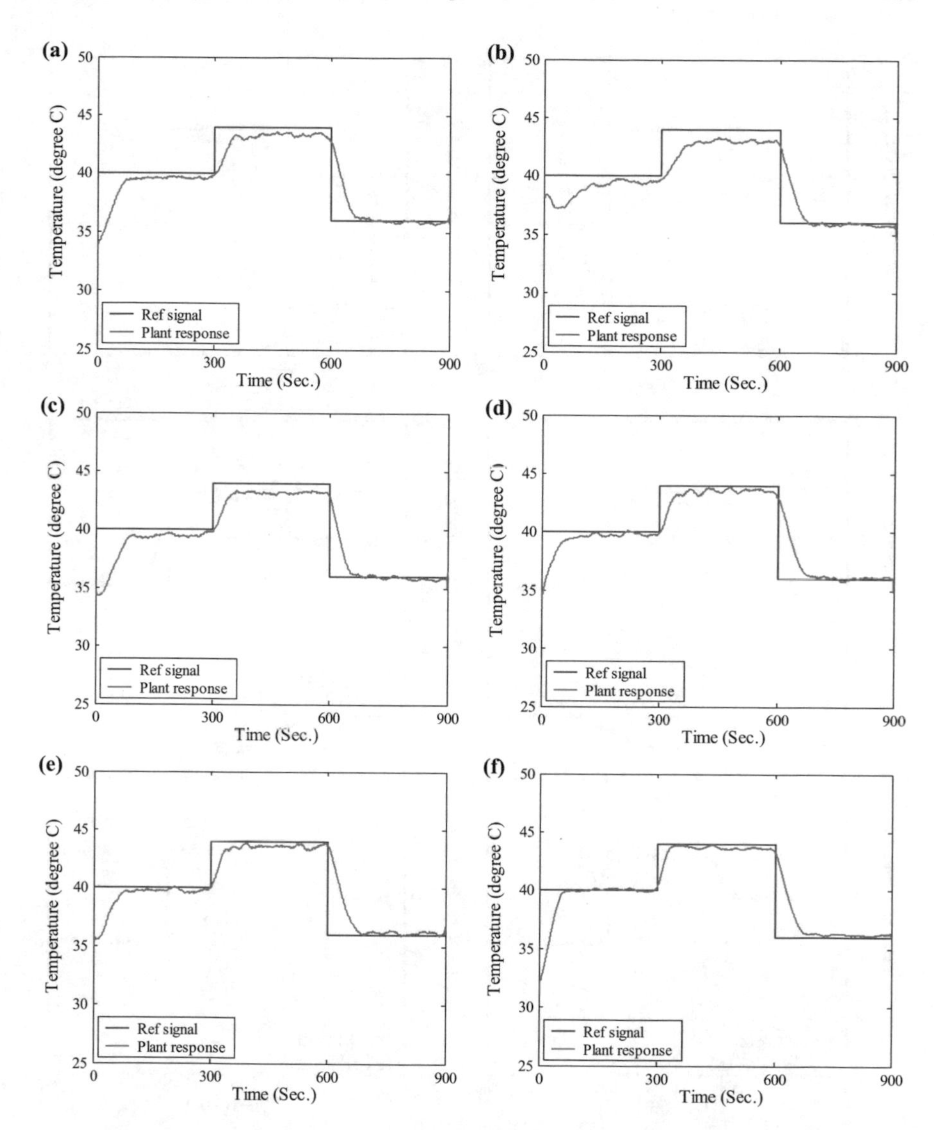

Fig. 8.20 Responses of the adaptive fuzzy controller in practical experimentation, for sensor #1 with fixed disturbance, of the air heater system: **a** fixed structure HSABA, **b** fixed structure HSA-Hybd-Con model, **c** fixed structure HSA-Hybd-Pref model, **d** variable structure HSABA, **e** variable structure HSA-Hybd-Con, **f** variable structure HSA-Hybd-Pref model

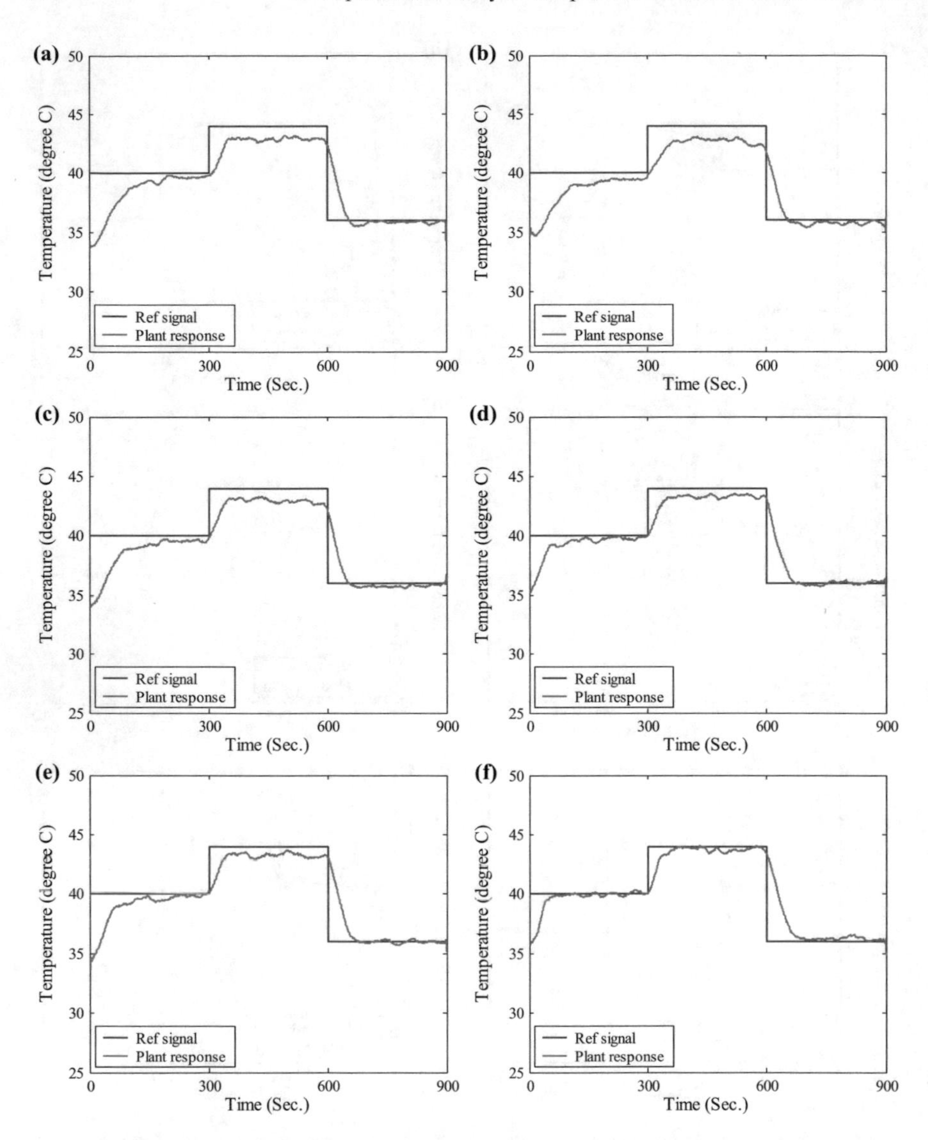

Fig. 8.21 Responses of the adaptive fuzzy controller in practical experimentation, for sensor #1 with randomly variable disturbance, of the air heater system: **a** fixed structure HSABA, **b** fixed structure HSA-Hybd-Con model, **c** fixed structure HSA-Hybd-Pref model, **d** variable structure HSABA, **e** variable structure HSA-Hybd-Con model, **f** variable structure HSA-Hybd-Pref model

Table 8.6 Comparison of experimental results for the air heater system for sensor #3: PSO-based design strategies

Disturbance	Control strategy	No. of MFs	IAE
Fixed disturbance	PSOBA	5	675.44
	PSO-Hybd-Con		635.52
	PSO-Hybd-Pref		638.83
	PSOBA-V	5	628.51
	PSO-Hybd-Con-V	3	584.51
	PSO-Hybd-Pref-V	4	861.49
Randomly variable disturbance	PSOBA	5	740.49
	PSO-Hybd-Con		594.37
	PSO-Hybd-Pref		612.82
	PSOBA-V	5	616.60
	PSO-Hybd-Con-V	3	800.14
	PSO-Hybd-Pref-V	4	790.39

Evaluation time = 600 s

8.4.3.2 HSA Algorithm-Based Design

In this case study too, the HS algorithm-based controllers are demonstrated to be robust enough to cope with the transportation delay present in the process. Table 8.7 shows that the HSA-Hybd-Pref model evolved as the superior control option in each test environment. Figures 8.24 and 8.25 show the performances of different HSA algorithm-based controllers in different test situations. In harmony search algorithm-based controller design strategies also HSA-Hybd-Pref model evolved as the best control option, on the whole, and the designed controllers are to a large extent adequately robust to provide the satisfactory performances.

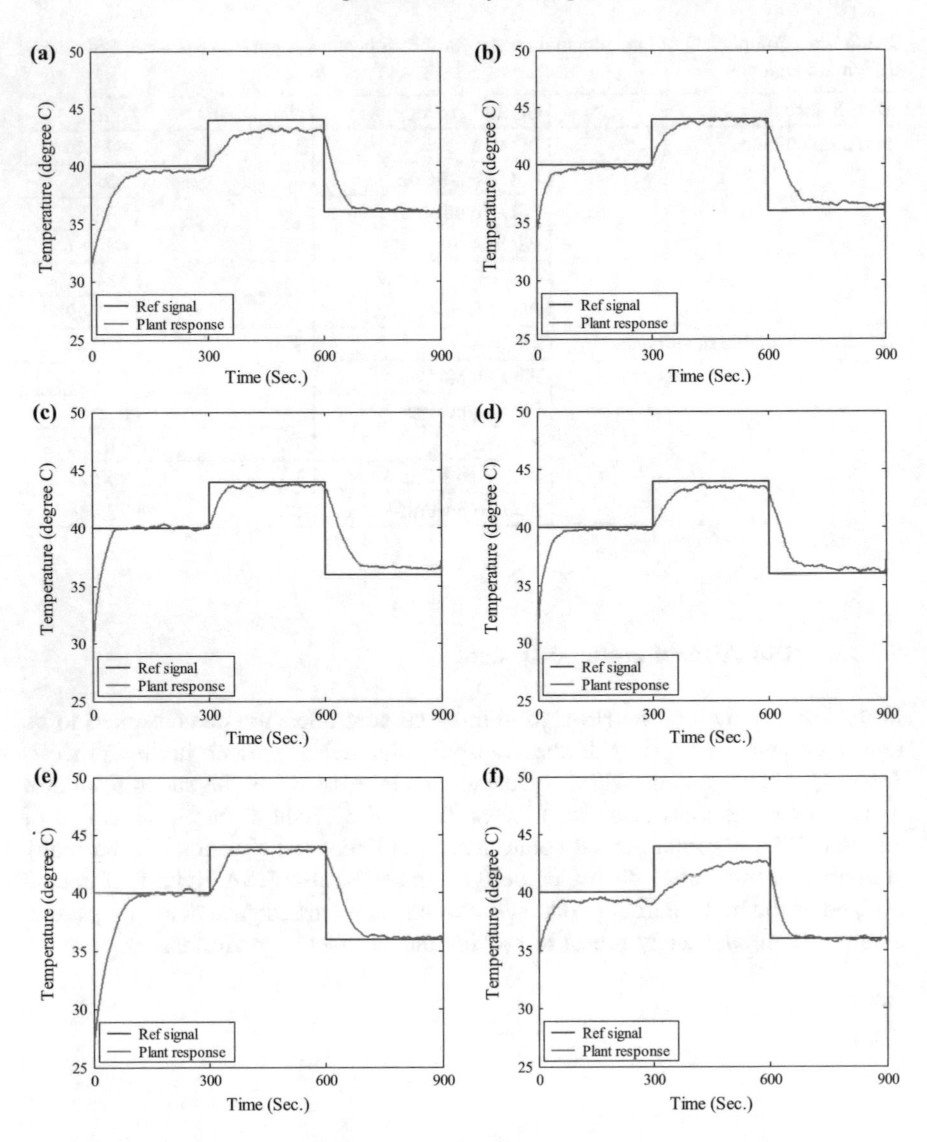

Fig. 8.22 Responses of the adaptive fuzzy controller in practical experimentation, for sensor #3 with fixed disturbance, of the air heater system: **a** fixed structure PSOBA, **b** fixed structure PSO-Hybd-Con model, **c** fixed structure PSO-Hybd-Pref model, **d** variable structure PSOBA, **e** variable structure PSO-Hybd-Con model, **f** variable structure PSO-Hybd-Pref model

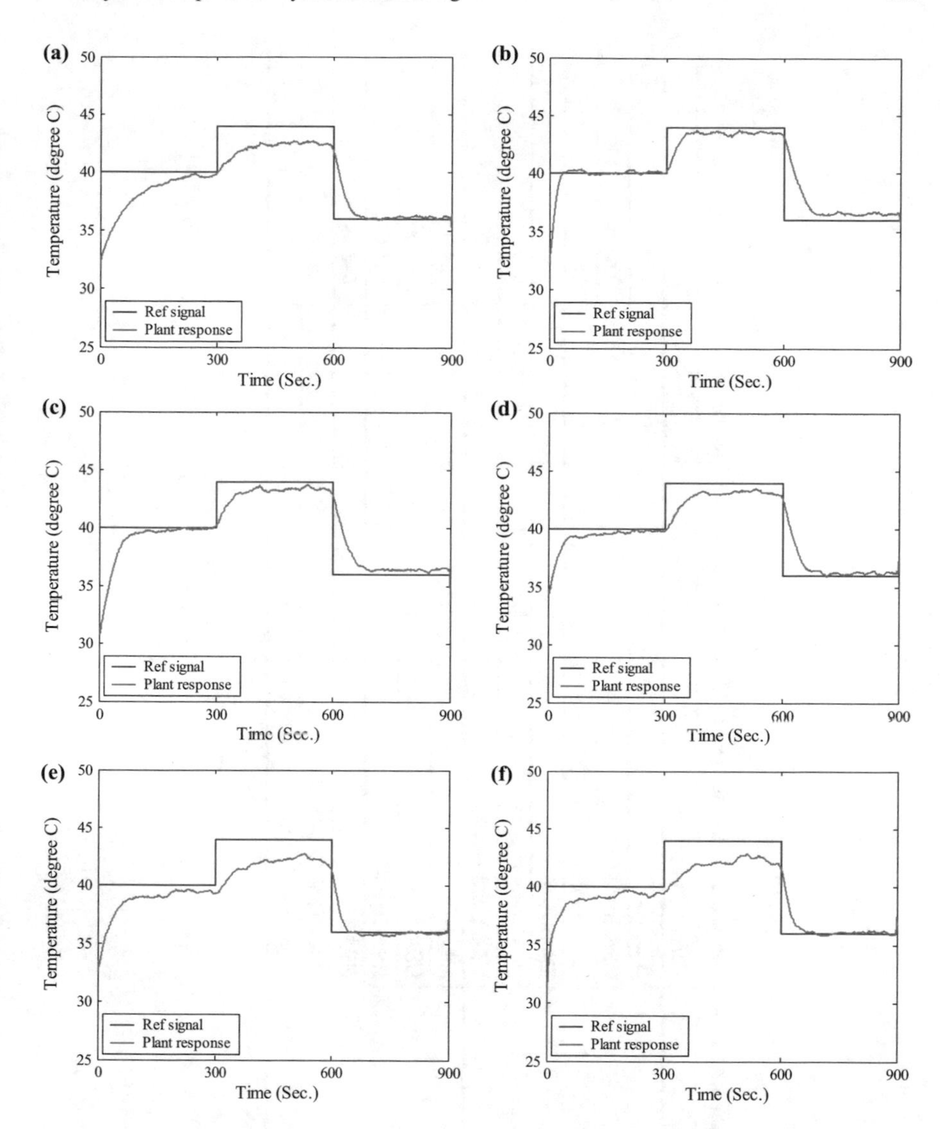

Fig. 8.23 Responses of the adaptive fuzzy controller in practical experimentation, for sensor #3 with randomly variable disturbance, of the air heater system: **a** fixed structure PSOBA, **b** fixed structure PSO-Hybd-Con model, **c** fixed structure PSO-Hybd-Pref model, **d** variable structure PSOBA, **e** variable structure PSO-Hybd-Con model, **f** variable structure PSO-Hybd-Pref model

Table 8.7 Comparison of experimental results for the air heater system for sensor #3: HSA-based design strategies

Disturbance	Control strategy	No. of MFs	IAE
Fixed disturbance	HSABA	5	769.25
	HSA-Hybd-Con		772.09
	HSA-Hybd-Pref		757.25
	HSABA-V	7	600.76
	HSA-Hybd-Con-V	4	612.24
	HSA-Hybd-Pref-V	5	510.28
Randomly variable disturbance	HSABA	5	1024.02
	HSA-Hybd-Con		1037.90
	HSA-Hybd-Pref		845.64
	HSABA-V	7	640.89
	HSA-Hybd-Con-V	4	664.51
	HSA-Hybd-Pref-V	5	592.71

Evaluation time = 600 s

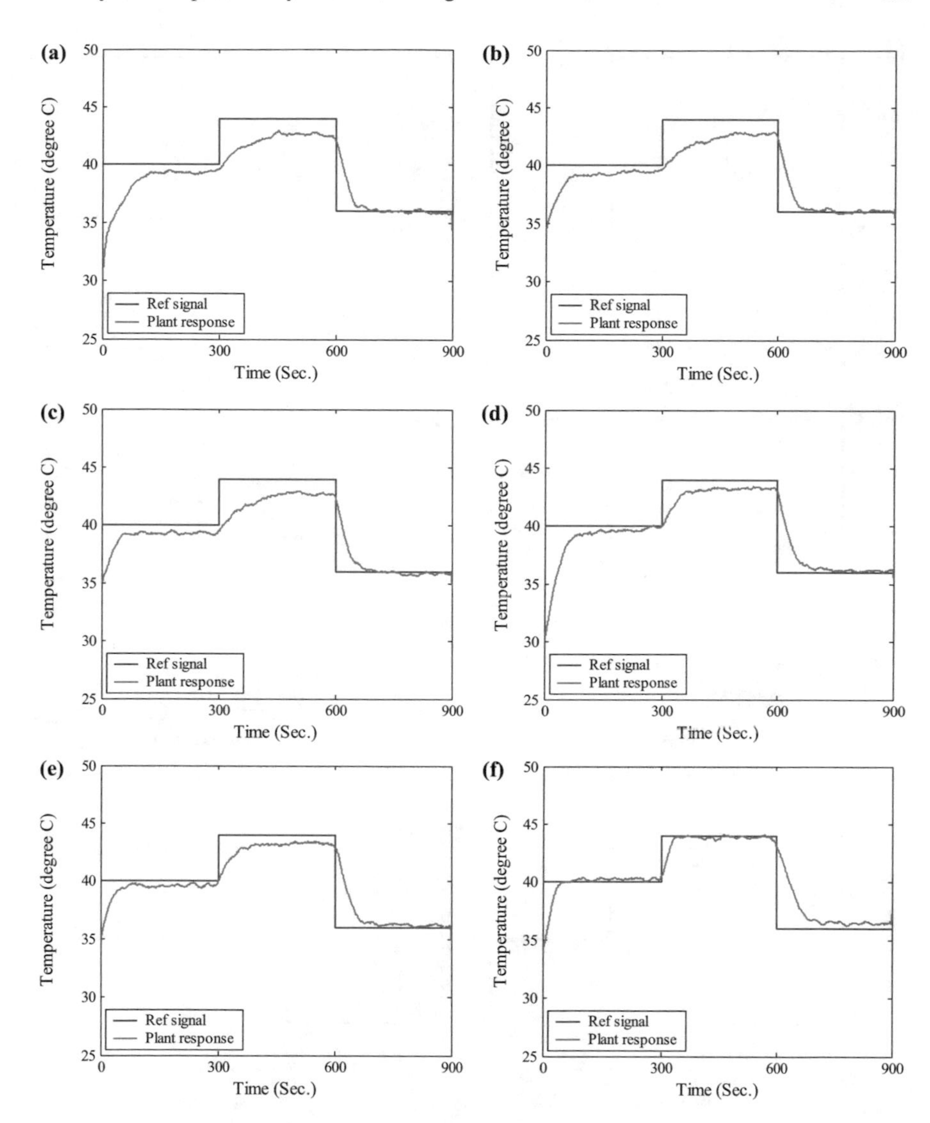

Fig. 8.24 Responses of the adaptive fuzzy controller in practical experimentation, for sensor #3 with fixed disturbance, of the air heater system: **a** fixed structure HSABA, **b** fixed structure HSA-Hybd-Con model, **c** fixed structure HSA-Hybd-Pref model, **d** variable structure HSABA, **e** variable structure HSA-Hybd-Con model, **f** variable structure HSA-Hybd-Pref model

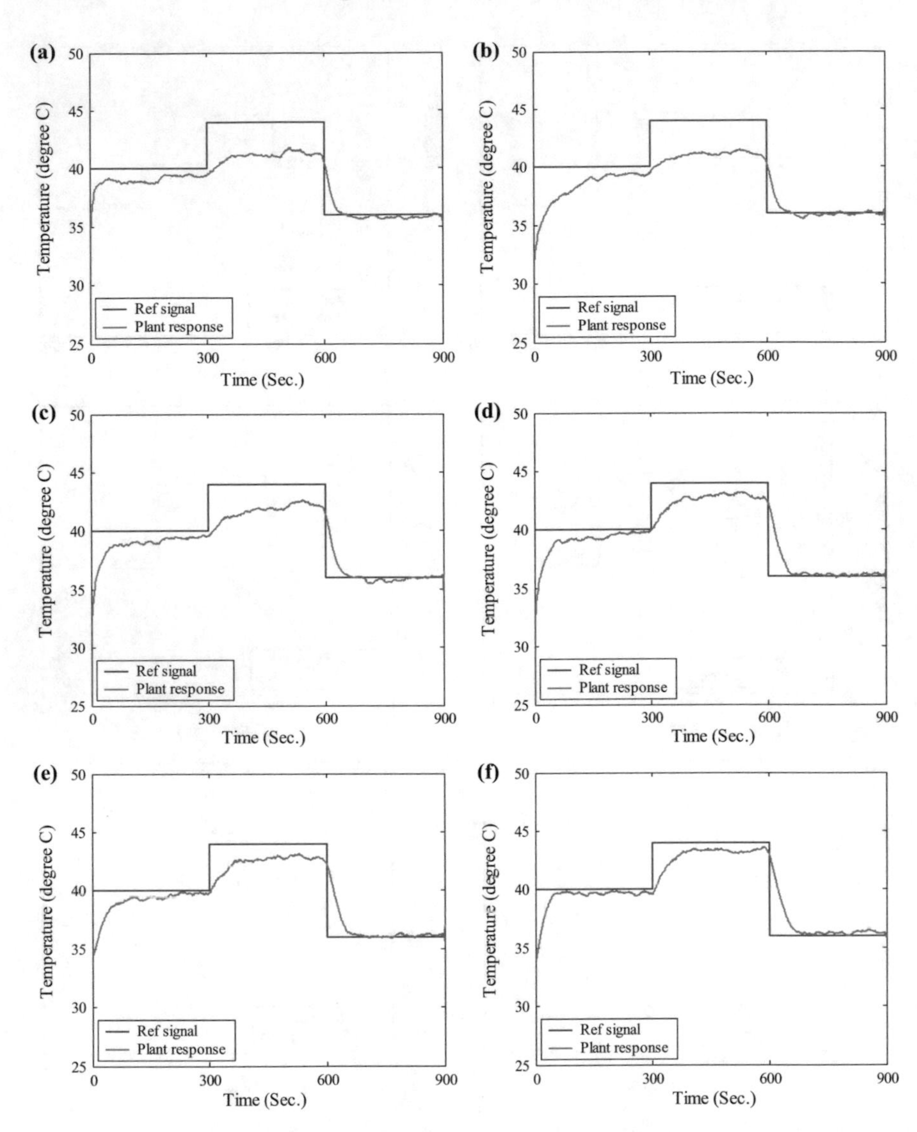

Fig. 8.25 Responses of the adaptive fuzzy controller in practical experimentation, for sensor #3 with randomly variable disturbance, of the air heater system: **a** fixed structure HSABA, **b** fixed structure HSA-Hybd-Con model, **c** fixed structure HSA-Hybd-Pref model, **d** variable structure HSABA, **e** variable structure HSA-Hybd-Con model, **f** variable structure HSA-Hybd-Pref model

8.5 Summary

This chapter demonstrated the successful implementations of the contemporary hybrid stable adaptive fuzzy controllers, based on the Lyapunov theory and PSO or HS algorithms, discussed in previous chapters, for controlling the temperature of the air heater process, at different cross section of air flow duct. The controllers discussed are designed incorporating the knowledge of heat transfer dynamics of the process and implemented satisfactorily in the presence of fixed and randomly variable disturbances. The same controllers designed for "no transportation delay" case are also utilized to control the temperature with the incorporation of transportation delay in the process. The designed controllers showed that they were still able to control the process, but with a reduced transient performance, which can be easily appreciated. Thus, it can be concluded that the proposed design methodologies are competent to provide satisfactory performances for the air heater process with variable heat transfer dynamics. In this real implementation experiment, the proposed preferential hybrid control strategy, for both PSO-based and HSA-based algorithms, overall emerge as the most suitable control option among the other discussed competing design techniques, for both fixed and variable structure configurations of the controller.

In the next chapter, another real implementation of vision-based mobile robot navigation is performed to demonstrate prospective wide range acceptance of the hybrid control approaches, discussed in this book.

References

1. U.-C. Moon, K.Y. Lee, Hybrid algorithm with fuzzy system and conventional PI control for the temperature control of TV glass furnace. IEEE Trans. Control. Syst. Tech. **11**(4), 548–554 (2003)
2. C.-F. Juang, C.-H. Hsu, Temperature control by chip-implemented adaptive recurrent fuzzy controller designed by evolutionary algorithm. IEEE Trans. Circuits Syst.-I **52**(11), 2376–2384 (2005)
3. R. Haber, J. Hetthessy, L. Keviczky, I. Vajk, A. Feher, N. Czeiner, Z. Csaazer, A. Turi, Identification and adaptive control of a glass furnace. Automatica **17**(1), 175–185 (1981)
4. S. Aoki, S. Kawachi, M. Sugeno, Application of fuzzy control logic for dead time process in a glass melting furnace. Fuzzy Sets Syst. **38**(3), 251–265 (1990)
5. E. Araujo, L.S. Coelho, Particle swarm approaches using Lozi map chaotic sequences to fuzzy modelling of an experimental thermal-vacuum system. Appl. Soft Comput. **8**(4), 1354–1364 (2008)
6. M.A. Denai, F. Palis, A. Zeghbib, Modeling and control of non-linear systems using soft computing techniques. Appl. Soft Comput. **7**(3), 728–738 (2007)
7. D. Filev, M. Lixing, T. Larsson, Adaptive control of nonlinear MIMO systems with transport delay: conventional, rule based or neural, in *Proceedings on 9th IEEE International Conference on Fuzzy Systems* (*FUZZ-IEEE 2000*), (2000), pp. 587–592
8. D. Simon, Kalman filtering for fuzzy discrete time dynamic systems. Appl. Soft Comput. **3** (3), 191–207 (2003)

9. C.-F. Juang, J.-S. Chen, Water bath temperature control by a recurrent fuzzy controller and its FPGA implementation. IEEE Trans. Indus. Electron. **53**(3), 941–949 (2006)
10. E. Kayacan, Y. Oniz, A.C. Aras, O. Kaynak, R. Abiyev, A servo system control with time-varying and nonlinear load conditions using type-2 TSK fuzzy neural system. Appl. Soft Comput. **11**(8), 5735–5744 (2011)
11. L.X. Wang, Stable adaptive fuzzy control of nonlinear system. IEEE Trans. Fuzzy Syst. **1**(2), 146–155 (1993)
12. D. Magazzeni, A framework for the automatic synthesis of hybrid fuzzy/numerical controllers. Appl. Soft Comput. **11**(4), 276–284 (2011)
13. A.R. Babaei, M. Mortazavi, M.H. Moradi, Classical and fuzzy-genetic autopilot design for unmanned aerial vehicles. Appl. Soft Comput. **11**(1), 365–372 (2011)
14. K. Das Sharma, A. Chatterjee, A. Rakshit, Harmony search algorithm and Lyapunov theory based hybrid adaptive fuzzy controller for temperature control of air heater system with transport-delay. Appl. Soft Comput. **25**, 40–50 (2014)
15. K. Das Sharma, *Hybrid Methodologies for Stable Adaptive Fuzzy Control*, Ph.D. Dissertation, Jadavpur University, Kolkata, India, 2012
16. K. Fischle, D. Schroder, An improved stable adaptive fuzzy control method. IEEE Trans. Fuzzy Syst. **7**(1), 27–40 (1999)
17. M. Mucientes, D.L. Moreno, A. Bugarin, S. Barro, Design of a fuzzy controller in mobile robotics using genetic algorithms. Appl. Soft Comput. **7**(2), 540–546 (2007)
18. Md.F Rahmat, Y.K. Hoe, S. Usman, N.A. Wahab, Modelling of PT326 hot air blower trainer kit using PRBS signal and cross-correlation technique. J. Teknologi **42**, 9–22 (2005)
19. K. Das Sharma, A. Chatterjee, A. Rakshit, A hybrid approach for design of stable adaptive fuzzy controllers employing Lyapunov theory and particle swarm optimization. IEEE Trans. Fuzzy Syst. **17**(2), 329–342 (2009)
20. K. Das Sharma, A. Chatterjee, A. Rakshit, Design of a hybrid stable adaptive fuzzy controller employing Lyapunov theory and harmony search algorithm. IEEE Trans. Control Syst. Tech. **18**(6), 1440–1447 (2010)
21. V. Giordano, D. Naso, B. Turchiano, Combining genetic algorithm and Lyapunov-based adaptation for online design of fuzzy controllers. IEEE Trans. Syst. Man, Cybern. B Cybern. **36**(5), 1118–1127 (2006)
22. Y. Shi, R.C. Eberhart, A modified particle swarm optimizer, in *Proceedings of IEEE International Conference Evolution Computational* (Alaska, May, 1998)
23. G. Venter, J.S. Sobieski, Particle swarm optimization, in *43rd AIAA Conference* (Colorado, 2002)
24. Z.W. Geem, J.H. Kim, G.V. Loganathan, A new heuristic optimization algorithm: harmony search. Simulation **76**(2), 60–68 (2001)
25. K.S. Lee, Z.W. Geem, A new structural optimization method based on the harmony search algorithm. Comp. Struct. **82**(9–10), 781–798 (2004)
26. LM35 Data sheet, National Semiconductor Inc., Nov. 2002, pp. 1–13
27. Microcontroller to sensor interfacing techniques, User Manual: ver 1.01, BiPOM Electronics Inc. (Texas, 2006), pp. 1–13
28. BTA/BTB12 and T12 series data sheet, ST Microelectronics (Italy, 2002), pp. 1–7
29. MOC3061/MOC3062/MOC3063—6-Pin DIP zero-cross optoisolators data sheet, Motorola Semiconductor Technical Data, 1995, pp. 1–6
30. 4-Wire pulse width modulation (PWM) controlled fans, Specification: Rev. 1.2, Intel Corporation, July 2004, pp. 1–23

Chapter 9
Experimental Study II: Vision-Based Navigation of Mobile Robots

9.1 Introduction

Autonomous mobile robot navigation has been regarded as a popular research area for quite sometimes now, which has still several open problems. There have been many attempts to solve the problems related to mobile robot navigation in both map-based and mapless environments [1–9]. In mobile robot navigation, simultaneous localization and mapping (SLAM) technique has also been popularly researched and several probable solutions to the problem have been proposed by the researchers [5, 6]. Earlier, robot navigation was mainly achieved by using different kinds of position sensors like IR, sharp IR, sonar sensors etc. [5, 7, 8]. The chief problem associated with these techniques is that the data fusion algorithms, used to synchronize the sensory data, cannot process the data at the same time stamp due to the out-of-sequence measurements (OOSM) problem. In order to make precise navigation, this OOSM problem should be solved. For this purpose, vision has become a popular alternative as a sensing mechanism [5, 6]. In some vision-based approaches, the navigation system is used to construct a high-level topological representation of the world. A robot using this system learns to recognize rooms or spaces and to navigate between them by building models of those spaces and their connections [8–10]. Many robot navigation systems also focus on producing a detailed metric map of the world using expensive hardware, such as a laser scanner. After this map is built, the robot then has to solve a complicated path planning algorithm [11–15]. These methods pose a computational burden to the processing unit and create difficulties during the real-life applications.

This chapter demonstrates a contemporary idea of formulating a mobile robot navigation problem as a tracking control problem, where vision is used as the primary sensing element for navigation [15, 16]. Here, vision is used to generate the reference path and adaptive state feedback fuzzy controllers are utilized to track that path. In this work, a single camera is used to capture the image of the environment with available floor area for navigation. Then, by employing an efficient path

K. Das Sharma et al., *Intelligent Control*, Cognitive Intelligence and Robotics,
https://doi.org/10.1007/978-981-13-1298-4_9

planning algorithm, a reference path is planned. The path is then decomposed to create two separate reference signals for x-direction and y-direction movements and two previously tuned adaptive fuzzy controllers are implemented, for x-direction and y-direction movements of the mobile robot. These controllers generate the control signals to produce the resultant drive signals, for left and right wheel actuators, and thus, the robot navigates. The major contribution of this work is that the controllers are tuned with some arbitrary reference signals generated from an alike environment where the robot will actually navigate and readily applied to the real applications. The system is so designed that, once the reference path is generated, the robot is commanded to track a specified fraction of this reference path. Then, a new image of the environment is acquired and the whole process of generation of new reference path, calculation of control actuations, and the robot navigation for a specified fraction of the reference path is repeated again and again in an iterative manner. This philosophy is adopted with the objective of coping the system with dynamically varying environments, so that robot navigation with dynamically varying positions of obstacles can still be carried out successfully.

The present chapter also demonstrates how the concepts, introduced in the earlier chapters, of designing stable adaptive fuzzy logic controllers (AFLCs) can be utilized for designing those x-direction and y-direction controllers for desired tracking performance. Here also, this is carried out with a systematic design procedure employing a hybridization of Lyapunov strategy-based approach (LSBA) and stochastic optimization, like particle swarm optimization (PSO) or harmony search algorithm (HSA)-based approach, i.e. PSOBA or HSABA, respectively [17–19]. This hybridization process attempts to combine the strong points of both methods to evolve a superior control strategy as mentioned before. In this design strategy, simultaneous adaptation of both the FLC structure and its free parameters are carried out. The stochastic global optimization-based method is employed to determine the most suitable completely optimized vector containing flags for determining controller structure and the free parameters of the AFLC, e.g., positions of input MFs, scaling gains and positions of output singletons. For each such global candidate solution of controller design, local adaptation by LSBA is carried out to further adapt the output singletons only.

In the present experimental case study, two types of hybridization of locally operative LSBA and global stochastic optimization have been utilized to design AFLCs for vision-based mobile robot navigation. As mentioned in earlier chapters, these hybrid adaptation strategy-based approaches (HASBA) discussed are called the concurrent and preferential hybrid approaches. This contemporary scheme for vision-based mobile robot navigation with the stable adaptive fuzzy controllers is implemented for both simulation case studies and two real implementation case studies, and the utility of the proposed approaches is aptly demonstrated. Thus, the novelty presented in these case studies is that the stable adaptive fuzzy controllers, designed by the stochastic optimization-based hybrid methodologies, are utilized for the tracking of the desired path for a vision based mobile robot and the tracking is guaranteed and theoretically proven with the help of Lyapunov theory.

9.2 Vision-Based Mobile Robot Navigation

The schematic of an autonomous mobile robot considered in this work is shown in Fig. 9.1. The robot consists of a vehicle with two driving wheels mounted on the same axis and two rear free wheels. The two driving wheels are independently driven in a differential manner by the two RC servo motors to achieve the motion and orientation. The robot is moving in a two-dimensional plane in which the world Cartesian coordinate system is defined. The present pose of the robot can be defined by the variable $p = [x \quad y \quad \varphi]^{\mathrm{T}}$ in the world Cartesian coordinate system. The coordinate (x, y) denotes the location of the center of mass of the robot, and the heading direction φ is taken counterclockwise from the x-axis. The entire locus of the point ($x(t)$, $y(t)$) is called the path or trajectory of the robot. If the time derivatives $\dot{x}$ and $\dot{y}$ exist, then $\varphi(t)$ can be expressed as follows:

$$\varphi(t) = \tan^{-1}\left[\frac{\dot{y}(t)}{\dot{x}(t)}\right] \tag{9.1}$$

The motion of the mobile robot in a plane can be described as follows [20]:

$$\begin{bmatrix} \dot{x} \\ \dot{y} \\ \dot{\varphi} \end{bmatrix} = \dot{p} = \begin{bmatrix} \cos\varphi & 0 \\ \sin\varphi & 0 \\ 0 & 1 \end{bmatrix} m \tag{9.2}$$

$$M(p)\dot{m} + V(p,\dot{p})m + G(p) = \tau \tag{9.3}$$

where $m = [v \quad \omega]^T$ is the vector of velocities, $m \in \Re^{n\times 1}$, v and ω are the linear and angular velocities, respectively, $\tau \in \Re^{n\times 1}$ is the input vector, $M(p) \in \Re^{n\times n}$ is a

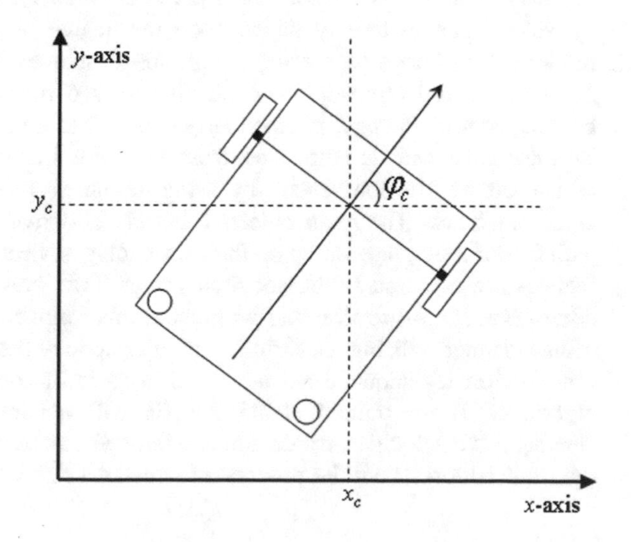

Fig. 9.1 Schematic of non-holonomic wheeled mobile robot

symmetric and positive definite inertia matrix, $V(p,\dot{p}) \in \Re^{n\times n}$ is the centripetal and Coriolis matrix, $G(p) \in \Re^{n\times 1}$ is the gravitational vector.

Equation (9.2) represents the kinematics or steering system of the mobile robot. Assuming that no-slip condition is satisfied as a non-holonomic constraint, it can be written as follows:

$$\dot{y}\cos\varphi - \dot{x}\sin\varphi = 0 \tag{9.4}$$

Thus, it is revealed from (9.4) that the mobile robot can only move in the direction normal to the axis of the driving wheels.

In this present work, the control objective is that, for a given reference trajectory $p_r(t)$ and orientation of mobile robot, a controller must be designed that applies suitable torque τ such that the current robot position $p_c(t)$ achieves the desired reference position $p_r(t)$:

$$\underset{t\to\infty}{Lt}\,(p_r(t) - p_c(t)) = 0 \tag{9.5}$$

To accomplish this control objective, the system should be so designed that a $\tau(t)$ is derived based on the differentially steered drive velocities to the left and right wheels of the robot according to the steering system in (9.2). This objective is satisfied by designing a stable adaptive fuzzy logic controller that can ensure asymptotic stability, achieve satisfactory transient performance, and can suitably configure this mobile robot navigation problem as a tracking control problem, where the controller guides the robot to track or navigate a desired tracking or navigation path. This navigation of the mobile robot can be logically carried out using different position sensors, e.g., ultrasonic sensors, IR sensors, global positioning system (GPS). In this case study, mobile robot navigation is carried out using vision as a primary sensing mechanism. While vision has emerged as a popular alternative as a robot sensor in recent times, the quality of solutions offered by vision sensing largely depends on the degree of sophistication involved in implementing image processing algorithms/techniques in such systems.

The proposed philosophy of the vision-based mobile robot navigation, using tracking control strategy, is described in Fig. 9.2 as a block diagram representation. The reference path is calculated from the color image acquired by the camera, which acts as a vision sensor, by using the image processing and path planning algorithm block. The main objective here is to derive a suitable, safe navigation path for the robot, depending on the surrounding environment, whose characteristic features are captured in the acquired image. Two stable adaptive T-S-type fuzzy controllers, for x-direction and y-direction robot movements, will be trained in an offline manner utilizing a hybridization of Lyapunov theory and a global stochastic optimization technique, such as particle swarm optimization or harmony search algorithm. These trained stable AFLCs will generate appropriate differential steering drive velocities, for the left and right wheels of the robot, in such a way that the required torque will be produced to track the reference path. A feedback signal

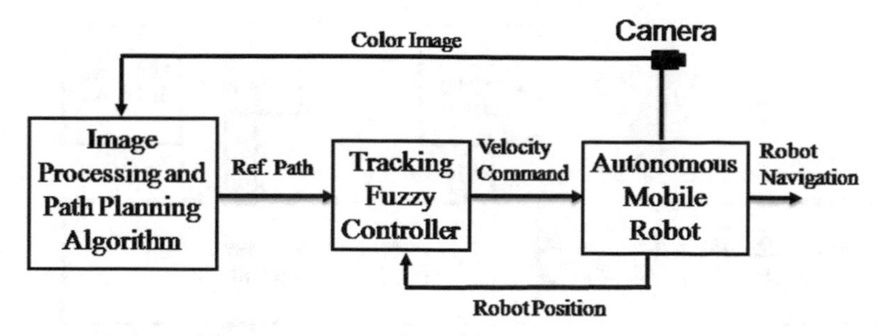

Fig. 9.2 Block diagram of the proposed vision-based mobile robot navigation scheme

containing the information about the current position of the robot is suitably utilized by the fuzzy controller to generate the left and right wheel velocities [21, 22].

9.3 Vision-Based Mobile Robot Experimental Platform

9.3.1 Mobile Robot Hardware Description

A differentially steered drive robot, indigenously designed and built in the Measurement and Instrumentation Laboratory, Electrical Engineering Department, Jadavpur University, Kolkata, India, is used as the experimental platform, as shown in Fig. 9.3. The proposed algorithms are implemented in Visual Basic on an ASUS EEE PC Laptop, with one GB solid-state HD, one GB RAM, Windows XP operating system, mounted on board [23]. The Laptop communicates with the robot base which comprises a PIC18F4550 processor, through USB link. The robot has one wheel on each front side, connected with RC servo motors, and two rear wheels, which are not motorized. The robot is equipped with encoder feedback from the left and right wheel encoders which generates four pulses/rotations and uses hall-effect switches in its operation. The Laptop camera with auto-focus facility serves as the mono-vision sensor. The robot base is energized (5 V, 1 A) from the Laptop through two USB cables. No separate power source is required for mobile robot operation, and all RC servos are power controlled for energy saving. Six IR proximity sensors and an IR range sensor processor PIC12F683 are also attached to the robot base, although they are not utilized in this work [21–23].

9.3.2 Image Processing Technique

A mobile robot using IR sensors can detect obstacles only over a limited range [20, 24]. However, a camera mounted on the robot can help it to extend its range and, hence, its applicability in the real world [2, 15, 16, 25].

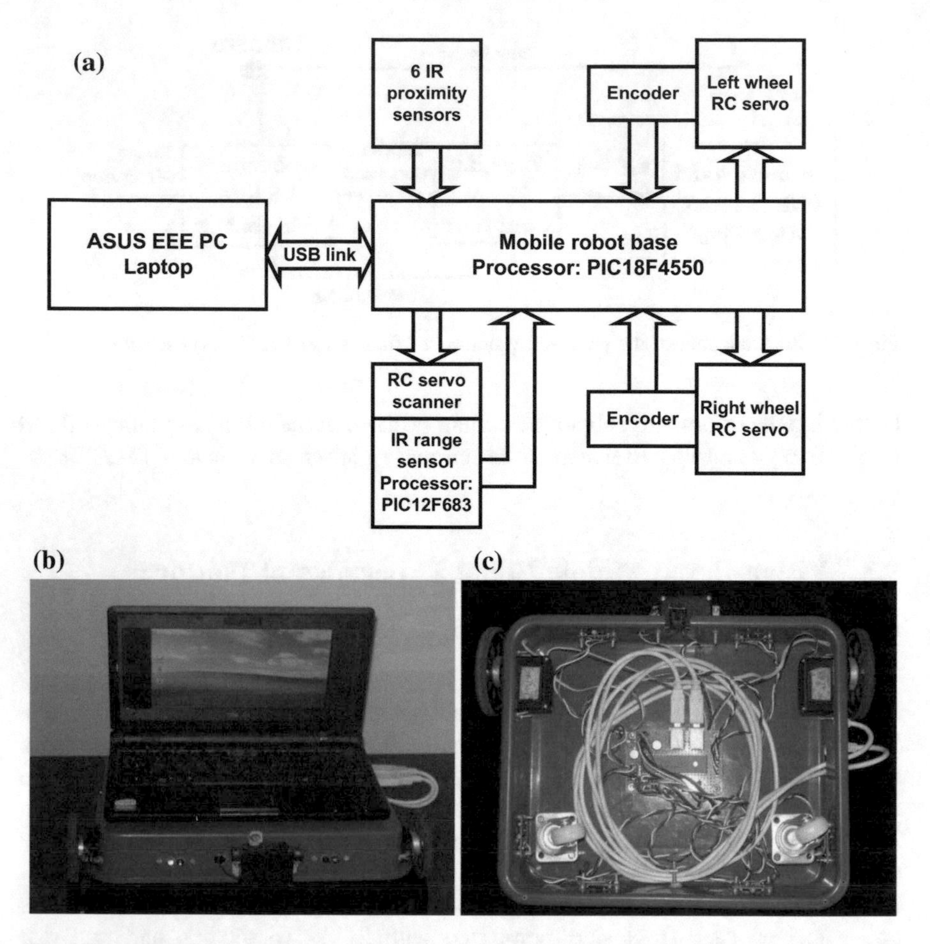

Fig. 9.3 **a** Block diagram of internal circuitry of the indigenously developed mobile robot, **b** front view and **c** bottom view of the actual robot built

During the process of navigation, the robot intermittently stops, the camera acquires image of the environment in front of the robot, image processing is performed on that image to determine the reference path, the controllers determine the suitable drive commands, and the robot navigates for a desired duration. Then, the robot stops again, acquires image, and goes through the before-mentioned steps once again. This process of navigating the robot for a desired duration and then stopping it and activating the vision sensor is carried out in an iterative fashion, until the robot completes the navigation job. In the image processing step, first the image is captured by the auto-focusing camera in-built in the Laptop. Then, this RGB color image is converted into a grayscale image for further processing. Gaussian blur filter is then used to de-noise this image. It is a common practice in computer vision-based applications that, to enhance the image structure scales, Gaussian smoothing is used as a preprocessing stage. Also, the Gaussian blur filter

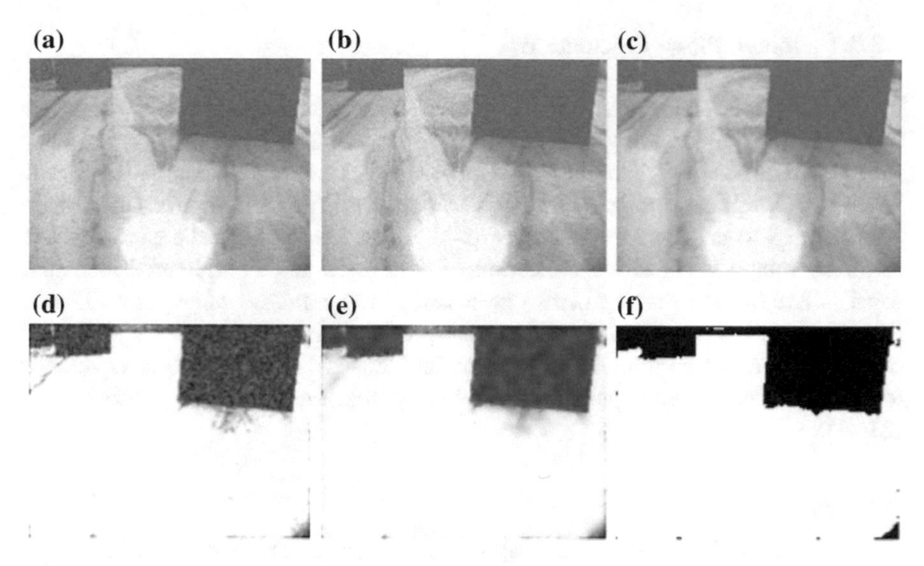

Fig. 9.4 Results of the image processing steps for a sample case: **a** captured RGB image, **b** grayscale image, **c** Gaussian blur filtered image, **d** contrast adjusted image, **e** geometric mean filtered image, and **f** final binary image after thresholding

can eliminate the high-frequency components from the image. Next, the image is processed by contrast adjustment to sharpen the distinguishable boundaries of the objects in the image. Then, geometric mean filter, a special type of nonlinear mean filter, is applied which is more efficient in removing Gaussian-type noise and preserving edge features, compared to the arithmetic mean filter. Once this series of image filtering steps is completed, then image segmentation technique is applied to the grayscale preprocessed image. For the segmentation purpose, thresholding technique is used, known as a popular, simple method to create a binary image [21–23]. Figure 9.4a–f shows a sample situation where a raw captured image undergoes these preprocessing steps and final processed binary image is produced.

9.3.3 Path Planning Algorithm

Several path planning techniques have so far been reported by the researchers during the last two decades [3, 8, 9, 13, 14, 26]. In this work, a simple but accurate path planning algorithm is proposed in the image plane and that path is subsequently projected in real-world Cartesian coordinate system. These proposed methodologies are detailed next.

9.3.3.1 Image Plane Calculations

The single Laptop camera is mounted on the robot in such a way that the captured image carries the necessary information about the available floor area in the front for navigation, along with the other objects in the environment, as shown in Fig. 9.4a. The binary image (Fig. 9.4f) as produced from the segmentation operation is used to calculate the reference path of the mobile robot at a given instant. A binary image is actually a collection of data in 0s and 1s structured in matrix form, where 0s represent the black pixels and 1s represent the white pixels. Due to the segmentation operation, the illuminated floor area mostly becomes white and most of the other portions of the image turn black. An algorithm is developed to calculate the reference path from the binary image as shown in Algorithm 9.1 [21–23].

BEGIN

1. Determine the size of the captured image as $(nrows \times ncols)$.
2. Initialize $i = 0$.
3. **FOR** $i = nrows$ to 1 (in steps of -1),
   3a. Store the pixel positions of white pixels along the complete image row in the array *w_pixel*(i,:).
   3b. Calculate the middle pixel position for ith row as *middle_w_pixel*(i).
   ENDFOR
4. Set $i = 0$.
5. **FOR** $i = nrows$ to 2 (in steps of -1),
   5a. Connect *middle_w_pixel*(i) and *middle_w_pixel*(i-1) by a straight line segment.
   ENDFOR

END

Algorithm 9.1: Generation of reference path in image plane.

In step 3(b), if there is an even number of white pixels in a row, then there is no single middle value and the middle pixel position is defined to be the mean of the two middle values. But as it is known that a pixel position is always an integer number, thus in that case, the mean value is floored to its next integer value. In step 5, connect the stored middle positions of the pixels by straight line segments to create the reference path, for that captured image. If there is any obstacle in the path, then the binary image will show two available paths, the algorithm will then decide which path is wider by measuring the white pixels as stored in step 3(a), and the reference path will be calculated along that path avoiding the obstacle.

Once the reference path is generated in the image plane, it is projected on the real space of the Cartesian coordinate system, and the fuzzy controller is used to navigate the robot, tracking the reference path. However, one important point is that, during the process of navigating the robot, the environment may get changed and the algorithm should be flexible to cope with the changes in the dynamic environment. For this, the robot should intermittently acquire images of the environment and calculate reference path, each time, and perform navigation accordingly. In this real implementation, for each reference path generated, 1/4th of this path is used by the robot for immediate tracking purpose, and once this path tracking is exhausted, a new image is acquired for subsequent generation of reference path. This process is continued in an iterative fashion. Figure 9.6a shows the reference path calculated in terms of the pixel positions, corresponding to the binary image as shown in Fig. 9.4f [21–23].

9.3.3.2 Image Plane to Real Space Conversion Technique

After calculating the reference path in the image plane, i.e., in terms of image pixel positions, it is required to project these locations to the real world coordinates, where the robot will actually navigate. The coordinate transformation is calculated by measuring four distances as shown in Fig. 9.5. These include the height of the camera from the floor, h. It is to be noted that there is always a blind area immediately in front of the robot. Hence, the distance of the nearest point that can be viewed by the camera is denoted as y_b.

The furthest point visible to the camera is at a distance $(y_b + y_l)$, and the fourth distance is the furthest point to the left or right of the camera lateral view, denoted as x_l. With these measurements, the three transformation angles are calculated as follows [25]:

$$\begin{aligned} \alpha &= \tan^{-1}\left[\frac{h}{y_b}\right], \\ \beta &= \tan^{-1}\left[\frac{h}{y_b + y_l}\right], \text{ and} \\ \gamma &= \tan^{-1}\left[\frac{x_l}{y_b + y_l}\right] \end{aligned} \tag{9.6}$$

The point $P(x, y)$ in the world coordinate, which is obtained by transforming from the point $P'(u, v)$ in the image plane, is obtained as follows [25]:

$$y = h \tan\left[\left(\frac{\pi}{2} - \alpha\right) + \left[\frac{u}{S_y}\right](\alpha - \beta)\right] \tag{9.7}$$

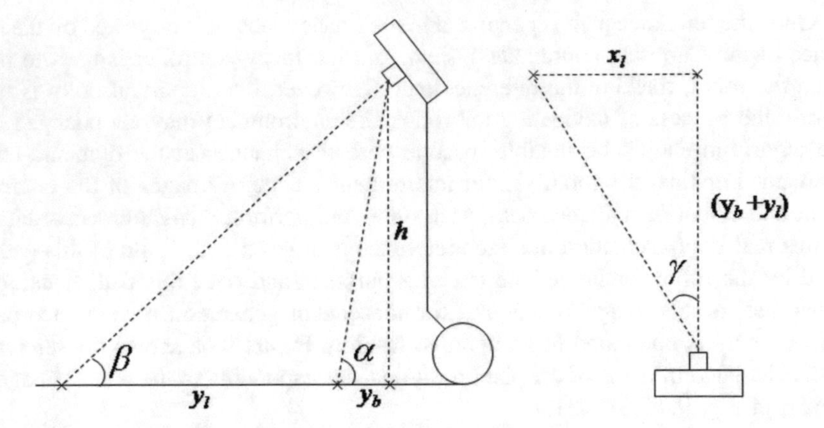

Fig. 9.5 Relationship between the image plane coordinates and the mobile robot real space coordinates

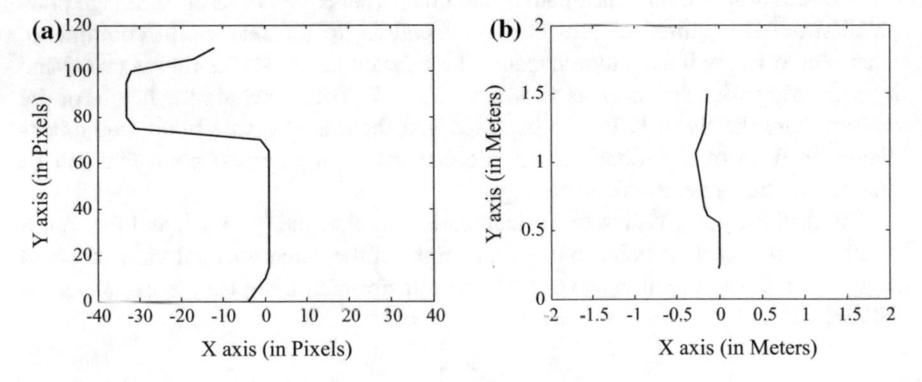

Fig. 9.6 Planned path in **a** image plane and **b** real space (world coordinate)

$$x = y \tan\left[\frac{2v}{S_x} \times \gamma\right] \tag{9.8}$$

where S_x and S_y are the number of image pixels in the vertical axis and horizontal axis, respectively [23, 25].

Thus, after transforming the data in image plane to the real space coordinates, the reference path is created in the real space, as shown in Fig. 9.6b. Hence, the image processing and the path planning technique can be summarized in form of the block diagram as shown in Fig. 9.7.

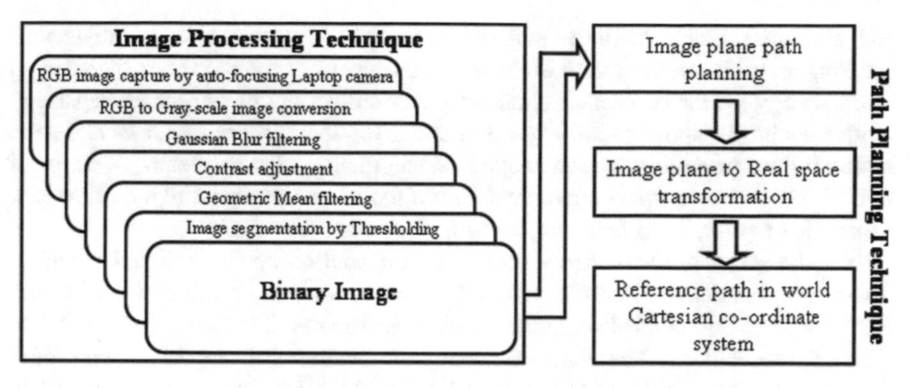

Fig. 9.7 Flowchart of image processing and path planning techniques

9.4 Hybrid Adaptive Fuzzy Tracking Controller Design

A contemporary methodology for autonomous mobile robot navigation utilizing the concept of tracking control will be detailed now. It will be discussed now how vision-based path planning and subsequent tracking can be performed by utilizing proposed stable adaptive state feedback fuzzy tracking controllers, designed using the Lyapunov theory and the stochastic global optimization technique, like particle swarm optimization or harmony search algorithm-based hybrid approaches, as discussed in detail in earlier chapters in this book. The objective is to design two self-adaptive fuzzy controllers, for *x*-direction and *y*-direction movements, optimizing both its structures and free parameters, such that the designed controllers

IF $|\Delta\varphi| \le 10°$,
  THEN $V_L = V$; $V_R = V$; % the robot will move straight
ELSE IF $((\Delta\varphi > 10°)$ and $(\Delta\varphi < 90°))$,
  THEN $V_L = 0$; $V_R = V$; % the robot will rotate left
ELSE IF $((\Delta\varphi > -90°)$ and $(\Delta\varphi < -10°))$,
  THEN $V_L = V$; $V_R = 0$; % the robot will rotate right
ENDIF

Algorithm 9.2 Calculation of left and right wheel velocities.

can guarantee desired stability, and simultaneously, they can provide satisfactory tracking performances for the vision-based navigation of mobile robot. The design methodology for the controllers simultaneously utilizes the global search capability of the stochastic optimization algorithm and Lyapunov theory-based local search method, thus providing a high degree of automation. Two different variants of hybrid approaches, namely concurrent hybrid model and preferential hybrid model, as detailed earlier, have been employed in this work [21–23].

The design of adaptive fuzzy controllers for controlling the x-direction and y-direction movements of the vision-based mobile robot is achieved by the PSO and HSA-based hybrid control strategies, as discussed in detail in Sects. 3.3 and 3.4 in Chap. 3, respectively. The concurrent and preferential hybrid models of controller design techniques, both PSO-based and HSA-algorithm based, as discussed in Sect. 5.2 in Chap. 5, are utilized along with the solely PSO-based and HSA algorithm-based design procedures, for navigation of the wheeled mobile robot with vision sensor.

To evaluate the effectiveness of the proposed control strategies, first the autonomous mobile robot navigation schemes in simulation are developed. During simulation, the fuzzy controllers according to different control schemes as stated before are trained. These trained controllers are implemented online to perform the real implementations. In simulation study, the process model is simulated with a sampling time of 1 s., keeping the hardware constraints for the robot in real implementations, in mind [23].

The steering equations of the mobile robot as in (9.2) can be rewritten as follows [21–23]:

$$
\begin{aligned}
\dot{x} &= v\cos\varphi = u_x \\
\dot{y} &= v\sin\varphi = u_y \\
\dot{\varphi} &= \omega
\end{aligned}
\tag{9.9}
$$

Now (9.9) is a special case of (2.1) as mentioned in Chap. 2, where $f(\cdot) = 0$ and $b = 1$. $u = u_x$ is the control signal from x-direction tracking controller and $u = u_y$ is the control signal from y-direction tracking controller. As stated before, φ can be obtained from (9.1).

Thus, the main idea of using state feedback tracking controller for vision-based mobile robot navigation is that, after acquiring an image and calculating a reference path for navigation in real Cartesian space (x-y plane), the path is decomposed into two separate reference paths, i.e., (x vs. t) for x-direction controller and (y vs. t) for y-direction controller. The two controllers are trained simultaneously to achieve the desired tracking of both x-direction and y-direction reference paths, and this should consequently produce resultant tracking of the reference path in x-y plane [22].

It is assumed that the reference point lies at the midpoint of the line joining the two drive wheels. Let V_L and V_R denote the velocities of the left and the right wheels, respectively. The linear velocity V and the orientation φ of the mobile robot can be described as follows [22, 23]:

$$\begin{aligned} V &= \sqrt{\left(u_x^2 + u_y^2\right)}, \\ \varphi &= \tan^{-1}\left(\frac{u_y}{u_x}\right) \end{aligned} \tag{9.10}$$

However, the theoretical expressions are suitably modified, to satisfy the hardware constraints of the robot system and to maintain robustness of the mechanical components of the robot developed for long-term operation. The V_L and V_R, left and the right wheel velocities, respectively, are individually determined from the linear velocity V and the orientation φ according to the algorithm 9.2 as follows:

The magnitudes of u_x and u_y are so generated from the respective fuzzy controllers that the linear velocity V and the corresponding $V_L$and V_R will drive the robot to track the generated path by the vision sensor.

The current position of the real robot can be determined form the position encoders as follows [21–23]:

$$\begin{aligned} \delta_{\text{new}} &= \delta_{\text{old}} + (\Delta p_L - \Delta p_R) * 10^\circ \\ x_{\text{new}} &= x_{\text{old}} + (\Delta p_L + \Delta p_R)\cos(\delta_{\text{new}}) \\ y_{\text{new}} &= y_{\text{old}} + (\Delta p_L + \Delta p_R)\sin(\delta_{\text{new}}) \end{aligned} \tag{9.11}$$

where Δp_L and Δp_R are the increments of the left and right encoders, respectively, and δ is the heading angle of the robot. x_{new} and y_{new} are used as inputs to the state feedback x-direction and y-direction controllers, respectively, for both simulation and real implementations.

In stochastic optimization-based design methodology, the optimum scaling gains and positions of output singletons, as well as the optimum controller structure are determined [27, 28]. Thus, in this design methodology, as described before, a candidate solution vector's position in solution space is a vector containing all required information to construct a fuzzy controller, e.g., (i) information about the number and positions of the input MFs, (ii) number of rules, (iii) values of scaling gains, (iv) positions of the output singletons. This candidate solution vector is formed as follows [18–23]:

$$\begin{aligned} \underline{Z} = [&\text{structural flags for input MFs}|\text{center locations of the input MFs}\ldots \\ &|\text{scaling gains}|\text{positions of the output singletons}] \end{aligned} \tag{9.12}$$

The structural flags are utilized to obtain the structure of the fuzzy controllers, i.e., the number and positions of the input MFs and subsequently number of rules and positions of output singletons. As the number of rules of a fuzzy controller has much importance from the point of view of the real implementation, thus in these case studies, structural flags are changed in such a fashion that the number of membership functions required to fuzzify each input variable of the fuzzy controller is varied from a minimum value of 2 to a maximum value of 7, so as to keep the computational burden for the processor as minimum as possible and also simultaneously achieve desired performance.

In hybrid methods, a candidate solution vector $\underline{Z}$ is divided into two sub-groups, given as follows[18, 19]:

$$\underline{Z} = [\underline{\psi} \mid \underline{\theta}] \tag{9.13}$$

where

$$\left.\begin{array}{l} \underline{\psi} = [\text{structural flags for MFs}|\text{center locations of the MFs}|\ldots|\text{scaling gains}] \\ \underline{\theta} = [\text{positions of the output singletons}] \end{array}\right\} \tag{9.14}$$

The partition of $\underline{Z}$ vector separates the parameters in hybrid design approaches accordingly as $\underline{\psi}$ has nonlinear influence and $\underline{\theta}$ has linear influence on the control effort taken by the fuzzy controller, respectively [18, 29–31].

9.4.1 Simulation Case Study

9.4.1.1 PSO-Based Study

During the simulation, each process model is simulated using a fixed step fourth-order Runge–Kutta method. In this work, all the parameters of $\underline{Z}$ are optimized in PSOBA algorithm by the global version of basic PSO [18, 32–34]. To perform this simulation, a population size of ten particles is chosen. For each simulation, 200 iterations of PSO are performed. In fixed structure controller design, a fixed number of evenly distributed input MFs are used for the initial population of PSO. PSOBA algorithm tunes the free parameters of the controller and the positions of the output singletons, to minimize the tracking error. In variable structure controller design, the structural flags are utilized to automatically vary the structure of the candidate controller in each iteration. The input MFs and the corresponding positions of the output singletons are set according to these structural flags. Integral absolute error (*IAE*) computed between the reference position p_r and the robot current position p_c, over a chosen time duration is chosen as the fitness function for optimization [21–23].

In concurrent hybrid model (PSO-Hybd-Con), the global version PSO-based optimization of parameters for $\underline{Z} = \left[\underline{\psi}|\underline{\theta}\right]$, and the Lyapunov-based algorithm, which only adjusts $\underline{\theta}$, are run concurrently. The upper and lower bounds of the parameters are kept same as in PSOBA. The adaptation gains $v_x = 0.01$ and $v_y = 0.1$ are chosen, for *x*-direction and *y*-direction controllers, respectively, providing a good tradeoff between fast learning and small amplitudes of the parameter oscillations due to inherent approximation error. The structural flags are set and reset in the same manner as it is implemented in PSOBA and utilized to automatically determine the structure of the candidate controller [23].

In preferential hybrid model (PSO-Hybd-Pref), both concurrent hybrid model and PSOBA are implemented for the same particle separately at the same time, and the updating of each particle's position and velocity is carried out accordingly. The initial values of the population, i.e., the initial position of the particles in the solution space, used in hybrid models, are the same as in PSOBA simulation, so that a fair comparison can be carried out among the strategies in an identical platform. The results of simulation studies are tabulated in Table 9.1. In all these simulations, the robot navigation is evaluated in implementation phase for different competing controllers adapted offline during separate training procedures [23].

The simulations are carried out for the fixed structures of 5 input MFs and each variable structure with a potentially biggest possible structure of 7 input MFs. Table 9.1 presents the comparison of two PSO-Hybd controllers vis-à-vis PSOBA controllers, each trained with its fixed and variable versions. For a given controller MF configuration, among the competing controllers in fixed structure configuration, the PSO-Hybd-Pref model shows the best performance, whereas in variable structure case, PSOBA evolved as a better solution, as shown in Table 9.1. In variable structure configuration, the number of input MFs is reduced to some extent, and hence, the number of rules. Such a simplified structure of an FLC will be very

Table 9.1 Comparison of simulation results for mobile robot navigation: PSO-based design strategies

Control strategy	No. of MFs in *x*-dir controller	No. of MFs in *y*-dir controller	IAE
PSOBA	5	5	1.0153
PSO-Hybd-Con			1.6289
PSO-Hybd-Pref			0.9880
PSOBA-V	4	4	0.9860
PSO-Hybd-Con-V	3	5	1.6280
PSO-Hybd-Pref-V	3	6	1.6004

useful from the point of view of implementation in real robot. It should be noted here that, during tabulation of results, *-V denotes the variable structure configurations of the controllers, i.e., PSOBA-V means PSOBA algorithm in variable structure configuration.

Figures 9.8a, 9.9a and 9.10a show the positions of the input MFs. Figures 9.8b, 9.9b and 9.10b show the positions of output singletons, for x-direction controller and y-direction controller, respectively. Figures 9.8c, 9.9c and 9.10c show the robot motion with adaptive fuzzy controller, optimized by PSOBA, PSO-Hybd-Con model, and PSO-Hybd-Pref model, for the fixed structure x-direction controller and y-direction controller, with 5 input MFs. Similarly, Figs. 9.11, 9.12, and 9.13 show similar results for the adaptive fuzzy controllers, optimized by PSOBA, PSO-Hybd-Con model and PSO-Hybd-Pref model, with the variable structure configuration.

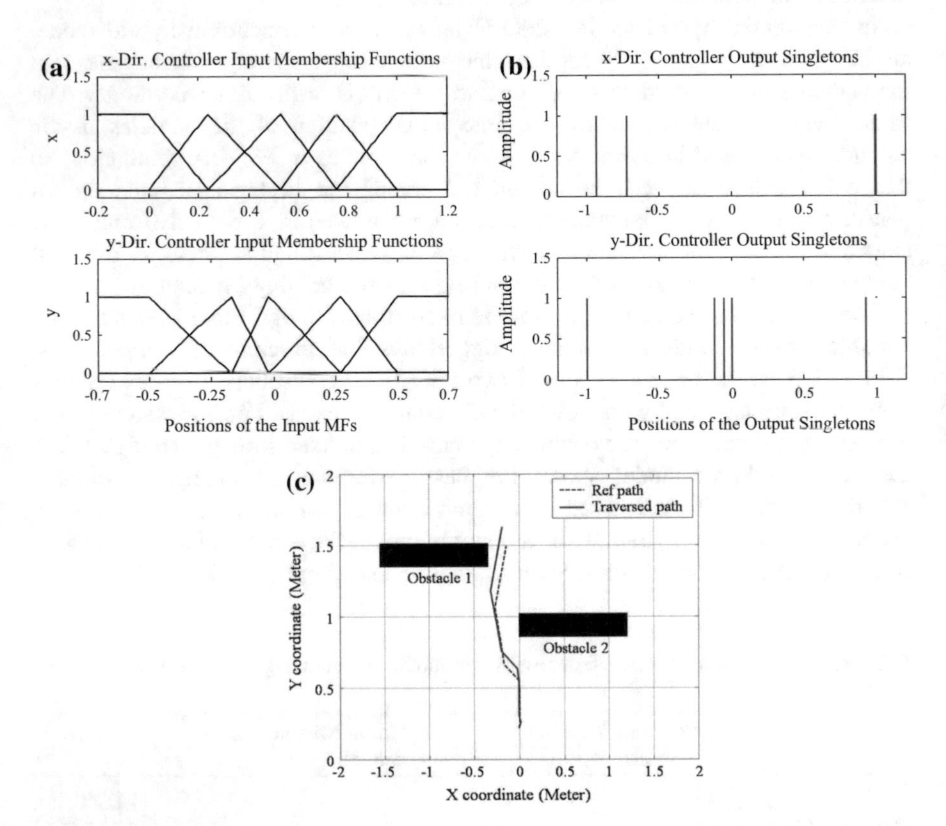

Fig. 9.8 Responses of the adaptive fuzzy controller with PSOBA, for fixed structure configuration of input MFs, of the mobile robot navigation during simulation study: **a** positions of input MFs for x-direction and y-direction controllers, **b** positions of output singletons for x-direction and y-direction controllers, **c** evaluation period response after 200 PSO generations

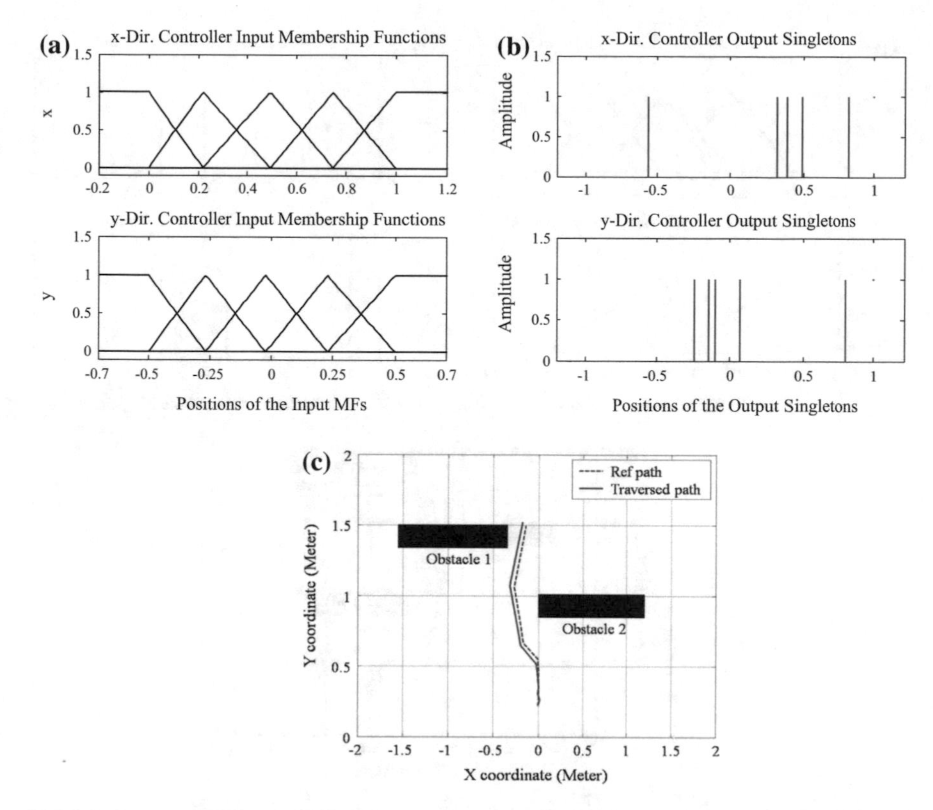

Fig. 9.9 Responses of the adaptive fuzzy controller with PSO-Hybd-Con model, for fixed structure configuration of input MFs, of the mobile robot navigation during simulation study: **a** positions of input MFs for x-direction and y-direction controllers, **b** positions of output singletons for x-direction and y-direction controllers, **c** evaluation period response after 200 PSO generations

9.4.1.2 HSA Based Study

This section describes how HS algorithm [35–37] is used to design the hybrid stable adaptive state feedback fuzzy controllers for autonomous mobile robot navigation, utilizing the concept of tracking control [22, 23]. Vision-based path planning and the design of tracking controllers for controlling the x-direction and y-direction movements are performed with the same objective as discussed in PSO-based design technique in Sect. 9.4.1.1. In this HS algorithm-based case study too, the concurrent and preferential hybrid model-based controllers are implemented to study the effectiveness of the proposed control strategies.

In HS algorithm-based design, the information needed to construct the direct adaptive fuzzy controllers for x-direction and y-direction movements of mobile robot is obtained in the same fashion as in the case of the PSO-based design

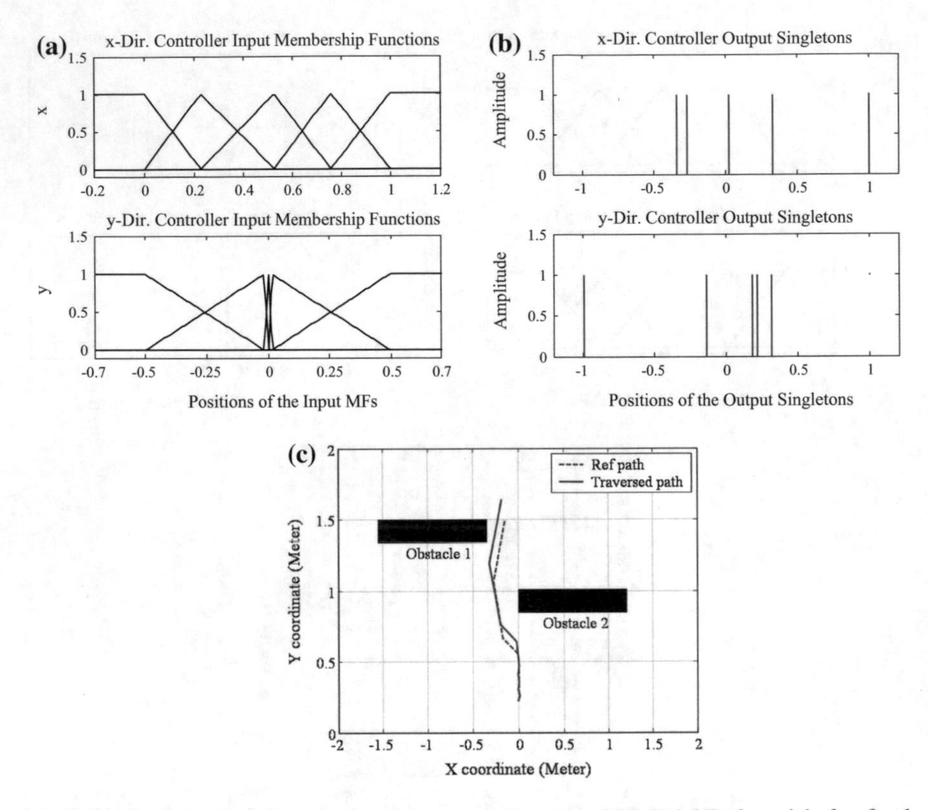

Fig. 9.10 Responses of the adaptive fuzzy controller with PSO-Hybd-Pref model, for fixed structure configuration of input MFs, of the mobile robot navigation during simulation study: **a** positions of input MFs for *x*-direction and *y*-direction controllers, **b** positions of output singletons for *x*-direction and *y*-direction controllers, **c** evaluation period response after 200 PSO generations

discussed before. In this design methodology too, the number of MFs in each input variable, the supports of input MFs, the number of rules, the values of the scaling gains, and the positions of output singletons are optimized to evaluate a candidate controller settings for better performance, in terms of the *IAE* value between the path planned using vision sensor and the actual path traversed by the mobile robot.

Both the process models are simulated using a fixed step fourth-order Runge–Kutta method in this case study too. All the parameters of $\underline{Z}$ are optimized in HSABA algorithm by the modified harmony search algorithm [19]. In this simulation study, a harmony memory of ten harmonies is chosen, and for each simulation, 200 improvisations of the harmony memory are performed. The fixed and variable configuration controllers, for both *x*-direction and *y*-direction, are designed in the same fashion as in PSO-based study given before. The *IAE* values are

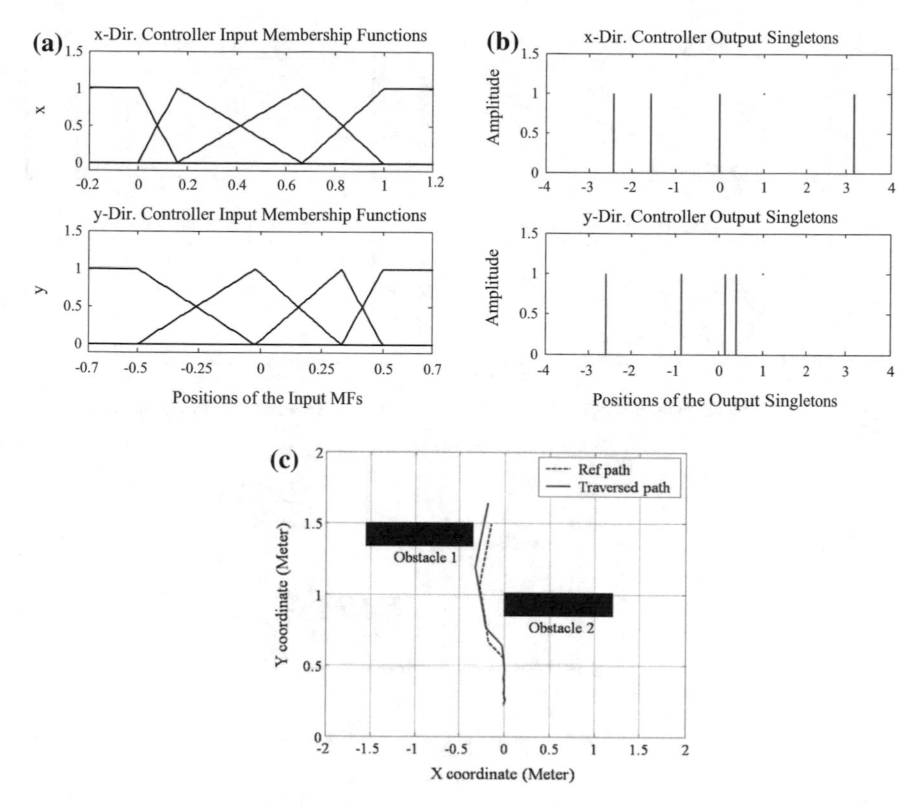

Fig. 9.11 Responses of the adaptive fuzzy controller with PSOBA, for variable structure configuration of input MFs, of the mobile robot navigation during simulation study: **a** positions of input MFs for x-direction and y-direction controllers, **b** positions of output singletons for x-direction and y-direction controllers, **c** evaluation period response after 200 PSO generations

computed to compare the performances of the navigation schemes with different HS algorithm-based control strategies. In this case study too, the concurrent hybrid model and the preferential hybrid model are implemented utilizing the locally adaptive Lyapunov theory and HS algorithm-based global optimization technique. The adaptation gains ($v_x = 0.01$ and $v_y = 0.1$ are taken for x-direction and y-direction controllers, respectively), and the other design parameters are chosen same, as in PSO-based study [23]. The results of HS algorithm-based simulation studies are given in Table 9.2 which shows the performances of robot navigation, in simulation, evaluated in implementation phase for different competing controllers,

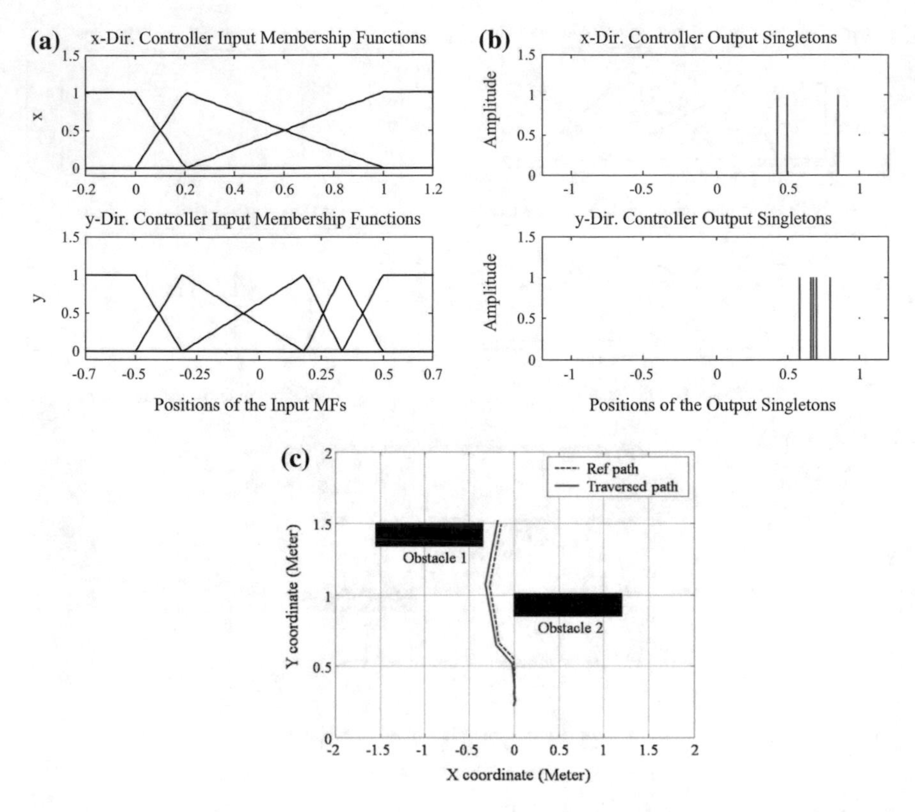

Fig. 9.12 Responses of the adaptive fuzzy controller with PSO-Hybd-Con model, for variable structure configuration of input MFs, of the mobile robot navigation during simulation study: **a** positions of input MFs for x-direction and y-direction controllers, **b** positions of output singletons for x-direction and y-direction controllers, **c** evaluation period response after 200 PSO generations

adapted during separate training procedure. Figures 9.14a, 9.15a, and 9.16a show the positions of the input MFs. Figures 9.14b, 9.15b, and 9.16b show the positions of output singletons for x-direction controller and y-direction controller. Figures 9.14c, 9.15c and 9.16c show the robot motion with adaptive fuzzy controller, optimized by HSABA, HSA-Hybd-Con model, and HSA-Hybd-Pref model, for the fixed structure x-direction controller and y-direction controller, with five input MFs. In a similar fashion, Figs. 9.17, 9.18, and 9.19 show the same items for the adaptive fuzzy controllers optimized by HSABA, HSA-Hybd-Con model, and HSA-Hybd-Pref model with the variable structure configuration. It should also be noted here that, during tabulation of results, *-V denotes the variable structure configurations of the controllers, i.e., HSABA-V means HSABA algorithm in variable structure configuration.

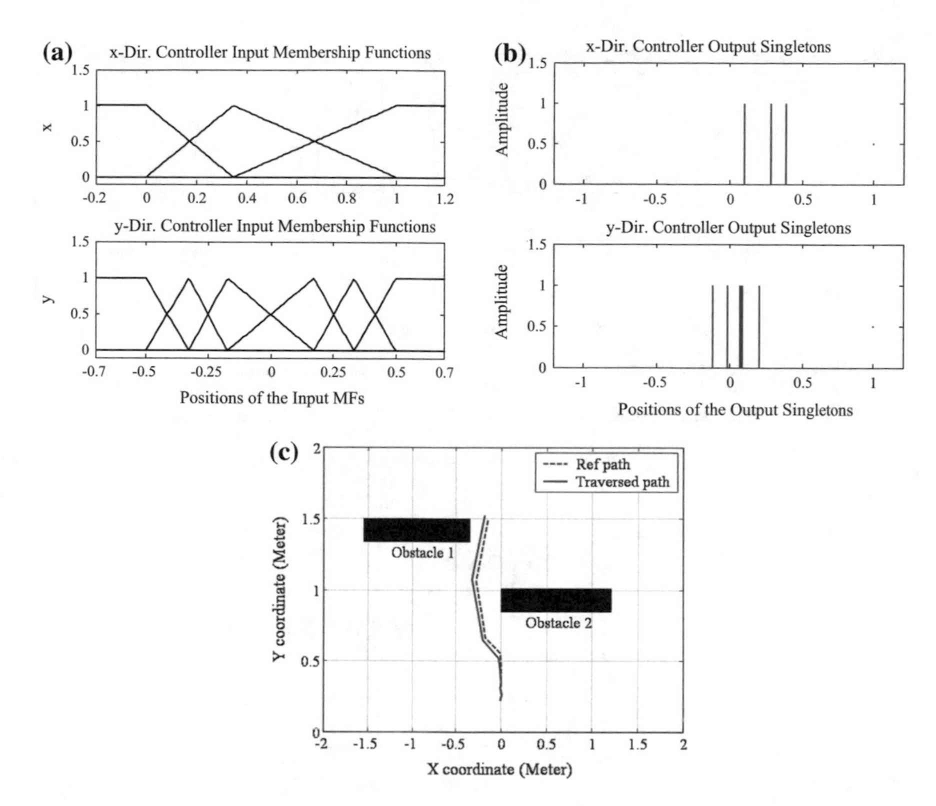

Fig. 9.13 Responses of the adaptive fuzzy controller with PSO-Hybd-Pref model, for variable structure configuration of input MFs, of the mobile robot navigation during simulation study: **a** positions of input MFs for x-direction and y-direction controllers, **b** positions of output singletons for x-direction and y-direction controllers, **c** evaluation period response after 200 PSO generations

Table 9.2 Comparison of simulation results for mobile robot navigation: HSA-based design strategies

Control strategy	No. of MFs in x-dir controller	No. of MFs in y-dir controller	IAE
HSABA	5	5	1.0176
HSA-Hybd-Con			0.9410
HSA-Hybd-Pref			0.9031
HSABA-V	3	5	1.0179
HSA-Hybd-Con-V	4	4	1.0016
HSA-Hybd-Pref-V	4	6	1.0432

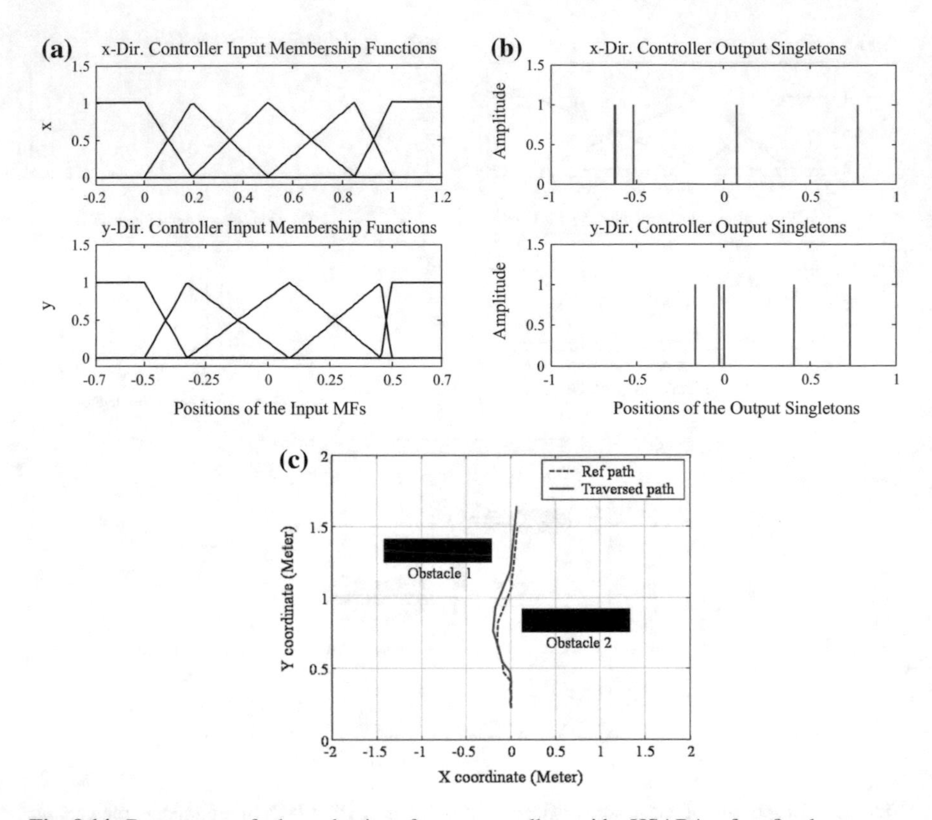

Fig. 9.14 Responses of the adaptive fuzzy controller with HSABA, for fixed structure configuration of input MFs, of the mobile robot navigation during simulation study: **a** positions of input MFs for x-direction and y-direction controllers, **b** positions of output singletons for x-direction and y-direction controllers, **c** evaluation period response after 200 HS generations

Among the competing controllers in fixed structure configuration, the HSA-Hybd-Pref model shows the best performance, whereas in variable structure case, HSA-Hybd-Con model evolved as a better solution, in terms of the *IAE* value computed between the planned path and the actual path traversed by the mobile robot, as shown in Table 9.2. In variable structure configuration, the number of input MFs and hence the number of rules are reduced to some extent which will be very useful from the point of view of real implementation. However, the best *IAE* performance is obtained with a fixed structure configuration.

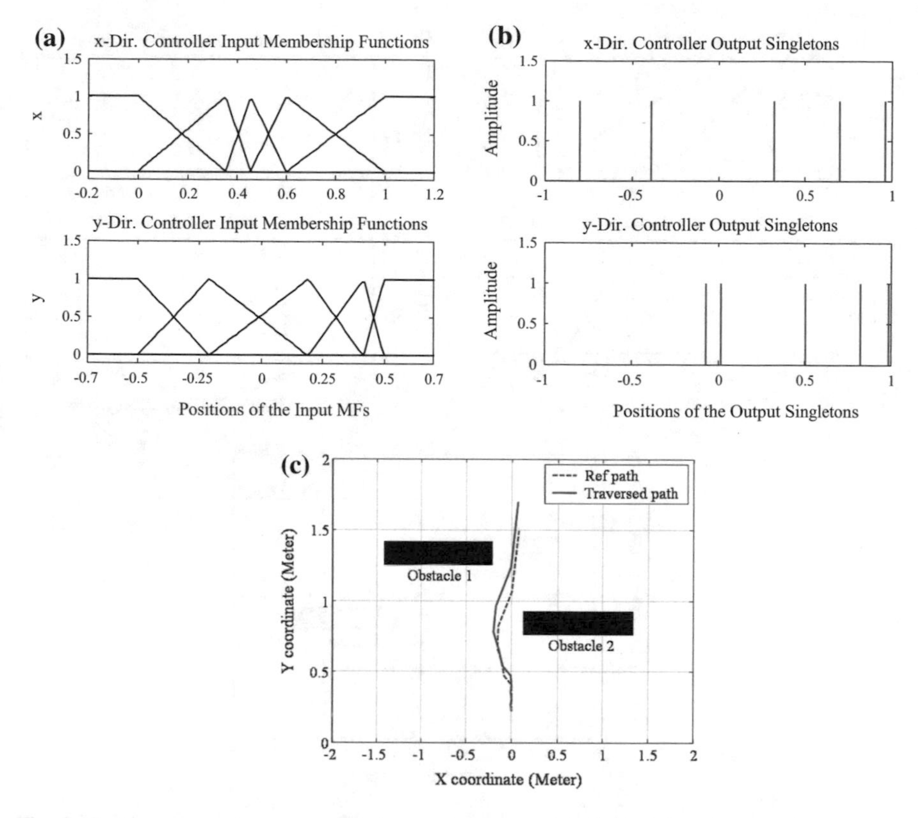

Fig. 9.15 Responses of the adaptive fuzzy controller with HSA-Hybd-Con model, for fixed structure configuration of input MFs, of the mobile robot navigation during simulation study: **a** positions of input MFs for x-direction and y-direction controllers, **b** positions of output singletons for x-direction and y-direction controllers, **c** evaluation period response after 200 HS generations

9.4.2 Experimental Case Study with Real Robot

9.4.2.1 PSO-Based Study

To demonstrate the usefulness of the proposed PSO-based design strategies in real implementations, the navigation of an autonomous mobile robot in two different test environments is performed. The adaptive fuzzy controllers trained during the simulation experiments are utilized to perform the navigation in real environments [21–23]. Figure 9.20 shows a series of snapshots of the mobile robot navigating through the obstacle path, which starts at the position of Fig. 9.20a, and finishes at

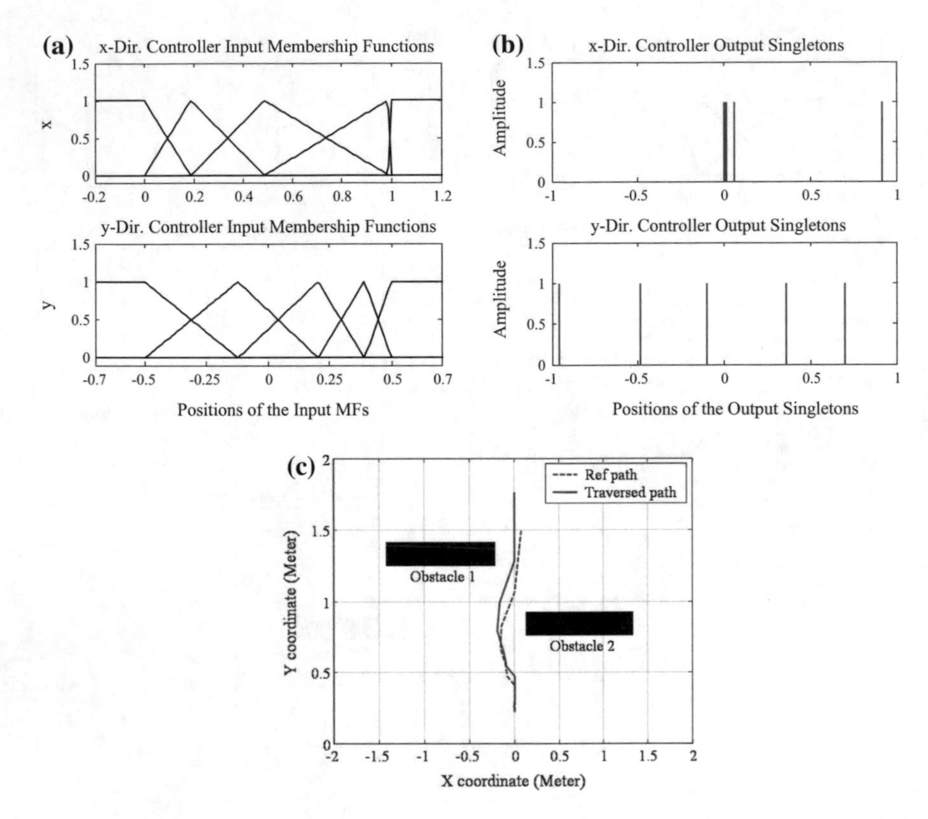

Fig. 9.16 Responses of the adaptive fuzzy controller with HSA-Hybd-Pref model, for fixed structure configuration of input MFs, of the mobile robot navigation during simulation study: **a** positions of input MFs for x-direction and y-direction controllers, **b** positions of output singletons for x-direction and y-direction controllers, **c** evaluation period response after 200 HS generations

Fig. 9.20f for a fixed structure preferential hybrid adaptive fuzzy controller in test environment-I. Figure 9.21 shows the robot navigation performance avoiding obstacles, for test environment-I, utilizing fixed and variable structure versions of the PSOBA, PSO-Hybd-Con model, and PSO-Hybd-Pref model controllers. Figure 9.22 shows another experimental case study for another sample environment, called test environment II, where robot moves from its starting position in Fig. 9.22a and finishes as in Fig. 9.22f, for the same fixed structure PSO-Hybd-Pref. Figure 9.23 shows the corresponding robot navigation performances for real test environment II, with fixed and variable structure versions of the

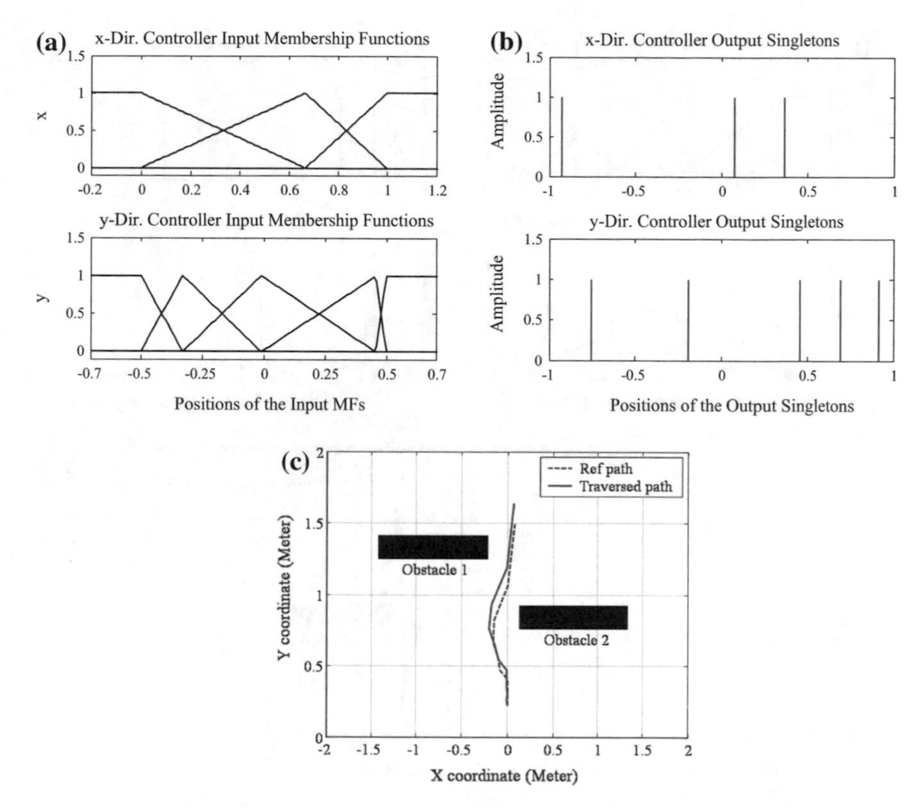

Fig. 9.17 Responses of the adaptive fuzzy controller with HSABA, for variable structure configuration of input MFs, of the mobile robot navigation during simulation study: **a** positions of input MFs for *x*-direction and *y*-direction controllers, **b** positions of output singletons for *x*-direction and *y*-direction controllers, **c** evaluation period response after 200 HS generations

PSOBA, PSO-Hybd-Con model, and PSO-Hybd-Pref model controller. Tables 9.3 and 9.4 present quantitative comparisons of competing controllers, in terms of *IAE*, in implementation phase, for both these real test environments chosen. The results reported in Tables 9.3 and 9.4 demonstrate that the proposed PSO-Hybd-Pref controller, using its fixed structure version, significantly outperforms all other competing controllers, implemented in both case studies. The variable structure versions of the controllers were successful in reducing the total number of input MFs, and hence fuzzy rules implemented, which are useful from the point of view of reducing computational loads in real implementations [21]. But these simplifications in controller structures were achieved at the cost of reduced accuracy in tracking performances. However, a closer look at Figs. 9.21 and 9.23 will reveal

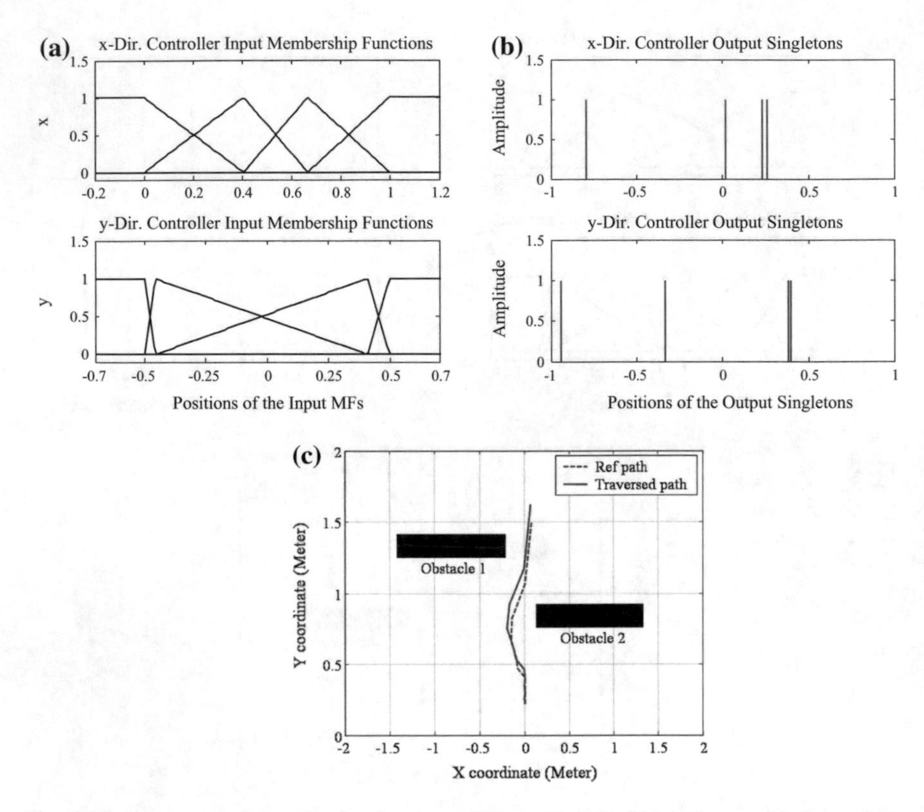

Fig. 9.18 Responses of the adaptive fuzzy controller with HSA-Hybd-Con model, for variable structure configuration of input MFs, of the mobile robot navigation during simulation study: **a** positions of input MFs for *x*-direction and *y*-direction controllers, **b** positions of output singletons for *x*-direction and *y*-direction controllers, **c** evaluation period response after 200 HS generations

that both fixed and variable structure versions of the PSO-Hybd-Pref model of AFLCs were successful in performing their basic duties, i.e., safe navigation of a mobile robot through an obstacle-filled environment, keeping good margins on both sides, successfully. Hence, depending on the problem in hand, one will have to relatively weigh the accuracy of tracking performance vis-à-vis the simplicity of the controller structure implementation and choose the best-suited control algorithm for that purpose.

Another concern for real implementation is the power management of the mobile robot during navigation. In this experimental mobile robot system, both the hardware and software are so synchronized that the power required to drive the RC servos is managed uniquely from the laptop mounted on the robot itself. Thus, no separate power device and associated connecting wire are required during the navigation. Figure 9.22 shows a series of snapshots where one can see that no power cable is connected to the robot during the navigation, in contrast to the earlier snapshots in Fig. 9.20. Thus, the tracking controllers designed can be flexibly

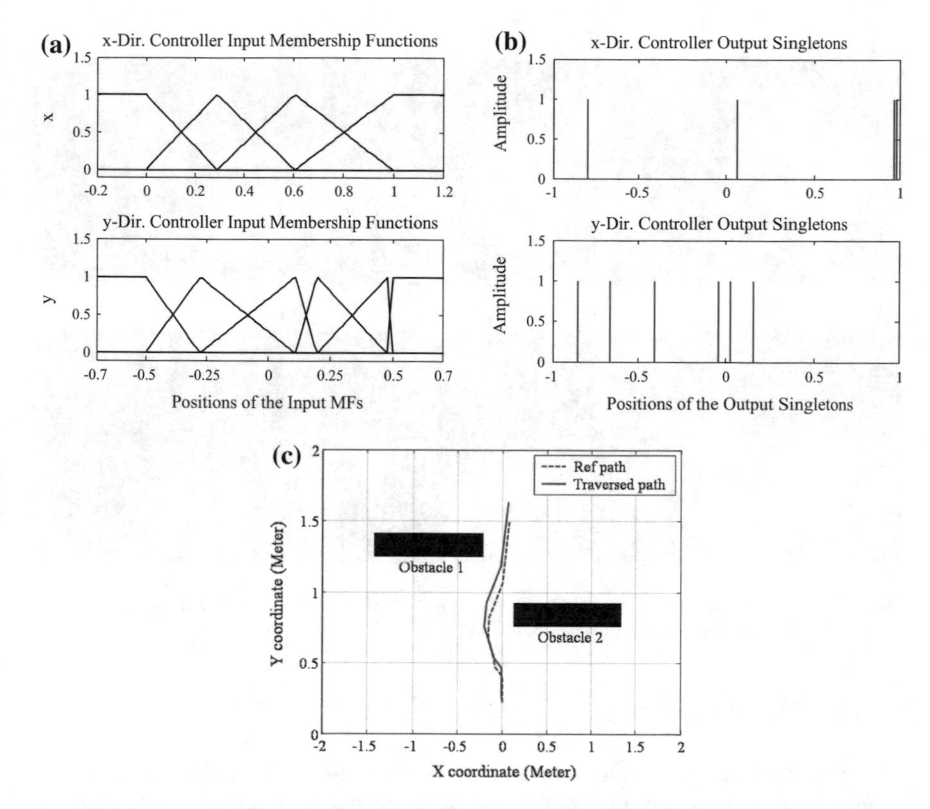

Fig. 9.19 Responses of the adaptive fuzzy controller with HSA-Hybd-Pref model, for variable structure configuration of input MFs, of the mobile robot navigation during simulation study: **a** positions of input MFs for x-direction and y-direction controllers, **b** positions of output singletons for x-direction and y-direction controllers, **c** evaluation period response after 200 HS generations

utilized by the mobile robot to navigate in arbitrarily oriented environments successfully, using vision as the primary sensing element.

Hence, from the results obtained in both simulation and real implementations, it can be inferred that, among the competing PSO-based controller design strategies for fixed and variable structure cases, hybrid preferential model i.e. PSO-Hybd-Pref model evolved as a superior design methodology, on the whole, for the application of vision-based mobile robot navigation schemes, using adaptive state feedback fuzzy tracking controllers.

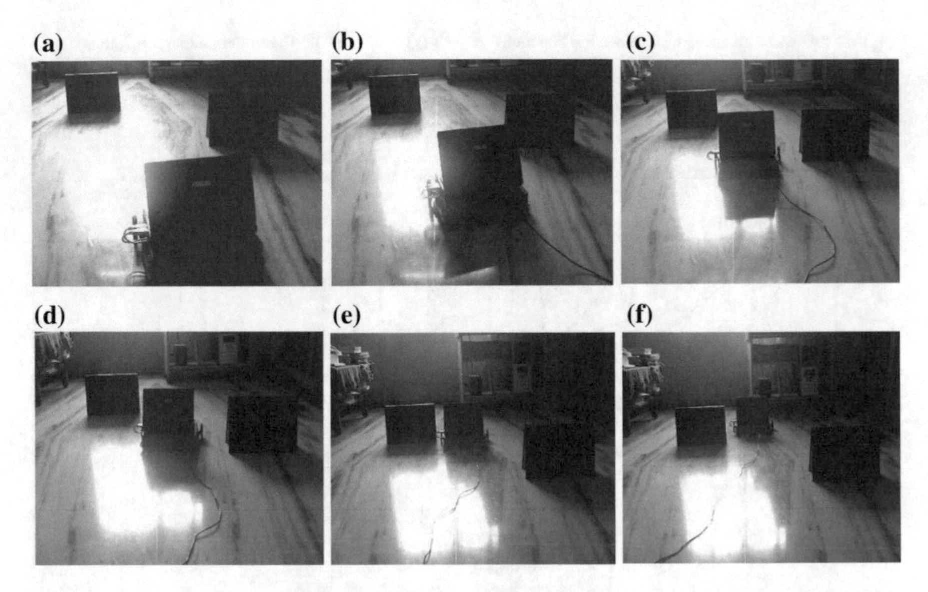

Fig. 9.20 Real robot path traversed in a sample test environment-I for fixed structure PSO-Hybd-Pref model adaptive fuzzy controller

9.4.2.2 HSA-Based Study

The proposed HSA algorithm-based design strategies are implemented for the navigation of the autonomous mobile robot in the two test environments with real robot, as used in PSO-based design [22, 23]. The mobile robot navigation in real environments is achieved by the adaptive fuzzy controllers trained during the HS algorithm based simulation experiments. A series of snapshots of the mobile robot navigating through the obstacle path is shown in Fig. 9.24a–f for a fixed structure preferential HSA-based hybrid adaptive fuzzy controller in test environment-I, as a representative case. Figure 9.25 shows the robot navigation performance avoiding obstacles, for test environment-I, utilizing fixed and variable structure versions of the HSABA, HSA-Hybd-Con model, and HSA-Hybd-Pref model controller. Figure 9.26 shows another experimental case study for sample test environment II where robot moves from its starting position in Fig. 9.26a and finishes as in Fig. 9.26f, whereas Fig. 9.27 shows the robot navigation performances for real test environment II with fixed and variable structure versions of the HSABA, HSA-Hybd-Con model, and HSA-Hybd-Pref model controller. Tables 9.5 and 9.6 present quantitative comparisons of competing controllers, in terms of *IAE*, in implementation phase, for both these real test environment chosen. In HS algorithm-based mobile robot navigation scheme, proposed HSA-Hybd-Pref model of controller design, using both fixed structure and variable structure configurations, evolved as superior control strategy, overall in terms of its *IAE* values as reported in Tables 9.5 and 9.6, in both real implementations.

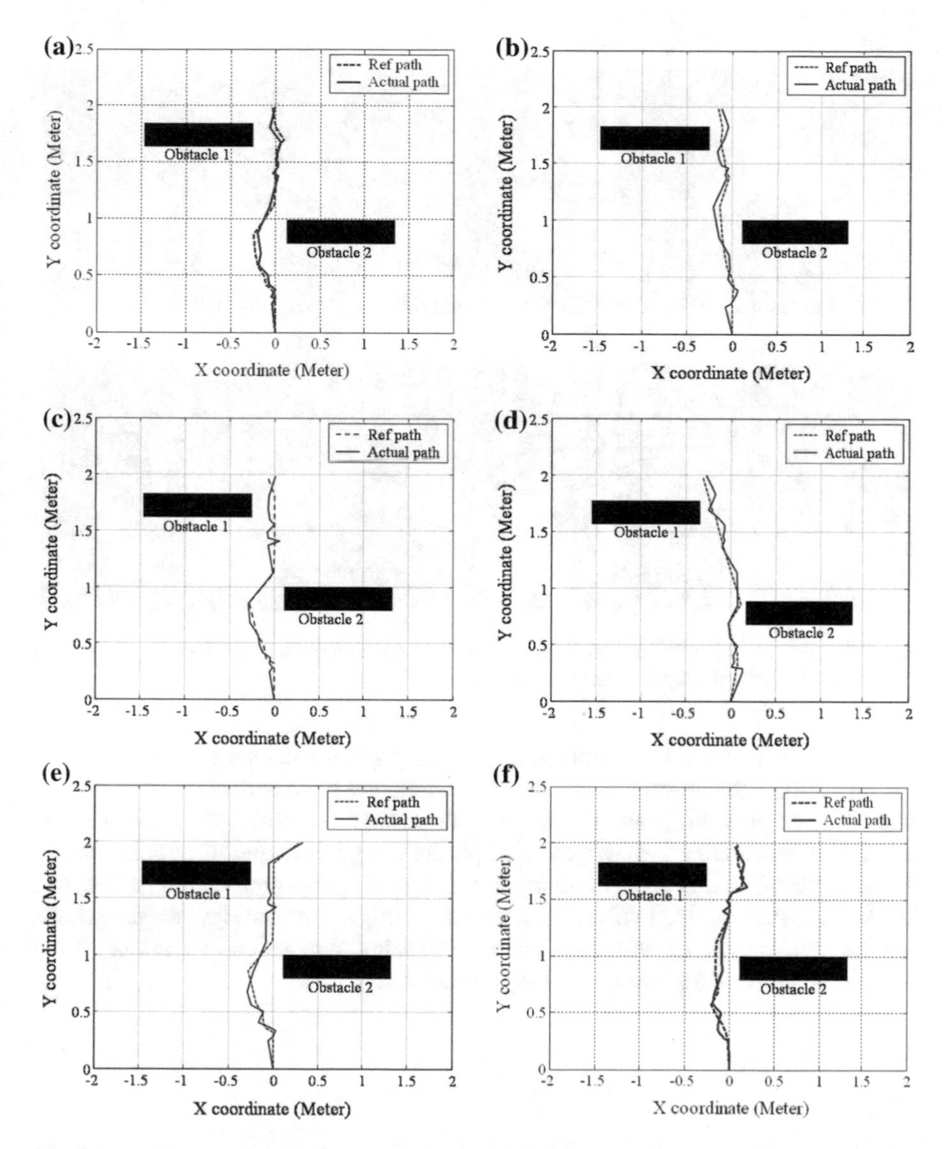

Fig. 9.21 Trajectory of robot movement for real implementation in sample test environment-I for: **a** fixed structure PSOBA, **b** fixed structure PSO-Hybd-Con model, **c** fixed structure PSO-Hybd-Pref model, **d** variable structure PSOBA, **e** variable structure PSO-Hybd-Con model, and **f** variable structure PSO-Hybd-Pref model, based adaptive fuzzy controller

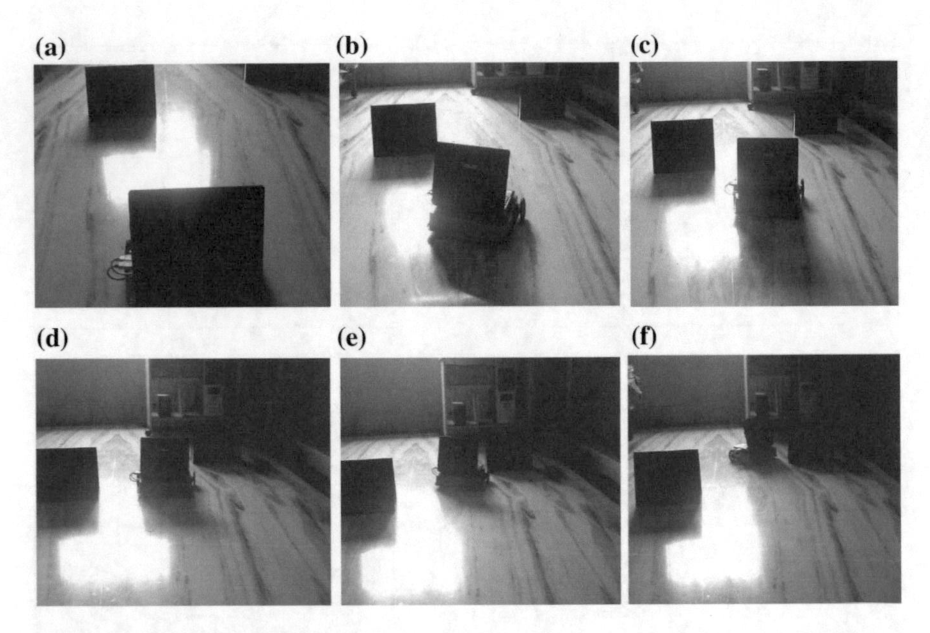

Fig. 9.22 Real robot path traversed in a sample test environment II for fixed structure PSO-Hybd-Pref model adaptive fuzzy controller

Therefore, in both the simulation and real implementations in two test environments, the HSA algorithm-based control strategies have performed their desired objectives of providing safe navigation path through the obstacles. It can also be concluded that, among the competing controller design strategies, for both PSO and HS algorithm-based design, harmony search-based hybrid preferential model, i.e., HSA-Hybd-Pref model, evolved as a superior design methodology, on the whole, for the application of vision-based mobile robot navigation, technique using adaptive state feedback fuzzy tracking controllers.

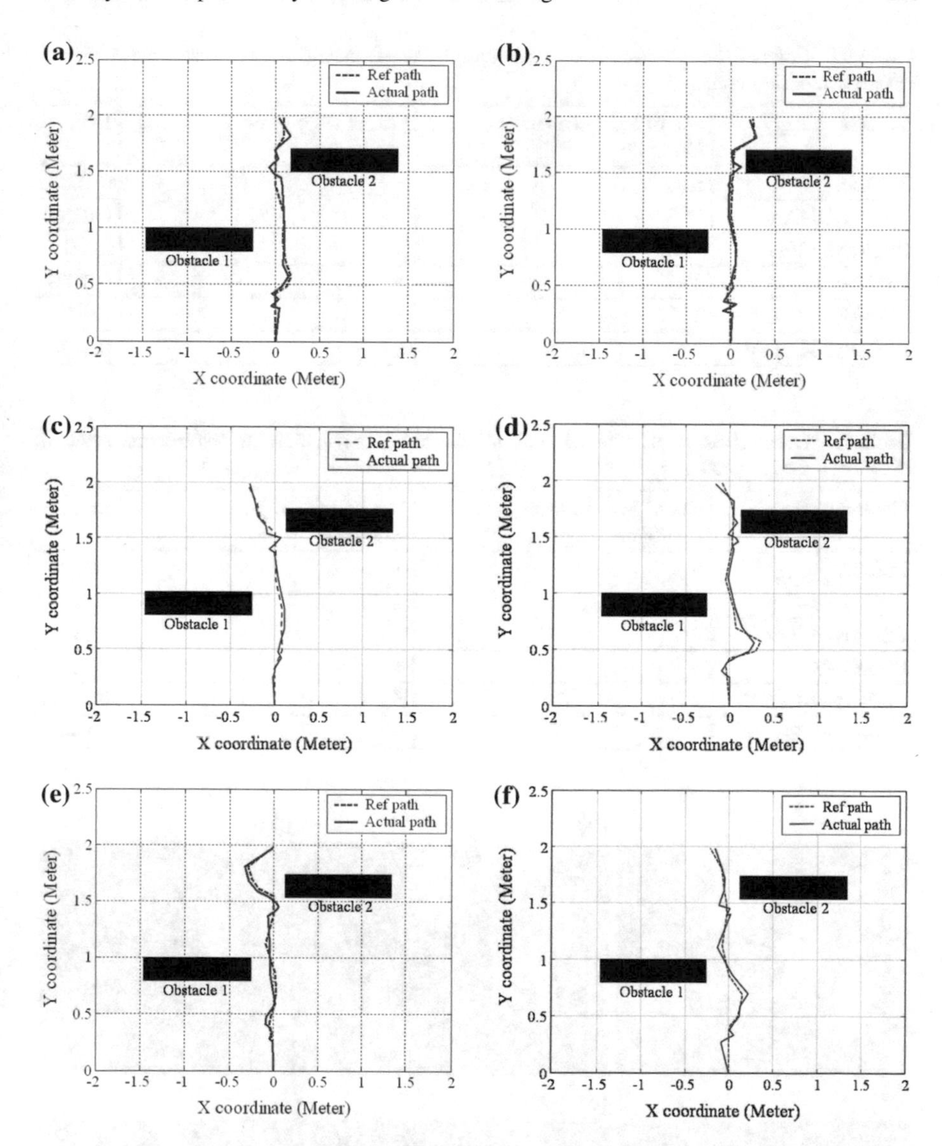

Fig. 9.23 Trajectory of robot movement for real implementation in sample test environment II for: **a** fixed structure PSOBA, **b** fixed structure PSO-Hybd-Con model, **c** fixed structure PSO-Hybd-Pref model, **d** variable structure PSOBA, **e** variable structure PSO-Hybd-Con model, and **f** variable structure PSO-Hybd-Pref model, based adaptive fuzzy controller

Table 9.3 Comparison of results of real mobile robot navigation in test environment-I: PSO-based design strategies

Control strategy	No. of MFs in x-dir Controller	No. of MFs in y-dir Controller	IAE
PSOBA	5	5	0.8488
PSO-Hybd-Con			0.8133
PSO-Hybd-Pref			0.6715
PSOBA-V	4	4	0.8260
PSO-Hybd-Con-V	3	5	0.9003
PSO-Hybd-Pref-V	3	6	0.8434

Table 9.4 Comparison of results of real mobile robot navigation in test environment II: PSO-based design strategies

Control strategy	No. of MFs in x-dir controller	No. of MFs in y-dir controller	IAE
PSOBA	5	5	0.8325
PSO-Hybd-Con			0.6983
PSO-Hybd-Pref			0.6168
PSOBA-V	4	4	0.9854
PSO-Hybd-Con-V	3	5	0.8887
PSO-Hybd-Pref-V	3	6	0.8444

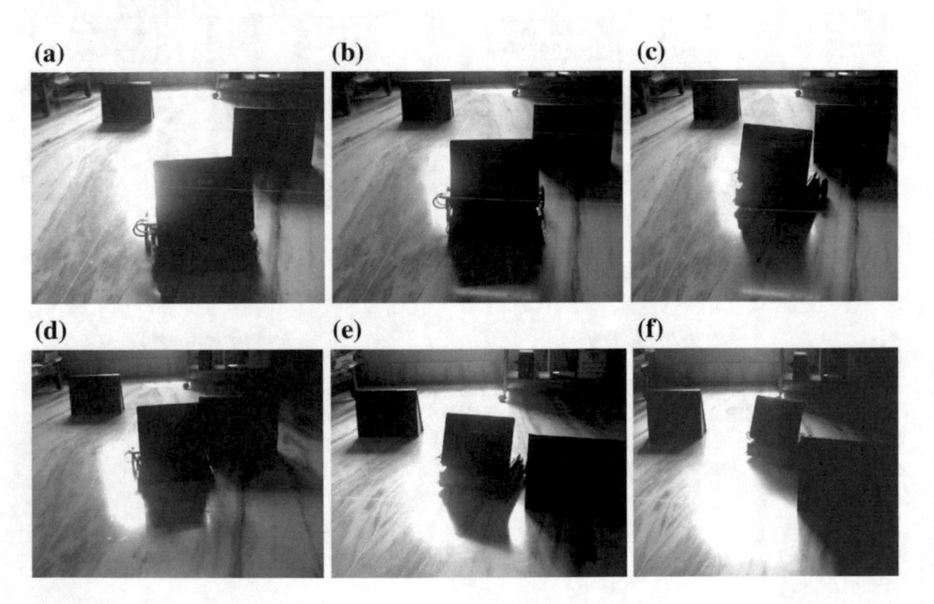

Fig. 9.24 Real robot path traversed in a sample test environment-I for fixed structure HSA-Hybd-Pref model adaptive fuzzy controller

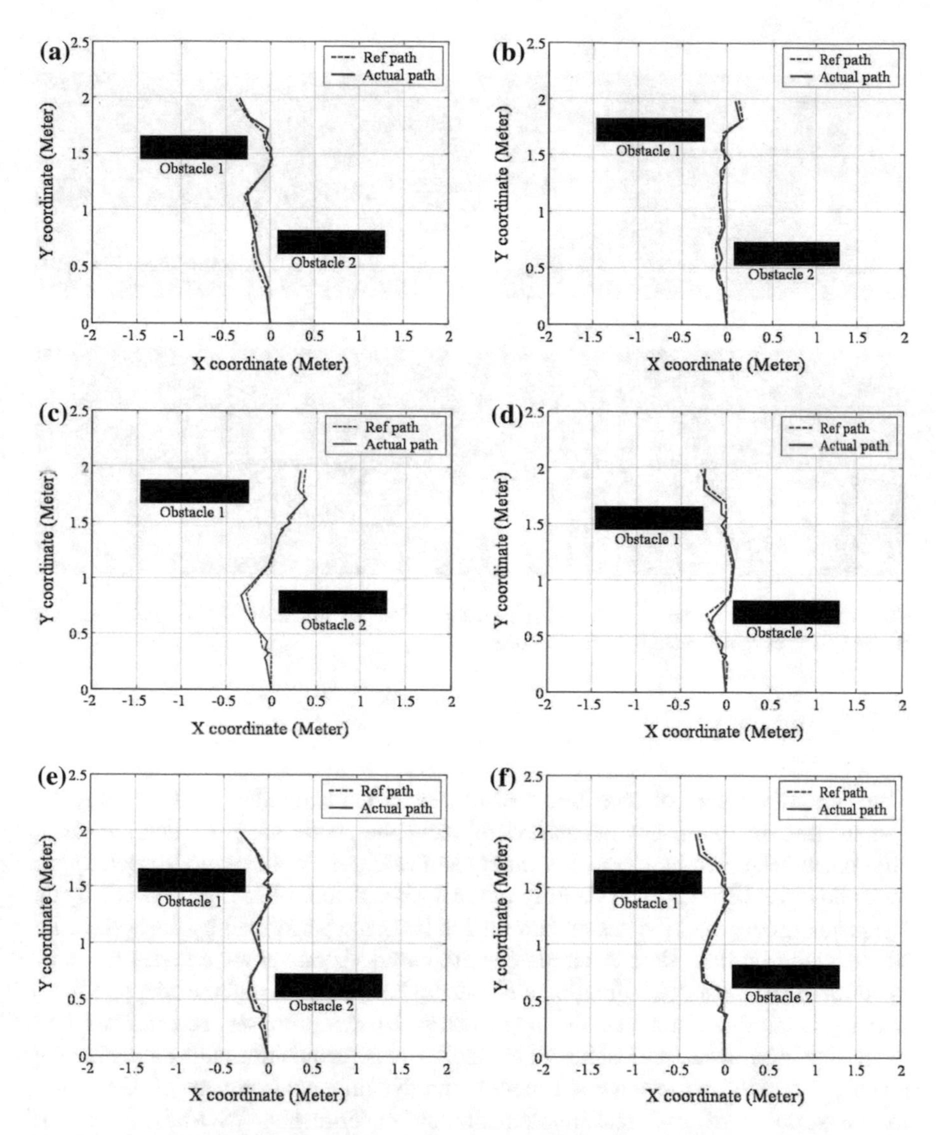

Fig. 9.25 Trajectory of robot movement for real implementation in sample test environment-I for: **a** fixed structure M-HSABA, **b** fixed structure HSA-Hybd-Con model, **c** fixed structure HSA-Hybd-Pref model, **d** variable structure HSABA, **e** variable structure HSA-Hybd-Con model, and **f** variable structure HSA-Hybd-Pref model, based adaptive fuzzy controller

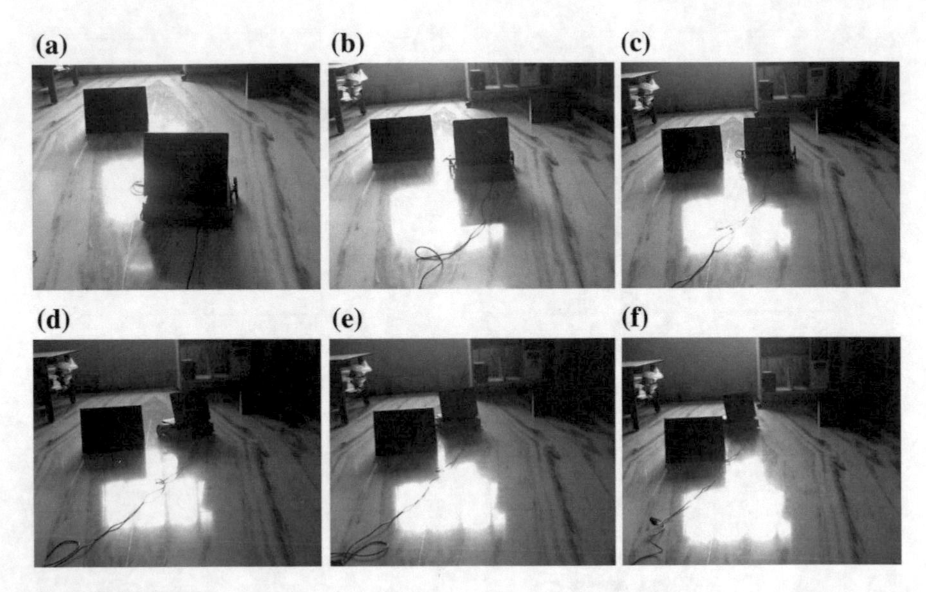

Fig. 9.26 Real robot path traversed in a sample test environment II for fixed structure HSA-Hybd-Pref model adaptive fuzzy controller

9.5 Summary

The present chapter of this book discussed a contemporary methodology for vision-based mobile robot navigation utilizing the stable adaptive fuzzy tracking controllers, designed by Lyapunov theory and stochastic optimization-based hybrid methodologies [21–23]. The stability of the designed controller is guaranteed by the Lyapunov theory, and the design automation has been performed by the application of stochastic optimization. A single camera-based vision sensor reduces the complexities of the sensor coordination and data fusion problems and the adaptive fuzzy tracking controllers enhance the applicability of the proposed scheme for both navigation and obstacle avoidance. Periodic image acquisition and path planning technique facilitates the scheme to tackle the dynamic environment of navigation. In the simulations and real implementation experiments, PSO-based and HS algorithm-based design approaches and two variants of hybrid design approaches,

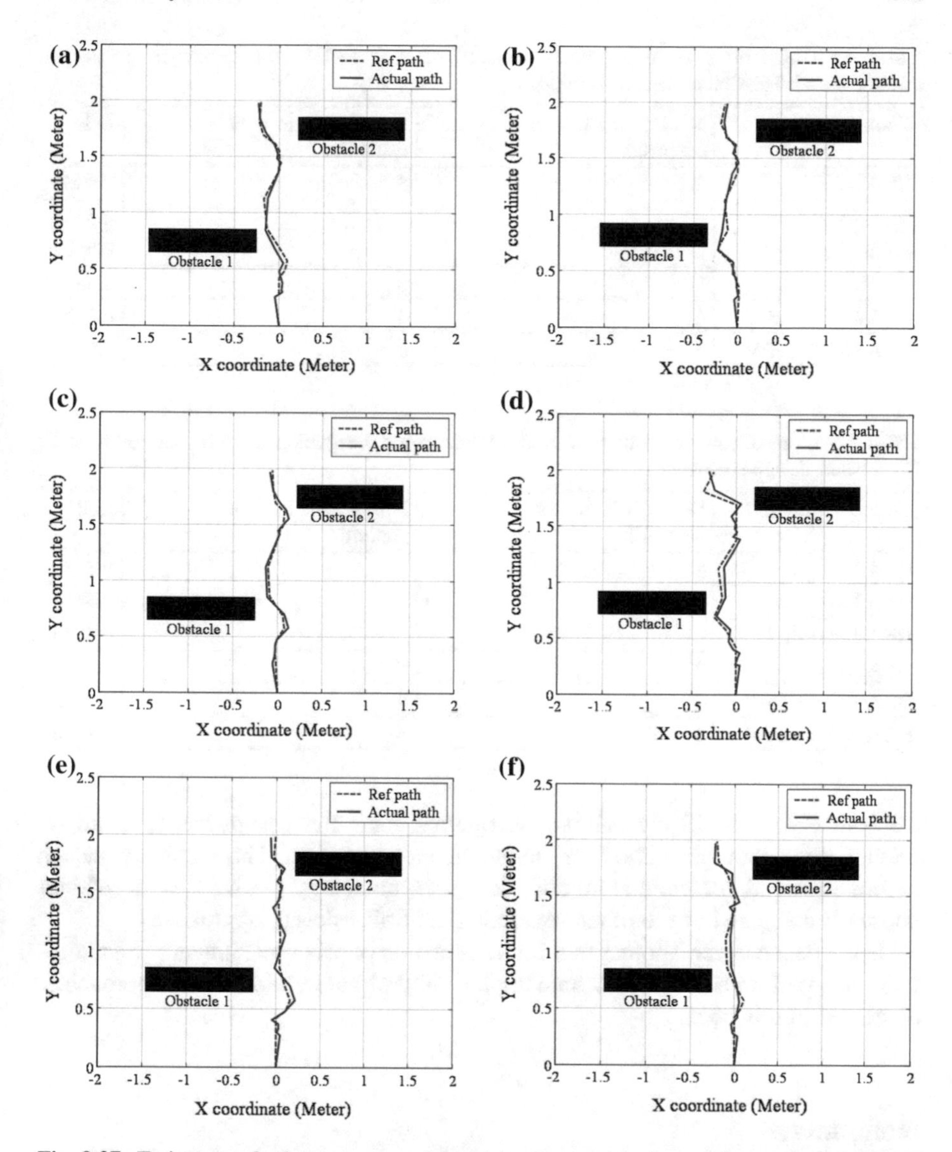

Fig. 9.27 Trajectory of robot movement for real implementation in sample test environment II for: **a** fixed structure HSABA, **b** fixed structure HSA-Hybd-Con model, **c** fixed structure HSA-Hybd-Pref model, **d** variable structure HSABA, **e** variable structure HSA-Hybd-Con model, and **f** variable structure HSA-Hybd-Pref model, based adaptive fuzzy controller

Table 9.5 Comparison of experimental results of real mobile robot navigation in test environment-I: HSA-based design strategies

Control strategy	No. of MFs in x-dir controller	No. of MFs in y-dir controller	IAE
HSABA	5	5	0.6356
HSA-Hybd-Con			0.5761
HSA-Hybd-Pref			0.5897
HSABA-V	3	5	0.6840
HSA-Hybd-Con-V	4	4	0.6737
HSA-Hybd-Pref-V	4	6	0.6174

Table 9.6 Comparison of results of real mobile robot navigation in test environment II: HSA-based design strategies

Control strategy	No. of MFs in x-dir controller	No. of MFs in y-dir controller	IAE
HSABA	5	5	0.5551
HSA-Hybd-Con			0.5818
HSA-Hybd-Pref			0.4646
HSABA-V	3	5	0.8219
HSA-Hybd-Con-V	4	4	0.7415
HSA-Hybd-Pref-V	4	6	0.7096

for both PSO and HS algorithms, for optimal controller design has been implemented for robot navigation in unknown environments. The harmony search algorithm based preferential hybrid design methodology evolved as a superior approach compared to the other competing controller design approaches.

In the next chapter, another real implementation of robot manipulator joint angle control is performed to demonstrate applicability of the hybrid control approaches, discussed in this book.

References

1. P. Rusu, E.M. Petriu, T.E. Whalen, A. Cornell, H.J.W. Spoelder, Behavior-based neuro-fuzzy controller for mobile robot navigation. IEEE Trans. Instrum. Measure. **52**(4), 1335–1340 (2003)
2. D. Santosh, S. Achar, C.V. Jawahar, Autonomous image-based exploration for mobile robot navigation, in *Proceedings of IEEE International Conference on Robotics and Automation, ICRA 2008,* pp. 2717–2722, May 2008
3. B.A. Merhy, P. Payeur, E.M. Petriu, Application of segmented 2-D probabilistic occupancy maps for robot sensing and navigation. IEEE Trans. Instrum. Measure. **57**(12), 2827–2837 (2008)

4. L. Astudillo, O. Castillo, P. Melin, A. Alanis, J. Soria, L.T. Aguilar, Intelligent control of an autonomous mobile robot using type-2 fuzzy logic. Eng. Lett. **13**(2), 93–97 (2006)
5. F. Bonin-Font, A. Ortiz, G. Oliver, Visual navigation for mobile robots: a survey. J. Intell. Robot. Syst. **53**(3), 263–296 (2008)
6. G.N. DeSouza, A.C. Kak, Vision for mobile robot navigation: a survey. IEEE Trans. Pattern Anal. Mach. Intell. **24**(2), 237–267 (2002)
7. N.N. Singh, *Vision based autonomous navigation of mobile robots*. Ph.D. dissertation, Jadavpur University, 2010
8. C.C. Chou, F.L. Lian, C.C. Wang, Characterizing indoor environment for robot navigation using velocity space approach with region analysis and look-ahead verification. IEEE Trans. Instrum. Measure. **60**(2), 442–451 (2011)
9. I.A.R. Ashokaraj, P.M.G. Silson, A. Tsourdos, B.A. White, Robust sensor-based navigation for mobile robots. IEEE Trans. Instrum. Measure. **58**(3), 551–556 (2009)
10. U. Larsson, J. Forsberg, A. Wernersson, Mobile robot localization: integrating measurements from a time-of-flight laser. IEEE Trans. Indust. Electron. **43**(3), 422–431 (1996)
11. W. Gueaieb, M.S. Miah, An intelligent mobile robot navigation technique using RFID technology. IEEE Trans. Instrum. Measure. **57**(9), 1908–1917 (2008)
12. C.C. Tsai, H.C. Huang, C.K. Chan, Parallel elite genetic algorithm and its application to global path planning for autonomous robot navigation. IEEE Trans. Ind. Electron. (2011, Early Access)
13. K.S. Hwang, M.Y. Ju, A propagating interface model strategy for global trajectory planning among moving obstacles. IEEE Trans. Indust. Electron. **49**(6), 1313–1322 (2002)
14. R. Lenain, E. Lucet, C. Grand, B. Thuilot, F.B. Amar, Accurate and stable mobile robot path tracking algorithm: an integrated solution for off-road and high speed context, in *Proceedings of IEEE/IROS International Conference on Intelligent Robots and Systems* (2010), pp. 196–201
15. S.Y. Hwang, J.B. Song, Monocular vision-based SLAM in indoor environment using corner, lamp, and door features from upward-looking camera. IEEE Trans. Indust. Electron. **58**(10), 4804–4812 (2011)
16. C.H. Ku, W.H. Tsai, Obstacle avoidance in person following for vision-based autonomous land vehicle guidance using vehicle location estimation and quadratic pattern classifier. IEEE Trans. Ind. Electron. **48**(1), 205–215 (2001)
17. K. Das Sharma, A. Chatterjee, F. Matsuno, A Lyapunov theory and stochastic optimization based stable adaptive fuzzy control methodology, in *Proceedings of SICE Annual Conference 2008, The University Electro-Communications,* Japan, pp. 1839–1844, Aug 2008
18. K. Das Sharma, A. Chatterjee, A. Rakshit, A hybrid approach for design of stable adaptive fuzzy controllers employing Lyapunov theory and particle swarm optimization. IEEE Trans. Fuzzy Syst. **17**(2), 329–342 (2009)
19. K. DasSharma, A. Chatterjee, A. Rakshit, Design of a hybrid stable adaptive fuzzy controller employing Lyapunov theory and harmony search algorithm. IEEE Trans. Control Syst. Tech. **18**(6), 1440–1447 (2010)
20. T. Das, I.N. Kar, Design and implementation of an adaptive fuzzy logic-based controller for wheeled mobile robots. IEEE Trans. Control Syst. Tech. **14**(3), 501–510 (2006)
21. K. Das Sharma, A. Chatterjee, A. Rakshit, A PSO-Lyapunov hybrid stable adaptive fuzzy tracking control approach for vision based robot navigation. IEEE Trans. Instrum. Measure. **61**(7), 1908–1914 (2012)
22. K. Das Sharma, A. Chatterjee, A. Rakshit, Harmony search-based hybrid stable adaptive fuzzy tracking controllers for vision-based mobile robot navigation. Mach. Vis. Appl. **25**, 405–419 (2014)
23. K. Das Sharma, *Hybrid methodologies for stable adaptive fuzzy control*. Ph.D. dissertation, Jadavpur University (2012)
24. M.W.M.G. Dissanayake, P. Newman, S. Clark, H.F. Durrant-Whyte, M. Csorba, A solution to the simultaneous localization and map building (SLAM) problem. IEEE Trans. Robot. Autom. **17**(3), 229–241 (2001)

25. P.G. Kim, C.G. Park, Y.H. Jong, J.H. Yun, E.J. Mo, C.S. Kim, M.S. Jie, S.C. Hwang, K.W. Lee, Obstacle avoidance of a mobile robot using vision system and ultrasonic sensor, in *Proceedings of 3rd International Conference on Advanced Intelligent Computing Theories and Applications, ICIC'07*, vol. 4681 (Springer, Berlin, 2007), pp. 545–553
26. A. Widyotriatmo, H. Keum-Shik, Navigation function-based control of multiple wheeled vehicles. IEEE Trans. Ind. Electron. **58**(5), 1896–1906 (2011)
27. O. Castillo, R. Martinez-Marroquin, J. Soria, Parameter tuning of membership functions of a fuzzy logic controller for an autonomous wheeled mobile robot using ant colony optimization, in *Proceedings of SMC* (2009), pp. 4770–4775
28. R. Martínez-Soto, O. Castillo, L.T. Aguilar, Optimization of interval type-2 fuzzy logic controllers for a perturbed autonomous wheeled mobile robot using genetic algorithms, in *Soft Computing for Hybrid Intelligent Systems* (2008), pp. 3–18
29. L.X. Wang, Stable adaptive fuzzy control of nonlinear system. IEEE Trans. Fuzzy Syst. **1** (2),146–155 (1993)
30. K. Fischle, D. Schroder, An improved stable adaptive fuzzy control method. IEEE Trans. Fuzzy Syst. **7**(1), 27–40 (1999)
31. V. Giordano, D. Naso, B. Turchiano, Combining genetic algorithm and Lyapunov-based adaptation for online design of fuzzy controllers. IEEE Trans. Syst. Man Cybern. B Cybern. **36**(5), 1118–1127 (2006)
32. R.C. Eberhart, J. Kennedy, A new optimizer using particle swarm theory, in *Proceedings of 6th International Symposium on Micromachine Human Science,* Nagoya, Japan (1995), pp. 39–43
33. Y. Shi, R.C. Eberhart, A modified particle swarm optimizer, in *Proceedings of IEEE International Conference on Evolutionary Computation*, Alaska, May 1998
34. A.A.A. Esmin, A.R. Aoki, G.L. Torres, Particle swarm optimization for fuzzy membership functions optimization, in *Proc. IEEE Intl. Conf. Syst., Man, Cybern.,* pp. 108–113 (2002)
35. Z.W. Geem, J.H. Kim, G.V. Loganathan, A new heuristic optimization algorithm: harmony search. Simulation **76**(2), 60–68 (2001)
36. M. Mahdavi, M. Fesanghary, E. Damangir, An improved harmony search algorithm for solving optimization problems. Appl. Math. Comput. **188**, 1567–1679 (2007)
37. K.S. Lee, Z.W. Geem, A new structural optimization method based on the harmony search algorithm. Comput. Struct. **82**, 781–798 (2004)

Chapter 10
Experimental Study III: Control of Robot Manipulators

10.1 Introduction

Nowadays, robot manipulators are widely used in industry to cater the various sophisticated tasks [1–7]. Particularly, in handling of poisonous materials, bio-hazardous substances, etc., the manipulators are employed to avoid any potential hazard to any human life [2]. In modern manufacturing industries, manipulators are utilized to reduce the human effort in large and complicated assembly line [3, 4]. Even in designing and manufacturing very intricate VLSI technology-based devices, manipulators are used for precise placement of electronic components and chips [3].

In all such applications of robot manipulators, a core component is the control scheme that is employed to achieve the desired movements of manipulator joint angles and thus the positioning of the end effector. The chief problem of applying the conventional control algorithms for controlling the movement of the end effector is the effects of nonlinearity present in the various parts of the manipulator, e.g., servomotors in joints, interactions in links [4–6]. These nonlinearities, which are very common in practice, can eventually degrade the performance of conventional controllers. The variable structure control, model reference adaptive control strategies, etc., are some of the many advanced control methodologies developed to reduce the effects of nonlinearity during the operation of the manipulator [7–11]. However, the performances of these methods depend on the accuracy of system identification technique applied to derive the model of the plant and estimation of parameters of it. In most of the cases, it is difficult to obtain an accurate nonlinear model of an actual manipulator and thus the parameter estimation from that plant model is only an approximation of actual values. To overcome the shortcomings of the conventional control strategies, many intelligent techniques have been developed and extensively used by the researchers [12–19]. Fuzzy logic controller is one of the encouraging solutions for this problem as it does not require a precise model of the plant to be controlled. Moreover, fuzzy modeling of the manipulator can

K. Das Sharma et al., *Intelligent Control*, Cognitive Intelligence and Robotics,
https://doi.org/10.1007/978-981-13-1298-4_10

further reduce the complex, nonlinear plant dynamics into a set of linear systems, each operating over a considerably small linear region [20, 21].

In this chapter, a manipulator of unknown parameters is chosen to demonstrate the usefulness of the stable adaptive control strategies presented in this book. In contrast to the simulation case studies discussed so far for benchmark processes, where the plant parameters were known, this situation provides a real-life challenge for the hybrid controllers, where no a priori knowledge is available regarding the plant under control. The model of the manipulator is hence, first, estimated with the help of T-S-type linear fuzzy modeling technique [20, 21] and a representative stochastic optimization technique, e.g., covariance matrix adaptation (CMA) [22–24]. Then the experiments have been performed to achieve the position control of the two-link manipulator. CMA algorithm is combined with the robust H^{∞} control strategy-based adaptation of controller parameters [25, 26], to develop the hybrid fuzzy controllers in such a fashion that can achieve the automation in the design process and also guarantee the stability of the designed controllers [27].

10.2 Hardware Description of the Experimental Platform

The experimental platform for the position control of a manipulator utilizing adaptive fuzzy controller consists of a laboratory-scale two-link manipulator hardware system, indigenously developed in the System Engineering Laboratory, Department of Applied Physics, University of Calcutta, Kolkata, India, interfaced to a PC through USB port. Figure 10.1 shows the two-link manipulator under consideration and the experimental arrangement where the PC-based controller controls the manipulator through an ATmega 2560 microcontroller-based Arduino-MEGA board [28]. The experimental scheme employs a 5 V, 2 A power source to drive the

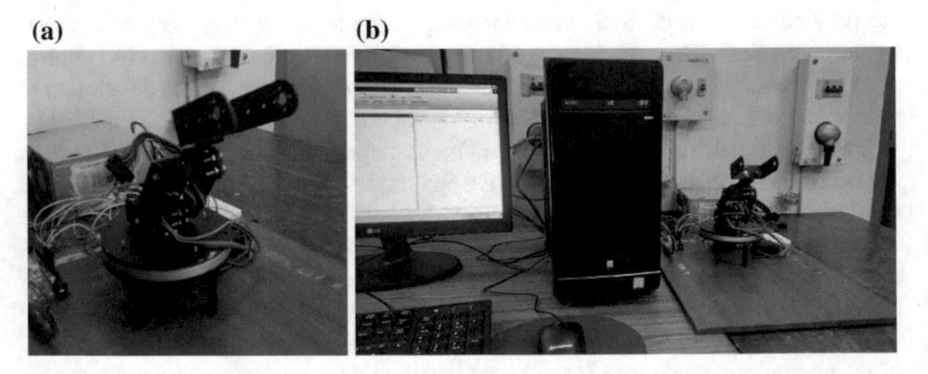

Fig. 10.1 **a** A two-link manipulator and **b** experimental arrangement for control of the manipulator

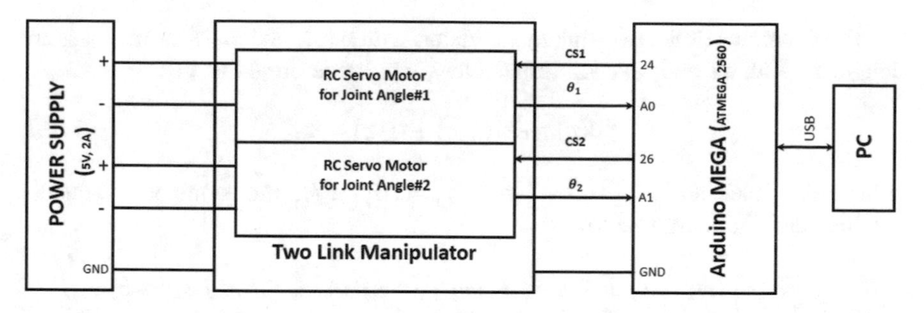

Fig. 10.2 Schematic diagram of experimental arrangement of the two-link manipulator. *CS*1: control signal for motor attached to joint #1, *CS*2: control signal for motor attached to joint #2

two RC servomotors, one each for each joint of the manipulator, to achieve the desired positional variation of the two joint angles simultaneously, as shown in Fig. 10.2.

During the experimentations, the measurement of the angular position, for each joint angle, is carried out using rotary potentiometers present inside RC servomotors [29]. The variable point of the rotary potentiometer is taken out for the measurement of the angular position of the RC servomotor. Then the two signals acquired for the angular positions are suitably filtered in the software developed, employing exponential averaging technique, to remove high-frequency noise pickup. The PWM cycle generated for each control input is of duration 100 ms (i.e., on-time + off-time). The controller outputs (*CS*1 and *CS*2 as shown in Fig. 10.2) are used to suitably control this on-time of each PWM cycle.

10.3 System Identification of the Two-Link Manipulator

The design methodology, for simultaneous control of two joint angle positions of a two-link manipulator, requires the knowledge of parameters of the manipulator for tuning the adaptive fuzzy controllers. Thus, before designing the controllers, the parameters of the manipulator are estimated, i.e., the system to be controlled is identified using CMA-tuned Takagi–Sugeno (T-S) fuzzy system modeling technique [20, 21, 30].

A system identification problem can be viewed as the construction of a mathematical model for a dynamic system using the knowledge and/or observations of that system [30–34]. Now, as the dynamics of the two-link manipulator is quite complex and highly nonlinear in nature, therefore, the present chapter utilizes T-S fuzzy modeling for the purpose of system identification. The T-S fuzzy modeling incorporates the set of fuzzy rules with linear consequents that can locally describe the system dynamics [4, 20, 21].

The dynamics of the two-link manipulator, with m_1, l_1 and m_2, l_2 as the mass and length for link #1 and link #2, respectively, can be described as [4]:

$$B(q)\ddot{q} + C(\dot{q}, q) + G(q) = \tau \tag{10.1}$$

where the joint angles $q^T(t) = [q_1 \quad q_2] = [\theta_1 \quad \theta_2]$, the symmetric bounded positive definite inertia matrix

$$B(q) = \begin{bmatrix} (m_1 + m_2)l_1^2 + m_2 l_2^2 + 2m_2 l_1 l_2 \cos\theta_2 & m_2 l_2^2 + m_2 l_1 l_2 \cos\theta_2 \\ m_2 l_2^2 + 2m_2 l_1 l_2 \cos\theta_2 & m_2 l_2^2 \end{bmatrix},$$

the centrifugal and Coriolis terms

$$C(\dot{q}, q) = \begin{bmatrix} -m_2 l_1 l_2 \sin\theta_2 (2\dot{\theta}_1 \dot{\theta}_2 + \dot{\theta}_2^2) \\ m_2 l_1 l_2 \sin\theta_2 \dot{\theta}_1 \dot{\theta}_2 \end{bmatrix},$$

the gravity vector

$$G(q) = \begin{bmatrix} -(m_1 + m_2) g l_1 \sin\theta_1 - m_2 g l_2 \sin(\theta_1 - \theta_2) \\ m_2 g l_2 \sin(\theta_1 - \theta_2) \end{bmatrix},$$

and the torque applied to the joints $\tau = \begin{bmatrix} f_{\theta_1} \\ f_{\theta_2} \end{bmatrix}$.

To perform the system identification task, the open-loop test data of the input voltages applied, corresponding to each joint angle motor torques, and its corresponding outputs in form of joint angles of the manipulator are acquired for 100 s, where each input–output data exemplar is measured at an interval of 0.1 s. The voltages that are to be applied to the RC servomotors, connected to each joint of the manipulator, for the identification of its parameters are chosen as sinusoidal waves. The measurement of the angular displacement is carried out using in-built rotary encoder of the RC servomotor [29], attached to each joint of the manipulator.

For the identification purpose of extremely nonlinear input–output characteristics of the two-link manipulator, we utilize the following form of the T-S fuzzy modeling rules [20, 21]:

Rule i :

$$\begin{aligned} &\text{IF } x_1 \text{ is } M_{11}^i \text{ and } x_2 \text{ is } M_{12}^i \\ &\text{THEN } \dot{x}_{12}(t) = A_{1i} x_{12}(t) + B_{1i} u_1(t) + W_{1i} w_{12}(t), \\ &\text{IF } x_3 \text{ is } M_{21}^i \text{ and } x_4 \text{ is } M_{22}^i \\ &\text{THEN } \dot{x}_{34}(t) = A_{2i} x_{34}(t) + B_{2i} u_2(t) + W_{2i} w_{34}(t), \end{aligned} \tag{10.2}$$

where $x^{\mathrm{T}}(t) = [x_{12} \quad x_{34}] = [x_1 \quad x_2 \quad x_3 \quad x_4] = [q_1 \quad \dot{q}_1 \quad q_2 \quad \dot{q}_2]$ is the state vector,

$q^{\mathrm{T}}(t) = [\,q_1 \quad q_2\,] = [\,\theta_1 \quad \theta_2\,]$
$u^{\mathrm{T}}(t) = [\,u_1 \quad u_2\,]$ is the control input vector and
$w^{\mathrm{T}}(t) = [\,w_{12} \quad w_{34}\,] = [\,x_3 \quad x_4 \quad x_1 \quad x_2\,]$ is the disturbance vector.

$i = 1, 2, \ldots, r$ represents the linearly distributed local areas of the system and thus the fuzzy rules, M_{jp}^{i} is the fuzzy membership of the premise variables, $A_{ji} \in \Re^{2\times 2}$, $B_{ji} \in \Re^{2\times 1}$ and $W_{ji} \in \Re^{2\times 1}$.

The defuzzified outputs can be written as:

$$\dot{x}_{12}(t) = \sum_{i=1}^{r} h_i(x_{12}(t))[A_{1i}x_{12}(t) + B_{1i}u_1(t) + W_{1i}w_{12}(t)] \tag{10.3}$$

$$\dot{x}_{34}(t) = \sum_{i=1}^{r} h_i(x_{34}(t))[A_{2i}x_{34}(t) + B_{2i}u_2(t) + W_{2i}w_{34}(t)] \tag{10.4}$$

where $h_i(\cdot) = \frac{\prod_{p=1}^{n} M_{jp}^{i}(\cdot)}{\sum_{i=1}^{r} \prod_{p=1}^{n} M_{jp}^{i}(\cdot)}, \quad p = 1, 2.$

In this case, CMA has been employed to evaluate the free parameters, i.e., the linear system matrices (A_{ji}), input matrices (B_{ji}), and the disturbance coupling matrices (W_{ji}) of the T-S fuzzy model employed for the two-link manipulator. A candidate solution vector's (CSV) position in solution space is chosen as a vector containing all information required to characterize the model of the manipulator.

The candidate solution vector $\underline{Z}$ is formed as [30–33]:

$$\begin{aligned}\underline{Z} = [&\text{center locations of the MFs} \mid \text{scaling gains} \mid \ldots \\ &\mid A_{ji} \text{ parameters} \mid B_{ji} \text{ parameters} \mid W_{ji} \text{ parameters}]\end{aligned} \tag{10.5}$$

The unknown parameters are identified utilizing the CMA algorithm, following a similar methodology as presented in Sect. 2.4 of Chap. 2. Here also, as discussed in the earlier system identification case discussed in Chap. 8, the CMA algorithm determines the best vector $\underline{Z}$ comprising the unknown parameters as shown in (10.5) for which discrepancy between the approximated outputs from the T-S fuzzy model and the actual experimental output data, for the same inputs, aggregated over the entire set of input–output real experimental data, is calculated to be minimum.

The plot of the open-loop input voltages, corresponding joint angles and the approximated joint angles of the manipulator joint #1 and joint #2 from the T-S fuzzy model are shown in Fig. 10.3a, b, respectively.

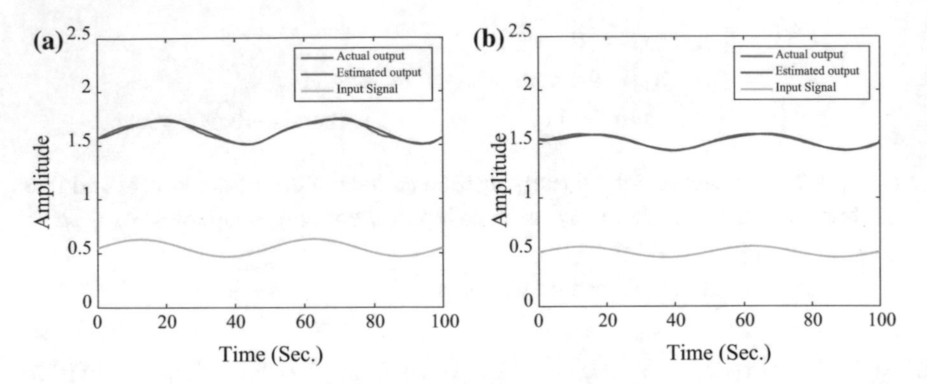

Fig. 10.3 Identification of manipulator **a** joint #1 dynamics and **b** joint #2 dynamics in open-loop configuration

10.4 Hybrid Adaptive Fuzzy Controller Design

The controller design methodology encompasses the simulation-based adaptation of fuzzy controllers with different adaptive control strategies utilizing CMA- and H^{∞}-based robust adaptive control law [27], as discussed in Chap. 7, for controlling the positions of the joint angles of a two-link manipulator using a fixed-step fourth-order Runge–Kutta method. Once the controller is successfully tuned in simulation, then it is employed in practice to control the movement of the real two-link manipulator [19]. It is intended to design these stable robust AFLCs with a fixed structure configuration of 5×5 input MFs, for this real experimental case study. Hence, during simulation, the same configuration, i.e., 5×5 input MFs, is used to optimize the controller settings. In this case study, the CMA technique is utilized to control the movement of the manipulator in individual and in hybrid configurations. Hence, CMABA, CMA-HASBA-Con, and CMA-HASBA-Pref control strategies [30], as discussed in Chap. 7, are utilized for designing the AFLCs in both simulation and real implementation. Along with these H^{∞} strategy-based robust approach, i.e., HSBRA, as discussed in Chap. 6, is also tested for designing the AFLC. To achieve a precise control over the two-link manipulator joint angles, the system dynamics is considered as the combined dynamics of each joint angle, having a coupling in between, as described in (10.3) and (10.4). Thus, two zero-order T-S-type AFLCs are utilized to control each joint angle movement with the control objective of simultaneous minimization of individual joint angle errors. The joint angle error is defined as the difference between the reference signal to each joint and the output of that joint.

It is to be noted that as T-S fuzzy model-based system identification approach has been utilized in this chapter, the design of the AFLCs for controlling the joint angle positions is modified accordingly from the methods adopted in Chaps. 8 and 9. Here, the structure and the antecedent part of the AFLCs are same as those employed for identifying the system; only the consequent part is to be tuned to achieve the desired control objectives for two joint angles of the manipulator. The candidate solution vector (CSV) $\underline{Z}^c$ can be formed as [27, 32]:

$$\underline{Z}^c = [\text{output scaling gains} \mid \text{position of the output singletons for joint angles}] \tag{10.6}$$

10.4.1 Case Study: Simulation and Experimental

To perform the simulation, a population size of ten CSV is chosen. For each simulation, 200 iterations of CMA are performed. The adaptive control strategies, as mentioned before, tune the output scaling gains of the two controllers, one each for each joint angles and its corresponding positions of the output singletons, to minimize the weighted average of individual integral absolute errors (*IAE*s) between the reference signals and the joint angles. The sampling time is taken as 100 ms in conformation with the requirement of the hardware driver circuit of the manipulator. The other robust control parameters are chosen as: prescribed attenuation level $\rho = 0.05$ and the positive scalar constant $r = 0.8$. The value of the adaptation gain is chosen as $v = 0.001$, which gives a good trade-off between the adaptation rate and small amplitudes of parameter oscillations, which are due to inherent approximation error. The simulation is performed for 100 s for each candidate controller configuration as it is planned that, during the real experimental study, the manipulator will be tested for the same duration. In this case study, CMA-HASBA-Pref control strategy produced best control configuration in both simulation and real implementation as indicated by the IAE values obtained, shown in Table 10.1. The locations of the center positions of input MFs and the positions of output singletons with HSBRA, CMABA, CMA-HASBA-Con, and CMA-HASBA-Pref control strategies are shown in (a) and (b) of Figs. 10.4, 10.5, 10.6, and 10.7, respectively. In (c) and (d) of Figs. 10.4, 10.5, 10.6, and 10.7, the evaluation period responses, in simulation and real implementation cases, respectively, are shown, for different control strategies, as mentioned before.

Table 10.1 Comparison of results for the two-link manipulator: CMA-based design strategies (evaluation time = 100 s)

Control strategy	No. of MFs	No. of rules	IAE	
			Simulation	Experimental
HSBRA	5 × 5	25	1.2318	2.6715
CMABA			1.0615	1.2818
CMA-HASBA-Con			0.6405	0.9002
CMA-HASBA-Pref			0.4447	0.7911

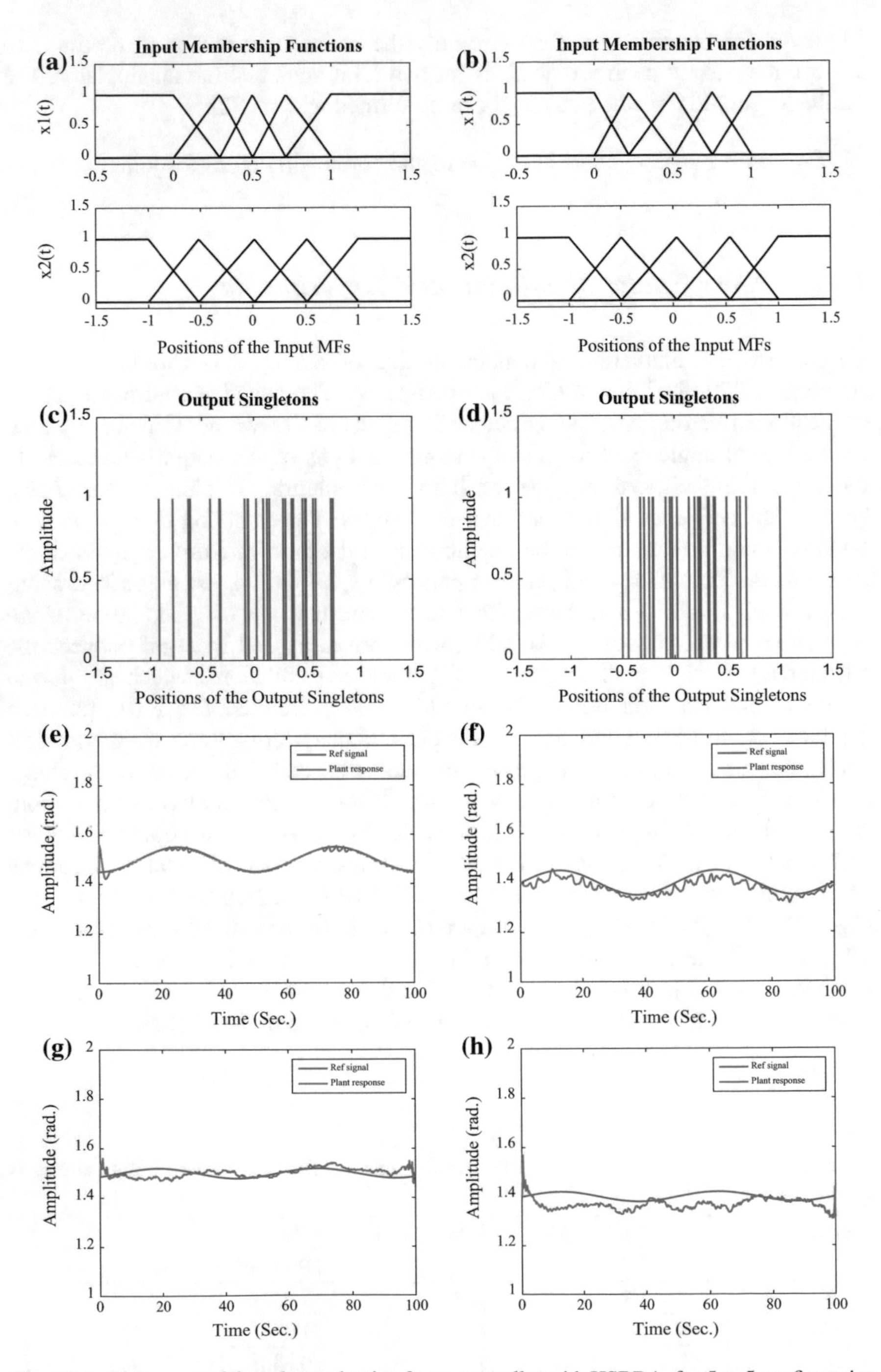

Fig. 10.4 Responses of the robust adaptive fuzzy controller with HSBRA, for 5 × 5 configuration of input MFs, of the two-link manipulator: **a** positions of input MFs, **b** positions of output singletons, **c** evaluation period response after 1000 s adaptation during simulation, **d** evaluation period response in real implementation experiment

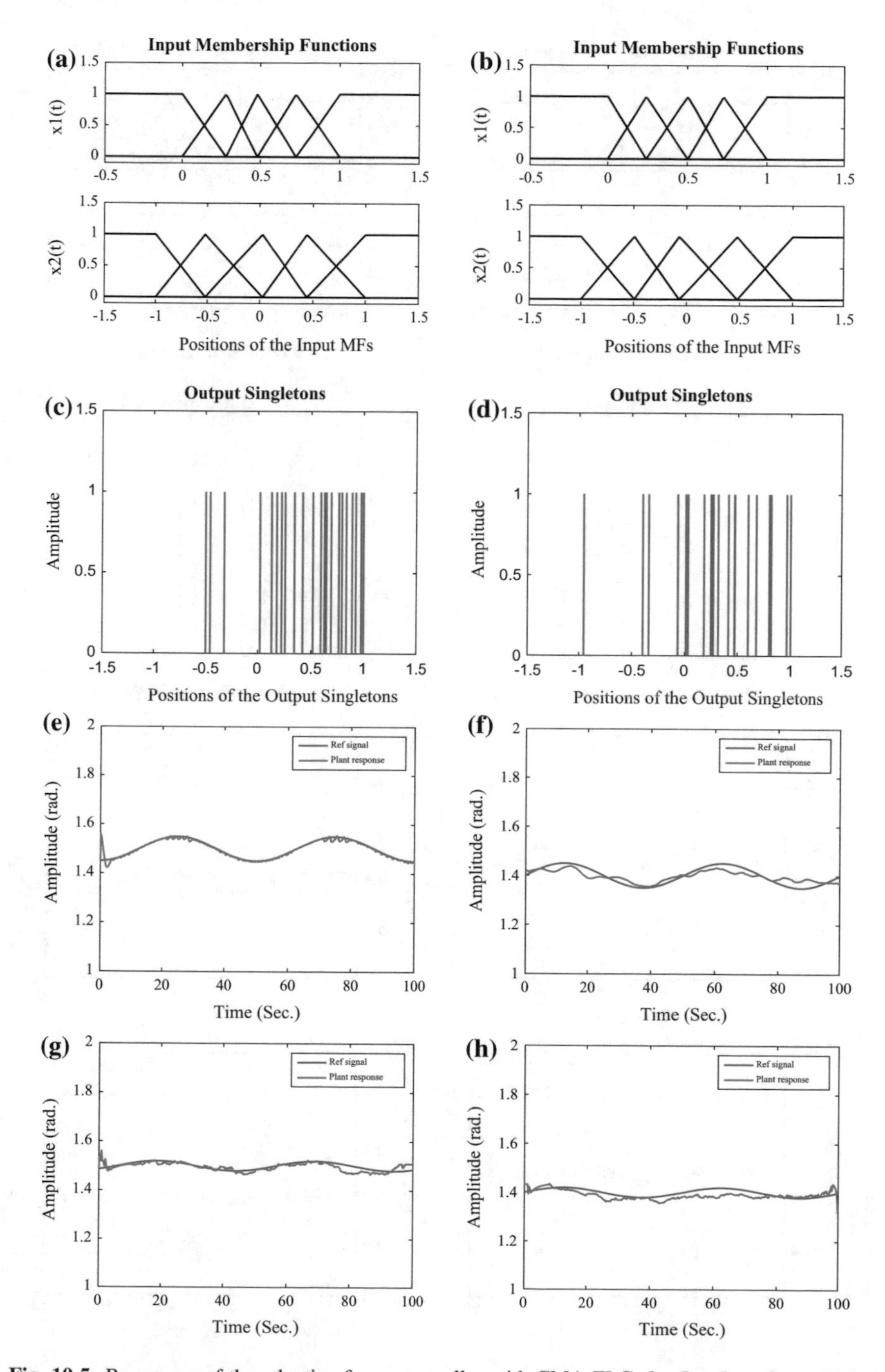

Fig. 10.5 Responses of the adaptive fuzzy controller with CMA-FLC, for 5 × 5 configuration of input MFs, of the two-link manipulator: **a** positions of input MFs, **b** positions of output singletons, **c** evaluation period response after 200 CMA generations during simulation, **d** evaluation period response in real implementation experiment

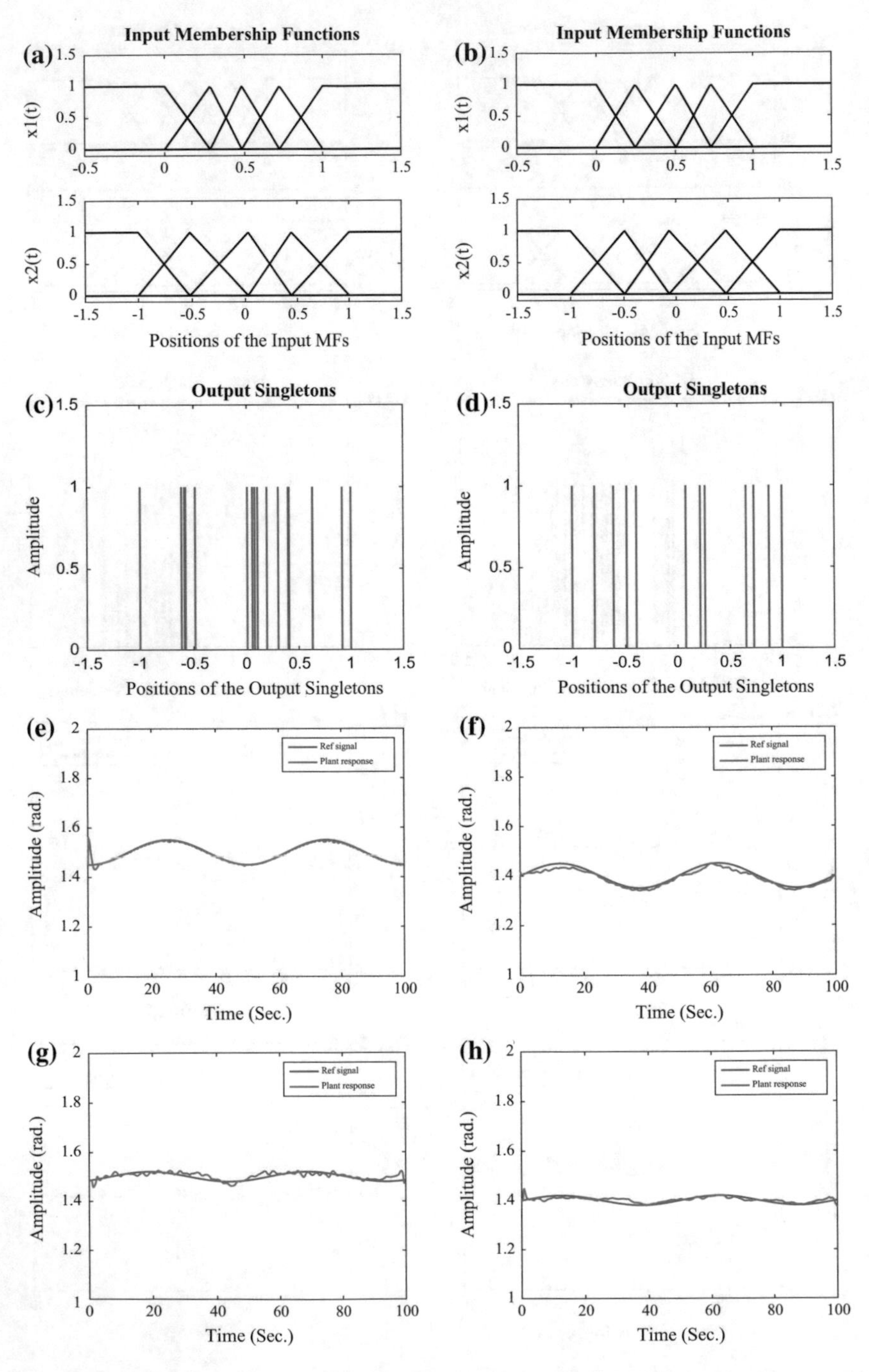

Fig. 10.6 Responses of the adaptive fuzzy controller with PSO-HASBA-Con model, for 5×5 configuration of input MFs, of the real DC motor: **a** positions of input MFs, **b** positions of output singletons, **c** evaluation period response after 200 CMA generations during simulation, **d** evaluation period response in real implementation experiment

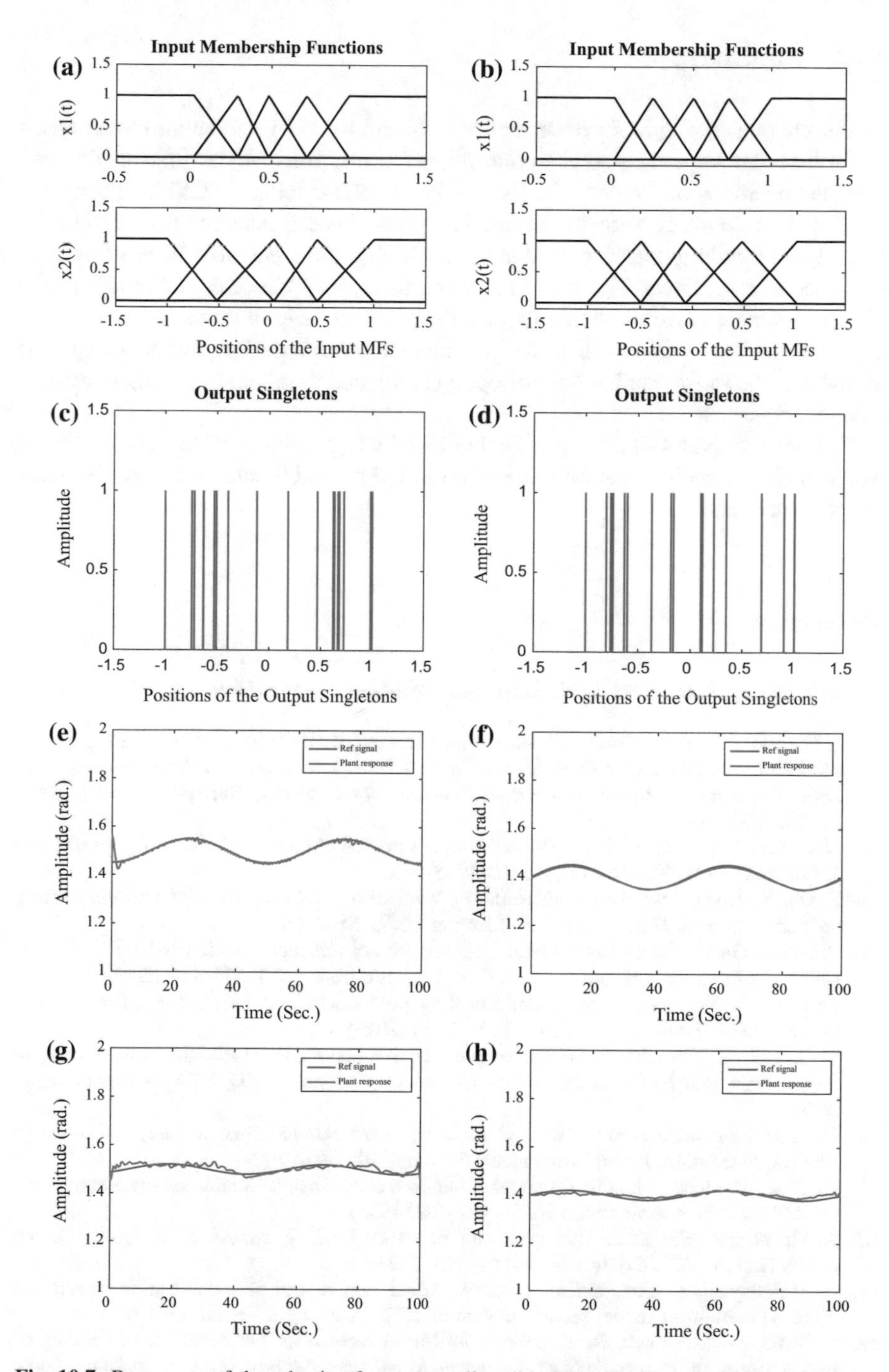

Fig. 10.7 Responses of the adaptive fuzzy controller with CMA-HASBA-Pref model, for 5 × 5 configuration of input MFs, of the real DC motor: **a** positions of input MFs, **b** positions of output singletons, **c** evaluation period response after 200 CMA generations during simulation, **d** evaluation period response in real implementation experiment

10.5 Summary

In this chapter, the hybrid concurrent model- and hybrid preferential model-based controllers are implemented for controlling the positions of the joint angles and thus, the positions of the end effector of a two-link manipulator. CMA algorithm is employed to combine with the robust H^{∞} control-based adaptation to design the hybrid methodology-based controllers. CMA algorithm has also been utilized in conjunction with linear T-S fuzzy modeling to identify the dynamic model of the highly nonlinear two-link manipulator system before designing the controllers. The experimentations showed that the preferential hybrid design methodology has evolved as the most suitable control option compared to other competing controller design techniques.

The next chapter will conclude the book and will provide a glimpse of the future research directions that can be undertaken in hybrid stable adaptive state feedback fuzzy controllers.

References

1. M.W. Spong, S. Hutchinson, M. Vidyasagar, *Robot Dynamics and Control*, 2nd edn. (Wiley, 2006)
2. A.G. Özkil, Z. Fan, S. Dawids, J. Klæstrup Kristensen, K.H. Christensen, H. Aanæs, Service robots for hospitals: a case study of transportation tasks in a hospital, in *Proceedings of 2009 IEEE International Conference on Automation and Logistics*, Shenyang, China, 2009, pp. 289–294
3. H.-C. Lin, P.J. Egbelu, C.-T. Wu, A two-robot printed circuit board assembly system. Int. J. Comput. Integr. Manuf. **8**(1), 21–31 (1995)
4. S. Oh, K. Kong, Two-degree-of-freedom control of a two-link manipulator in the rotating coordinate system. IEEE Trans. Ind. Electron. **62**(9), 5598–5607 (2015)
5. S. Cao, L. Guo, X. Wen, Robust fault diagnosis with disturbance rejection and attenuation for systems with multiple disturbances. J. Syst. Eng. Electron. **22**(1), 135–140 (2011)
6. D. Pucci, F. Romano, F. Nori, Collocated adaptive control of under actuated mechanical systems. IEEE Trans. Robot. **31**(6), 1527–1536 (2015)
7. S. Ulrich, J.Z. Sasiadek, Modified simple adaptive control for a two-link space robot, in *Proceedings of 2010 American Control Conference*, Baltimore, MD, USA, 2010, pp. 3654–3659
8. H. Jamali, *Adaptive control methods for mechanical manipulators: a comparative study*, Masters dissertation, Naval Postgraduate School, California, 1989
9. C.-Y. Su, Y. Stepanenko, T.-P. Leungs, Combined adaptive and variable structure control for constrained robots. Automatica **31**(3), 483–488 (1995)
10. A. Green, J.Z. Sasiadek, Dynamics and trajectory tracking control of a two-link robot manipulator. J. Vibr. Control **10**, 1415–1440 (2004)
11. A. Zhakatayev, M. Rubagotti, H.A. Varol, Closed-loop control of variable stiffness actuated robots via nonlinear model predictive control. IEEE Access **3**, 235–248 (2015)
12. J. Tervo, M. Mustonen, R. Korhonen, Intelligent techniques for condition monitoring of rolling mills, in *Proceedings of European Symposium on Intelligent Techniques*, 2000, pp. 1009–1012

13. S.A. Abdul Kareem, R. Muhida, R. Akmeliawati, Fuzzy control algorithm for educational light tracking system, in *Proceedings of 2nd International Conference on Engineering Education (ICEED)*, 2010, pp. 22–27
14. B. Singh, S. Prakash, A.S. Pandey, S.K. Sinha, Intelligent PI controller for speed control of DC motor. Int. J. Electron. Eng. Res. **2**(1), 87–100 (2010)
15. G. Madhusudhana Rao, B.V. Shanker Ram, A neural network based speed control for DC motor. Int. J. Recent Trends Eng. **2**(6), 121–124 (2009)
16. Z. Moravej, M. Co, Speed control DC motor based on neural net and fuzzy logic, in *Proceedings of 5th International Conference on System Theory Scientific Computation*, Malta, September 2005, pp. 186–193
17. R.J. Roesthuis, S. Misra, Steering of multisegment continuum manipulators using rigid-link modeling and FBG-based shape sensing. IEEE Trans. Robot. **32**(2), 372–382 (2016)
18. M. Sun, S.S. Ge, I.M.Y. Mareels, Adaptive repetitive learning control of robotic manipulators without the requirement for initial repositioning. IEEE Trans. Robot. **22**(3), 563–568 (2006)
19. D. Biswas, K. Das Sharma, G. Sarkar, Stable adaptive NSOF domain FOPID controller for a class of non-linear systems. IET Control Theory Appl. https://doi.org/10.1049/iet-cta.2017.0732
20. K. Tanaka, T. Hori, H.O. Wang, A multiple Lyapunov function approach to stabilization of fuzzy control systems. IEEE Trans. Fuzzy Syst. **11**(4), 582–589 (2003)
21. M. Polanský, Advanced robust PDC fuzzy control of nonlinearsystems. World Acad. Sci. Eng. Technol. **11**, 859–864 (2007)
22. N. Hansen, A. Ostermeier, A.Gawelczyk, On the adaptation of arbitrary normal mutation distributions in evolution strategies: the generating set adaptation, in *Proceedings of 6th International Conference on Genetic Algorithms*, San Francisco, USA, 1995, pp. 57–64
23. A. Auger, N. Hansen, A restart CMA evolution strategy with increasing population size, in *Proceedings of Congress on Evolutionary Computation*, Edinburgh, UK, 2005, pp. 1769–1776
24. P.N. Suganthan, N. Hansen, J.J. Liang, K. Deb, Y.-P.Chen, A. Auger, S. Tiwari, Problem definitions and evaluation criteria for the cec 2005 special session on real-parameter optimization. Technical Report, Nanyang Tech. University, Singapore, 2005
25. B.S. Chen, C.H. Lee, Y.C. Chang, H^{∞} tracking design of uncertain nonlinear SISO systems: adaptive fuzzy approach. IEEE Trans. Fuzzy Syst. **4**(1), 32–43 (1996)
26. B.S. Chen, C.L. Tsai, D.S. Chen, Robust H-infinity and mixed H2/H∞ filters for equalization designs of nonlinear communication systems: fuzzy interpolation approach. IEEE Trans. Fuzzy Syst. **11**(3), 384–398 (2003)
27. K. Das Sharma, A. Chatterjee, P. Siarry, A. Rakshit, CMA—H^{∞} hybrid design of robust stable adaptive fuzzy controllers for non-linear systems, in *Proceedings of 1st International Conference on Frontiers in Optimization: Theory and Applications (FOTA-2016)*, Kolkata, India, 2016 (in press)
28. Arduino Mega Data sheet, link: arduino.cc, 20 February 2018
29. RKI 1204: Metal geared standard servo motor (Economy class) Datasheet, Robokits India, 2014
30. K. Das Sharma, *Hybrid methodologies for stable adaptive fuzzy control*. PhD dissertation, Jadavpur University, 2012
31. K. Das Sharma, A. Chatterjee, F. Matsuno, A Lyapunov theory and stochastic optimization based stable adaptive fuzzy control methodology, in *Proceedings of SICE Annual Conference 2008*, Japan, 20–22 August, pp. 1839–1844
32. K. Das Sharma, A. Chatterjee, A. Rakshit, A hybrid approach for design of stable adaptive fuzzy controllers employing Lyapunov theory and particle swarm optimization. IEEE Trans. Fuzzy Syst. **17**(2), 329–342 (2009)
33. K. Das Sharma, A. Chatterjee, A. Rakshit, Design of a hybrid stable adaptive fuzzy controller employing Lyapunov theory and harmony search algorithm. IEEE Trans. Contr. Syst. Technol. **18**(6), 1440–1447 (2010)
34. L. Ljung, *System Identification: Theory for the User* (Prentice Hall, 1987)

Part V
Epilogue

Chapter 11
Emerging Areas in Intelligent Fuzzy Control and Future Research Scopes

In this book, a detailed description has been carried out for a particular class of nonlinear systems, as mentioned in Chap. 3 [1–4]. However, the presented hybrid stable adaptive fuzzy control methodologies can be potentially applied to other classes of the system as well. In this book, the hybrid design methodologies utilize the Lyapunov theory-based adaptation, as presented in Chap. 3 [5], and H^{∞}-based robust adaptation, as presented in Chap. 6 [6]. Other formulations of the adaptive control strategy [7–11] may also be utilized to design the adaptive fuzzy controllers using the presented hybrid strategies, and their relative performance evaluations may be evaluated in future.

The entire discussion of the book is based on the direct adaptive fuzzy control scheme [3]. In both the cases of Lyapunov theory [5, 12] and robust H^{∞} [6, 13, 14] based formulations of adaptation rules are focused on the framework of direct adaptive control scheme. The complete set of hybrid control methodologies presented in this book can potentially be applied for the indirect adaptive fuzzy control scheme [5, 6]. There are numerous numbers of practical applications exist where the direct state feedback is not possible due to physical constraints. An indirect adaptive fuzzy control is the probable solution to this kind of situation and hybrid control schemes may improve the performance of the system response.

The hybrid design strategies described in this book utilize a single-objective fitness function in stochastic optimization algorithms [3]. The presented design strategies can be further extended to explore the solution space with multi-objective fitness evaluations [15–19]. For example, weighted combinations of the integral absolute error (*IAE*), the number of fuzzy rules and the smoothness of control actuation can be chosen as the fitness function in multi-objective optimization based design of adaptive fuzzy controllers [15]. Hence, to explore the solution space in an efficient manner and to overcome the premature convergence of a multi-objective optimization process, the concept of Pareto optimal design technique may also be exploited.

K. Das Sharma et al., *Intelligent Control*, Cognitive Intelligence and Robotics,
https://doi.org/10.1007/978-981-13-1298-4_11

The other population-based stochastic optimization algorithms, like simulated annealing (SA) [20], differential evolution (DE) [21], tabu search [22], ant colony optimization (ACO) [23], fire fly algorithm [24], gravitational search algorithm (GSA) [25], gray wolf optimization (GWO) [26], can be potential candidates to design the hybrid strategies. In future, the performance of such hybrid strategy-based controllers can also be compared with results obtained in this book and, armed with such extensive comparisons [27], it may be inferred in future that which kind of global stochastic optimization technique is best suitable to design the hybrid strategies presented in this book.

The concept of hybrid strategies, namely cascade model, concurrent model, and preferential model of adaptive fuzzy controllers, can be developed for the fractional-order systems [28, 29]. The main challenge is to accommodate the fractional-order dynamics in the framework of hybrid adaptive fuzzy control law. A new design methodology has to be found out that eliminates the complexity of solving fractional-order system dynamics.

Other very popular adaptive fuzzy control schemes like model reference adaptive control [30–32], sliding mode control [33, 34] can be very easily hybridized with the contemporary stochastic optimization algorithms just like the hybridization models present in this book. This kind of formulation of stable hybrid fuzzy control schemes may lead to superior performance of the fuzzy controller in different practical applications.

All the discussion of this book is oriented with the conventional fuzzy sets (sometimes referred as type-I fuzzy sets). Recent development of type-II fuzzy sets [35, 36] opens the arena of new potentials and this type of fuzzy sets may be applied to design the adaptive fuzzy controllers optimized by the stochastic algorithms. The imprecision present in the measured data from the practical systems can be modeled very efficiently with this type-II fuzzy sets. Thus, the performances of the controllers designed with this type of fuzzy sets are expected to be superior to that of the conventional fuzzy controllers. Optimal design of type-II adaptive fuzzy controllers is of particular concern of the researchers.

References

1. K. Das Sharma, A. Chatterjee, F. Matsuno, A Lyapunov theory and stochastic optimization based stable adaptive fuzzy control methodology, in *Proceedings of SICE Annual Conference 2008*, Japan, August 20–22, pp. 1839–1844
2. K. Das Sharma, A. Chatterjee, A. Rakshit, A hybrid approach for design of stable adaptive fuzzy controllers employing Lyapunov theory and particle swarm optimization. IEEE Trans. Fuzzy Syst. **17**(2), 329–342 (2009)
3. K. Das Sharma, *Hybrid methodologies for stable adaptive fuzzy control*. PhD dissertation, Jadavpur University, 2012
4. K. Das Sharma, A. Chatterjee, A. Rakshit, Design of a hybrid stable adaptive fuzzy controller employing Lyapunov theory and harmony search algorithm. IEEE Trans. Contr. Syst. Technol. **18**(6), 1440–1447 (2010)

5. L.X. Wang, Stable adaptive fuzzy control of nonlinear system. IEEE Trans. Fuzzy Syst. **1**(2), 146–155 (1993)
6. B.S. Chen, C.H. Lee, Y.C. Chang, H-infinity tracking design of uncertain nonlinear SISO systems: adaptive fuzzy approach. IEEE Trans. Fuzzy Syst. **4**(1), 32–43 (1996)
7. K. Tanaka, Design of model-based fuzzy controller using Lyapunov's stability approach and its application to trajectory stabilization of a model car, in *Theoretical Aspects of Fuzzy Control*, ed. by H.T. Nguyen, M. Sugeno, R. Tong, R.R. Yager (Wiley, New York, 1995), pp. 31–50
8. G. Feng, Stability analysis of discrete time fuzzy dynamic systems based on piecewise Lyapunov functions. IEEE Trans. Fuzzy Syst. **12**(1), 22–28 (2004)
9. K. Tanaka, T. Hori, H.O. Wang, A multiple Lyapunov function approach to stabilization of fuzzy control systems. IEEE Trans. Fuzzy Syst. **11**(4), 582–589 (2003)
10. D.L. Tsay, H.Y. Chung, C.J. Lee, The adaptive control of nonlinear systems using the Sugeno-type of fuzzy logic. IEEE Trans. Fuzzy Syst. **7**(2), 225–229 (1999)
11. T. Egami, H. Morita, Efficiency optimized model reference adaptive control system for a DC motor. IEEE Trans. Ind. Electron. **37**(1), 28–33 (1990)
12. K. Fischle, D. Schroder, An improved stable adaptive fuzzy control method. IEEE Trans. Fuzzy Syst. **7**(1), 27–40 (1999)
13. B.S. Chen, C.L. Tsai, D.S. Chen, Robust H^{∞} and mixed H^2/H^{∞} filters for equalization designs of nonlinear communication systems: fuzzy interpolation approach. IEEE Trans. Fuzzy Syst. **11**(3), 384–398 (2003)
14. K. Das Sharma, A. Chatterjee, P. Siarry, A. Rakshit CMA—H^{∞} hybrid design of robust stable adaptive fuzzy controllers for non-linear systems, in *Proceedings of 1st International Conference on Frontiers in Optimization: Theory and Applications (FOTA-2016)*, Kolkata, India, 2016 (in press)
15. V. Giordano, D. Naso, B. Turchiano, Combining genetic algorithm and Lyapunov-based adaptation for online design of fuzzy controllers. IEEE Trans. Syst. Man Cybern. B Cybern. **36**(5), 1118–1127 (2006)
16. C.A.C. Coello, A short tutorial on evolutionary multi-objective optimization, in *Proceedings of 1st International Conference on Evolutionary Multi-Criterion Optimization, (EMO 2001)*, vol. 1993 (LNCS, 2001), pp. 21–40
17. K. Deb, A. Pratap, S. Agarwal, T. Meyarivan, A fast and elitist multi-objective genetic algorithm: NSGA-II. IEEE Trans. Evol. Comput. **6**(2), 182–197 (2002)
18. K. Deb, *Multi-Objective Optimization Using Evolutionary Algorithms* (Wiley, New York, NY, USA, 2001)
19. T. Murata, H. Ishibuchi, M. Gen, Specification of genetic search directions in cellular multi-objective genetic algorithms, in *Proceedings of 1st International Conference on Evolutionary Multi-Criterion Optimization, (EMO 2001)*, vol. 1993 (LNCS, 2001), pp. 82–95
20. S. Kirkpatrick, C.D. Gelatt, M.P. Vecchi, Optimization by simulated annealing. Science **220** (4598), 671–680 (1983)
21. R. Storn, K. Price, Differential evolution—a simple and efficient adaptive scheme for global optimization over continuous spaces, reported in Technical report, Int. Comp. Sci. Inst., Berkley, 1995
22. F. Glover, Future paths for integer programming and links to artificial intelligence. Comput. Oper. Res. **13**(5), 533–549 (1986)
23. A. Colorni, M. Dorigo, V. Maniezzo, Distributed optimization by ant colonies, in *Proceedings of 1st European Conference on Artificial Life* (Elsevier Publishing, Paris, France, 1991), pp. 134–142
24. X. Yang, *Firefly algorithms for multimodal optimization. Stochastic algorithms: foundations and applications*, Springer Berlin Heidelberg. pp. 169–178, 2009
25. E. Rashedi, H. Nezamabadi-pour, S. Saryazdi, GSA: a gravitational search algorithm. Inf. Sci. **179**(13), 2232–2248 (2009)
26. S. Mirjalili, S.M. Mirjalili, A. Lewis, Grey wolf optimizer. Adv. Eng. Softw. **69**(3), 46–61 (2014)

27. K. Das Sharma, in *A Comparison among Multi-Agent Stochastic Optimization Algorithms for State Feedback Regulator Design of a Twin Rotor MIMO System*. Handbook of Research on Natural Computing for Optimization Problems (IGI Global, 2016), pp. 409–448
28. D. Valério, J. S. Da Costa, Time-domain implementation of fractional order controllers. IEE Proc. Control Theory Appl. **152**(5), 539–552 (2005)
29. D. Biswas, K. Das Sharma, G. Sarkar, Stable adaptive NSOF domain FOPID controller for a class of non-linear systems. IET Control Theory Appl. https://doi.org/10.1049/iet-cta.2017.0732
30. K.J. Astrom, B. Wittenmark, *Adaptive Control*, 2nd edn. (Dover Publications, New York, 2001)
31. K. Narendra, A. Annaswamy, *Stable Adaptive Systems* (Prentice Hall, USA, 1986)
32. E. Lavretsky, Combined/composite model reference adaptive control. IEEE Trans. Autom. Control **54**(11) (2009)
33. V.I. Utkin, Sliding mode control design principles and applications to electric drives. IEEE Trans. Ind. Electron. **40**(1), 23–36 (1993)
34. A. Pisano, On the multi-input second-order sliding mode control of nonlinear uncertain systems. Int. J. Robust Nonlinear Control (2011). https://doi.org/10.1002/rnc.1788. Wiley Online Library
35. N.N. Karnik, J.M. Mendel, Operations on type-2 fuzzy sets. Fuzzy Sets Syst. **122**, 327–348 (2001)
36. J.M. Mendel, Type-2 fuzzy sets and systems: an overview. IEEE Comput. Intell. Mag. **2**(2), 20–29 (2007)

Index

K. Das Sharma et al., *Intelligent Control*, Cognitive Intelligence and Robotics,
https://doi.org/10.1007/978-981-13-1298-4

Printed in the United States
By Bookmasters